Robotics and Parallel Kinematic Machines

Robotics and Parallel Kinematic Machines

Guest Editors

Swaminath Venkateswaran
Jong-Hyeon Park

MDPI

Basel • Beijing • Wuhan • Barcelona • Belgrade • Novi Sad • Cluj • Manchester

Guest Editors
Swaminath Venkateswaran
Department of Science & Engineering
De Vinci Research Center
Courbevoie
France

Jong-Hyeon Park
School of Mechanical Engineering
Hanyang University
Seoul
South Korea

Editorial Office
MDPI AG
Grosspeteranlage 5
4052 Basel, Switzerland

This is a reprint of the Special Issue, published open access by the journal *Robotics* (ISSN 2218-6581), freely accessible at: https://www.mdpi.com/journal/robotics/special_issues/RPKM.

For citation purposes, cite each article independently as indicated on the article page online and as indicated below:

Lastname, A.A.; Lastname, B.B. Article Title. *Journal Name* **Year**, *Volume Number*, Page Range.

ISBN 978-3-7258-5291-8 (Hbk)
ISBN 978-3-7258-5292-5 (PDF)
https://doi.org/10.3390/books978-3-7258-5292-5

Cover image courtesy of Swaminath Venkateswaran

Contents

About the Editors

Swaminath Venkateswaran

Swaminath Venkateswaran obtained his Doctor of Philosophy with honours in Mechanical Engineering from Ecole Centrale de Nantes, France, in 2020. After his PhD, Venkateswaran worked as a Research Associate at the G-SCOP laboratory of Grenoble Institute of Technology (Grenoble-INP) from 2020 to 2021. Currently, Venkateswaran works as an Assistant Professor in Mechatronics and serves as the head of the Industry & Robotics program at the Leonardo da Vinci engineering school (ESILV), Paris. His research areas/expertise include product design and analysis, kinematics of mechanisms, control of robots, cobots for the circular economy, and Industry 4.0. His teaching activities are centred around the domain of mechatronics and industrial engineering for the Bachelor's and Master's levels. His research activities are affiliated with the group "New Materials, Intelligent Systems, and Innovative Companies" of the Da Vinci Research Center (DVRC). Venkateswaran has published in and contributed to around 30 international journals (*Mechanism and Machine Theory*, *Journal of Mechanisms and Robotics*) and conferences (ASME-IDETC, ASME-IMECE, and IFToMM). He has over 200 citations with an h-index of 8 (Google Scholar and Scopus).

Jong-Hyeon Park

Jong-Hyeon Park (Member, IEEE) received his B.S. degree in mechanical engineering from Seoul National University, Seoul, South Korea, in 1981, and his M.S. and Ph.D. degrees from the Massachusetts Institute of Technology (MIT), Cambridge, MA, USA, in 1983 and 1991, respectively. Since 1992, he has been with the School of Mechanical Engineering, Hanyang University, Seoul, where he is currently a Professor. He was with the Korea Science and Engineering Foundation and the Japan Society for the Promotion of Science. He was a Visiting Researcher with Waseda University, Tokyo, Japan, in 1999; a KOSEF-CNR Visiting Researcher with Scuola Superiore Sant'Anna, Pisa, Italy, in 2000; a Visiting Scholar with MIT, from 2002 to 2003; and a Visiting Scholar with the University of Stuttgart, Germany, from 2019 to 2020. He was also associated with Brooks Automation Inc., Chelmsford, MA, USA, from 1991 to 1992 and from 2001 to 2002. His research interests include biped robots, robot dynamics and control, haptics, and biorobots. He is a member of the Korean Society of Mechanical Engineers, the Korean Society of Automotive Engineers, the Korean Society of Precision Engineering, and the Institute of Control, Robotics and Systems. From 2012 to 2017, he served as a Senior Editor for the *Journal of Mechanical Science and Technology*.

robotics MDPI

Editorial

Special Issue: Robotics and Parallel Kinematic Machines

Swaminath Venkateswaran

De Vinci Higher Education, De Vinci Research Center, 92916 Paris, France; swaminath.venkateswaran@devinci.fr

Parallel kinematic machines (PKMs) are a class of robots that have long been recognized for their higher stiffness and high payload-to-weight ratio and precision compared to their serial counterparts [1,2]. Their applications are suitable in various areas such as high-speed machining, medical robotics, and space. Despite their advantages and widespread applications, PKMs suffer from various inherent challenges in design and control owing to complexities in kinematics, limited workspaces, and intricate singularity conditions [3]. Recent industrial developments and research work have focused on improving modeling techniques, the analysis of singular configurations, and reconfigurable architecture. Significant attention has been paid to domains such as the optimization of workspaces [4], the singularity avoidance approach [5], robust design procedures [6], and the integration of compliant components for PKM structures [7] in order to meet evolving application demands. However, several theoretical and practical aspects remain underexplored. For example, cuspidal configurations whereby a robot can transition between multiple inverse kinematic solutions without encountering singularities remain less documented [8,9]. Another interesting phenomenon includes self-motion conditions where the end-effector exhibits mobility even when all actuators are locked, thereby posing design challenges and difficulties in control [10,11]. Furthermore, emerging topics such as modular PKM architectures [12], dynamic performance evaluation [13], and control-aware design optimization [14] continue to attract the attention of the research community, especially for high-precision applications under uncertain or varying load conditions. The following Special Issue features eight articles in the field of PKMs and their potential applications to meet the present era's growing industrial demands.

In paper [15], the authors propose an approach for the 3D modeling of spatial manipulators using Maple 2023 software. This approach enables the creation of three-dimensional computer models of manipulators with clear visual representation of links, cross sections, kinematic pairs, and loads, differing in structure and degrees of freedom, while ensuring a comprehensive view from spatial perspectives. Using the Denavit–Hartenberg method for motion control of 3D models and the recursive Newton–Euler algorithm for calculating distributed dynamic loads, algorithms were developed for generating distribution diagrams of all dynamic loads in each link of a moving manipulator. This approach proves vital, especially during the design of new, innovative manipulators.

In paper [16], the author proposes a computationally efficient approach for dimensional synthesis using multi-objective particle swarm optimization with hierarchical constraints. The approach demonstrates the broad applicability of combined structural and dimensional synthesis for symmetric parallel robots with rigid links and serial-kinematic leg chains. The approach also allows for extension as an open-source MATLAB toolbox 2024b.

In paper [17], the authors present the design and analysis of a novel three-degrees-of-freedom parallel robot with self-sensing Nitinol actuators. The method is a continuation of a previous work and was extended to a 20 mm actuator length. The manipulator was realized, and the control strategy was validated experimentally. The work highlights the importance of self-sensing Nitinol actuators for the design of small parallel manipulators.

In paper [18], the authors propose a flexible surgical stapler mechanism for laparoscopic rectal cancer surgery. By performing workspace analysis and synthesizing kinematic equations, precise control of the mechanism was possible during surgical procedures in the rectal region. The singularities were also examined, considering the influential eyelet friction parameter. A preliminary design was presented, which facilitates the identification of the friction parameter.

In paper [19], the authors propose a neural network model to approximate the task time function of a generic multi-DOF robot. The method proposed a uniform interface that can be adapted to generic robots and tasks. The results presented an accurate model with evaluation times compatible with real-time process optimization.

In paper [20], the authors present the analysis of a spherical parallel mechanism of type 3-SPS-U. The usual singularity approach involves using Euler angles. However, this method is computationally expensive, especially when working with stacked models. Using the Tilt and Torsion approach, the singularity analysis was carried out successfully, which also enabled the work to be extended for stacked models. An experimental platform also aided in validating the approach.

In paper [21], the authors present a novel walking hybrid-kinematics robot that can be reconfigured to have three, five, or six degrees of freedom. A symmetric 3PRPR/3PRRR parallel manipulator with three translational DOFs was used to perform machining tasks. A serial module with two rotational DOFs was added to the 3T parallel mechanism to provide five DOFs. Similarly, a parallel 3SPR or 3SU mechanism can also be combined with the 3T parallel mechanism to provide six DOFs. The mobility, pose, kinematics, differential kinematics, singularities, and workspace of the parallel configuration are discussed in detail. The analysis showed that by making the moving platform of 3SPR or 3SU smaller than the base, the singularities can be easily avoided.

Lastly, in paper [22], the authors present a detailed review covering the fields of rehabilitation, assistive technologies, and humanoid robots. The authors focus on the study of parallel robots designed for human body joints with three degrees of freedom, particularly the neck, shoulder, wrist, hip, and ankle. The review authors discuss the timeline and advancements of parallel robots, focusing on technology readiness levels (TRLs), degrees of freedom, kinematic structure, workspace analysis, performance evaluation methods, and material selection for the design.

The Guest Editor would like to sincerely thank all of the contributors for taking a keen interest in our Special Issue. All of the articles included herein were subjected to rigorous peer review to ensure the production of a high-quality publication. We also thank all of the reviewers for their valuable comments and revisions. In addition, we would like to thank the editors from MDPI for their support in the organization and publication of this Special Issue.

It is our belief that this Special Issue will help fellow researchers and industrial experts to gain an in-depth understanding of the analysis of parallel kinematic machines and their applications based on their architecture.

Conflicts of Interest: The author declares no conflicts of interest.

References

1. Merlet, J.P. *Parallel Robots*; Springer: Dordrecht, The Netherlands, 2006.
2. Tsai, L.W. *Robot Analysis: The Mechanics of Serial and Parallel Manipulators*; John Wiley & Sons: Hoboken, NJ, USA, 1999.
3. Miller, K. Optimal design and modeling of spatial parallel manipulators. *Int. J. Robot. Res.* **2004**, *23*, 127–140. [CrossRef]

4. Soltanov, S.; Roberts, R. Design of a Novel Bio-Inspired Three Degrees of Freedom (3DOF) Spherical Robotic Manipulator and Its Application in Human–Robot Interactions. *Robotics* **2025**, *14*, 8. [CrossRef]
5. Di Gregorio, R. Dimensional synthesis of a novel 3-URU translational manipulator implemented through a novel method. *Robotics* **2022**, *11*, 10. [CrossRef]
6. Riabtsev, M.; Guilhem, J.M.; Petuya, V.; Urizar, M.; Laribi, M.A. A Cartesian Parallel Mechanism for Initial Sonography Training. *Robotics* **2025**, *14*, 95. [CrossRef]
7. Feller, D.; Siemers, C. Mechanical design and analysis of a novel three-legged, compact, lightweight, omnidirectional, serial–parallel robot with compliant agile legs. *Robotics* **2022**, *11*, 39. [CrossRef]
8. Wenger, P. Cuspidal robots. In *Singular Configurations of Mechanisms and Manipulators*; Springer International Publishing: Cham, Switzerland, 2019; pp. 67–99.
9. Wenger, P.; Chablat, D. A review of cuspidal serial and parallel manipulators. *J. Mech. Robot.* **2023**, *15*, 040801. [CrossRef]
10. Briot, S.; Arakelian, V.; Bonev, I.A.; Chablat, D.; Wenger, P. Self-motions of general 3-r pr planar parallel robots. *Int. J. Robot. Res.* **2008**, *27*, 855–866. [CrossRef]
11. Chablat, D.; Ottaviano, E.; Venkateswaran, S. Self-motion conditions for a 3-PPPS parallel robot with delta-shaped base. *Mech. Mach. Theory* **2019**, *135*, 109–114. [CrossRef]
12. Cammarata, A.; Maddio, P.D.; Sinatra, R.; Tian, Y.; Zhao, Y.; Xi, F. Elastostatic analysis of a module-based shape morphing snake-like robot. *Mech. Mach. Theory* **2024**, *194*, 105580. [CrossRef]
13. Elsamanty, M.; Faidallah, E.M.; Hossameldin, Y.H.; Rabbo, S.A.; Maged, S.A.; Yang, H.; Guo, K. Workspace analysis and path planning of a novel robot configuration with a 9-DOF serial-parallel hybrid manipulator (SPHM). *Appl. Sci.* **2023**, *13*, 2088. [CrossRef]
14. Maloisel, G.; Knoop, E.; Thomaszewski, B.; Bächer, M.; Coros, S. Singularity-Aware Design Optimization for Multi-Degree-of-Freedom Spatial Linkages. *IEEE Robot. Autom. Lett.* **2021**, *6*, 6585–6592. [CrossRef]
15. Utenov, M.; Sobh, T.; Temirbekov, Y.; Zhilkibayeva, S.; Patel, S.; Baltabay, D.; Zhumasheva, Z. Analysis of Distributed Dynamic Loads Induced by the Own Mass of Manipulator Links and Their Visualization on Interactive 3D Computer Models. *Robotics* **2025**, *14*, 46. [CrossRef]
16. Schappler, M. Dimensional Synthesis of Parallel Robots Using Bilevel Optimization for Design Optimization and Resolution of Functional Redundancy. *Robotics* **2025**, *14*, 29. [CrossRef]
17. Durante, F.; Raparelli, T.; Beomonte Zobel, P. Three-Degree-of-Freedom Cable-Driven Parallel Manipulator with Self-Sensing Nitinol Actuators. *Robotics* **2024**, *13*, 93. [CrossRef]
18. Khanzode, D.; Jha, R.; Thomieres, A.; Duchalais, E.; Chablat, D. Surgical staplers in laparoscopic colectomy: A new innovative flexible design perspective. *Robotics* **2023**, *12*, 156. [CrossRef]
19. Righettini, P.; Strada, R.; Cortinovis, F. Neural network mapping of industrial robots' task times for real-time process optimization. *Robotics* **2023**, *12*, 143. [CrossRef]
20. Venkateswaran, S.; Chablat, D. Mapping the Tilt and Torsion Angles for a 3-SP SU Parallel Mechanism. *Robotics* **2023**, *12*, 50. [CrossRef]
21. Rosyid, A.; Stefanini, C.; El-Khasawneh, B. A reconfigurable parallel robot for on-structure machining of large structures. *Robotics* **2022**, *11*, 110. [CrossRef]
22. Abarca, V.E.; Elias, D.A. A review of parallel robots: Rehabilitation, assistance, and humanoid applications for neck, shoulder, wrist, hip, and ankle joints. *Robotics* **2023**, *12*, 131. [CrossRef]

robotics

MDPI

Article

Analysis of Distributed Dynamic Loads Induced by the Own Mass of Manipulator Links and Their Visualization on Interactive 3D Computer Models

Muratulla Utenov [1], Tarek Sobh [2], Yerbol Temirbekov [3], Saltanat Zhilkibayeva [4,*], Sarosh Patel [5,*], Dauren Baltabay [6] and Zhadyra Zhumasheva [1]

1 Department of Mechanics, al-Farabi Kazakh National University, Almaty 050040, Kazakhstan; umu53@mail.ru (M.U.); zhadyra_14@mail.ru (Z.Z.)
2 Department of Electrical and Computer Engineering, Lawrence Technological University, Southfield, MI 48075, USA; tsobh@ltu.edu
3 Laboratory of Strength and Rigidity of Mechanical Structures, Institute of Mechanics and Engineering Science, Almaty 050000, Kazakhstan; temirbekove@mail.ru
4 Department of Mechanics and Engineering, M. Auezov South Kazakhstan University, Shymkent 160012, Kazakhstan
5 Department of Computer Science and Engineering, University of Bridgeport, Bridgeport, CT 06604, USA
6 Department of Mechanics, L.N. Gumilyov Eurasian National University, Astana 010008, Kazakhstan; dauren.baltabay.95@gmail.com
* Correspondence: saltanat.zhilkibaeva@auezov.edu.kz (S.Z.); saroshp@bridgeport.edu (S.P.); Tel.: +7-(775)-582-82-28S (S.Z.); +1-(203)-583-93-21 (S.P.)

Abstract: This study proposes an approach to 3D modeling of spatial manipulators in the Maple 2023 software environment. Algorithms and program codes have been developed to create computer 3D models of manipulators controlled by generalized coordinates. The implementation of these algorithms and program codes has enabled the creation of three-dimensional computer models of manipulators with clear visual representations of links, their cross-sections, kinematic pairs, grippers, and loads, differing in structure and degrees of freedom while ensuring a comprehensive view from all spatial perspectives. During the motion of the manipulator, complex distributed dynamic loads arise in its links due to their intrinsic masses. These dynamic loads create several challenges: for instance, excessive dynamic loads or significant deformation of the links may lead to failure of the manipulator or a loss of precision in the positioning of the gripper. Such loads significantly impact the design, operation, and reliability of manipulators. The study and understanding of dynamic loads in manipulators are crucial areas in mechanics and robotics, enabling the development of more reliable and efficient systems. The Denavit–Hartenberg method was applied to control the motion of the created computer 3D models of manipulators using generalized coordinates. Using the recursive Newton–Euler equations, the necessary kinematic characteristics of the manipulator's links were determined for calculating the distributed dynamic loads arising from the intrinsic masses of the links at each cross-section, relative to the local coordinate systems rigidly attached to the links. Algorithms and program codes were developed for controlling the motion of 3D models of manipulators, as well as for constructing visual diagrams of distributed dynamic loads in mutually perpendicular planes, formed by the principal axes of the link cross-sections and the axes passing along the longitudinal axes of the links. The implementation of these algorithms and program codes enabled the generation of distribution diagrams of all dynamic loads in all links of the moving manipulator. These diagrams visually illustrate the changes in direction and magnitude of the distributed dynamic loads in all cross-sections of the links throughout the full cycle of the manipulator's operation. This allows for the consideration of the identified dynamic loads in the strength and stiffness calculations of the manipulator links, which is essential for the design of new innovative manipulators.

Keywords: manipulators; 3D modeling; kinematics; dynamic loads; stress–strain diagram; Maplesoft

1. Introduction

Modern technological advancements have led to increased customer demands for designed products. A high level of competition and diverse operating conditions necessitate the development of products with enhanced characteristics, such as high functionality and quality. In such conditions, engineers must design machines that exhibit flexibility and broad applicability. However, the fabrication of prototypes for all potential devices subject to experimental research is constrained by both economic and time-related factors.

Consequently, modeling becomes a crucial stage in both the design of new structures and the modernization of existing ones. Modern CAD systems not only enable the creation of geometric models but also facilitate kinematic, dynamic, and strength analyses. However, integrating control systems into such applications often presents a complex or even unfeasible challenge. For this reason, Matlab, in combination with the Simulink module, is widely used for modeling control systems.

The study in [1] emphasizes that offline programming (OLP) is an effective method for controlling industrial robots. However, its widespread implementation is hindered by the high cost of commercial OLP systems, which often exceeds the price of the robotic systems themselves. This creates a demand for accessible and efficient OLP solutions based on widely used CAD platforms.

This paper presents the development of the robotic OLP system RobSim, created using SolidWorks Professional 2019 and Microsoft Visual Studio 2010. RobSim is integrated into the SolidWorks environment as an additional tool, providing users with the ability to create, import, and edit objects, as well as conveniently define and modify robot movement trajectories.

In [2], the authors outline the stages of constructing simulation models developed using SolidWorks and Matlab/Simulink. Examples of simulation models include a laboratory hoist and a forestry crane. These models allow for motion visualization, trajectory tracking, and monitoring of velocity and acceleration at any system point.

In [3], a computer model of a three-link vertically walking robot is developed in the Matlab environment, illustrating the basic working principles of such robots. The modeling of the vertical walking robot's movement was conducted using the SimMechanics library in Simulink within the Matlab environment.

Studies [4–6] present a methodology for analytically determining internal forces in the links of planar linkage mechanisms and manipulators, considering distributed dynamic loads arising from the self-weight of the links, for both statically determinate and indeterminate structures. Discrete models were developed, and equilibrium equations were derived to establish the relationship between the force vector and the dynamic characteristics of the links. Additionally, algorithms and software codes were created to generate visual plots of dynamic load distribution, transverse and longitudinal forces, and bending moments, with corresponding results provided.

In works [7,8], it is assumed that the masses are concentrated at the center of gravity of the manipulator links. The inertia forces and the inertia matrix of the i-th link are determined relative to its center of mass. To illustrate the application of the recursive Newton-Euler equations, a two-link manipulator with rotational kinematic pairs is considered.

In the dynamic analysis of a manipulator, gravitational and inertial forces, as well as moments of inertia, are referenced to the center of mass of each link. However, they can be transferred to the axis of rotation of the link using Steiner's theorem. When analyzing the dynamics of the entire system, forces and moments are referenced to the axes of rotation of each link. Bringing all forces and moments to the rotation axes allows for the formulation of manipulator dynamics in a convenient form (typically a matrix form) that is directly related to actuators.

In [9], a method for modeling the dynamics of manipulators with flexible links is presented. This method decomposes the manipulator into components, with the dynamics of these components determined using Newton–Euler equations and linearized based on state vectors. Transfer matrices are developed to establish the system equations, preventing an increase in the size of the dynamic equations when adding joints or beam elements. The method focuses on computing small-scale dynamic equations of components, simplifying the process.

In work [10], a physically justified approach is presented that accounts for the flexibility of robot links and reducers. The article aims to establish a foundation for modeling flexible robotic structures to enhance accuracy. The elasticity of components is introduced into the discrete-parameter model using pairs of springs and dampers, the number of which is determined by the need to optimize computational costs and simulation time for integrating the model into industrial environments.

In article [11], two novel arm exoskeletons based on pantograph linkages are introduced, capable of supporting loads of up to 9 kg. The pantograph linkage adapts to the arm's geometry, ensuring support in any position. Gravity compensation is achieved through shoulder support and a counteracting force behind the back. A passive exoskeleton with a gas spring and an active exoskeleton with a motor are described. The forces generated by the gas spring are analyzed and various mechanism configurations are investigated. The equations governing the pantograph linkage are derived, and the exoskeleton forces are modeled with a comparison to measured values.

Article [12] presents a method for analyzing the performance of serial robotic manipulators with gravity balancing under dynamic loads. Torque, power, and energy reduction indices are employed to assess performance in both unbalanced and balanced states. The proposed method enables a comprehensive evaluation of the manipulator while considering dynamic effects. As an example, a three-degree-of-freedom manipulator is examined, in which gravity balancing is achieved using a compact module with a geared spring on each link.

Article [13] introduces an advanced architecture of an internally balanced mechanism based on a four-bar 4R linkage—referred to as the 4R Grand Architecture. The mechanism consists of 24 links and exhibits 26-fold redundant mobility while maintaining mobility with a fixed center of mass. It is demonstrated that all center-of-mass tracking theories for four-bar linkages are contained within this architecture. Additionally, it is shown how new normally constrained two-degree-of-freedom balanced mechanisms can be derived from it by removing specific links.

In article [14], an approach to designing gravity compensators based on planar four-bar linkages and linear springs is presented. A class of compensators comprising 42 types with high efficiency and kinematic simplicity is developed. The compensators are created by combining four-bar linkages with rotating masses and linear springs, whose parameters are optimized to minimize torque within a specified balancing range.

Work [15] investigates the influence of manipulator motion on the base posture of a free-floating space robot. When a target is grasped, a closed system is formed, resulting in more complex dynamic interactions. The interaction is analyzed in both open and closed systems.

In manuscript [16], a mathematical procedure for determining the inertial parameters of maternal manipulators is first described. These parameters are essential for accurate positioning and force control in an electromechanical system designed to assist in surgical operations. The method for determining the mass and center of mass of the manipulator is based on solving static equilibrium equations for rigid bodies. Based on the estimated inertial parameters of the manipulator, the article demonstrates how to determine the force exerted by the manipulator.

Work [17] examines the effect of time-varying inertial loads on the dynamics of parallel manipulators in joint space, particularly during acceleration and deceleration phases. A comprehensive analysis method based on a two-mass model is proposed.

In work [18], with the development of parallel manipulators, the issue of inertia matching has become increasingly significant. However, a clear inertia indicator and a method for its reconciliation have not yet been proposed. This study addresses these aspects using the Stewart manipulator as an example.

Work [19] considers a scenario in which multiple manipulators control a single object. A dynamic model of the system is constructed that explicitly represents the relationship between the object's dynamics and joint space dynamics, taking into account the object's dynamics, the dynamics of individual manipulators, and the actuator torques in the joints. Methods are then presented that directly distribute the dynamics of the object and manipulators within the joint space according to criteria based on joint input forces.

In work [20], a computational methodology for determining the maximum payload capacity of a robotic manipulator with joint elasticity is described, considering constraints related to accuracy and actuator characteristics. The maximum payload achievable by the manipulator along a given trajectory is limited by several factors.

One of the key aspects of manipulator design is ensuring the strength and rigidity of its links throughout the entire operational process. The analysis of the stress–strain state of the links is complicated by the fact that the manipulator is in motion, leading to distributed dynamic loads of a complex nature, which are caused by the mass of the links at each of their cross-sections. These loads vary in magnitude and direction depending on the kinematic characteristics of the links.

Since the manipulator is in motion, it is impossible to determine in advance at which cross-section of a link and at what position the maximum internal force or deformation may occur. Therefore, it is necessary to investigate the stress–strain state of the manipulator over the entire operating cycle.

Moreover, it is crucial to visualize the stress–strain state of all links throughout the complete working cycle so that all types of dynamic loads, internal forces, and deformations are clearly visible across all link cross-sections. This allows the designer to analyze the stress–strain state of each link and make well-founded decisions when calculating the strength and rigidity of the manipulator's links.

Maple is a powerful mathematical software for symbolic and numerical computations, as well as for engineering modeling. It does not require integration with CAD software or other applications to create computer-based 3D models of manipulators controlled by generalized coordinates. In the Maple environment, three-dimensional manipulator models have been developed with clear visual representations of links, their cross-sections, kinematic pairs, grippers, and loads, which vary in structure and degrees of freedom, ensuring a comprehensive spatial view controlled by generalized coordinates. This environment enables the determination of link positions and orientations, the calculation of all kinematic characteristics required for assessing dynamic loads, and their visualization.

The topic of this study is highly relevant and significant in the field of robotics, particularly in the context of manipulator dynamics and design. The study of dynamic loads on manipulators plays a crucial role in the development of more reliable and efficient robotic systems.

2. Materials and Methods

2.1. Algorithm for 3D Modeling of Manipulators in the Maple 2023 Software Environment

The Maple 2023 software environment includes three-dimensional primitives from the plottools package, which enable the construction of 3D shapes and surfaces, such as cones, cylinders, parallelepipeds, and polyhedra. These primitives are used to create links, stands, grippers, and kinematic pairs of manipulators. The generated 3D components in Maple 2023 can be translated along the three spatial axes and rotated around them. Using these links and kinematic pairs, the designer can assemble a 3D model of a manipulator.

Subsequently, all elements of the manipulator are integrated into a single system by introducing fundamental connections (kinematic pairs) from the gripper to the stand. This allows the construction of a complete visual moving model controlled by the generalized coordinates of the manipulator. Maple 2023 provides the capability to connect links both rigidly and while accounting for their relative motion. This feature enables designers to create manipulators with the required structure and specified degrees of freedom. Using the described algorithms in the Maple 2023 environment, program codes were developed for constructing 3D models of manipulators, whose movement is governed by generalized coordinates. The implementation of the developed algorithms and program codes facilitated the creation of three-dimensional computer models of manipulators with distinct representations of links, their cross-sections, kinematic pairs, grippers, and loads, exhibiting various structures and degrees of freedom. All models can be viewed from different angles due to the viewing capabilities of Maple 2023 (Figures 1–3). By employing this algorithm, designers can construct computer models of manipulators with any desired structure and varying degrees of freedom. By defining specific patterns of generalized coordinates for the created manipulators, their motion in space can be observed.

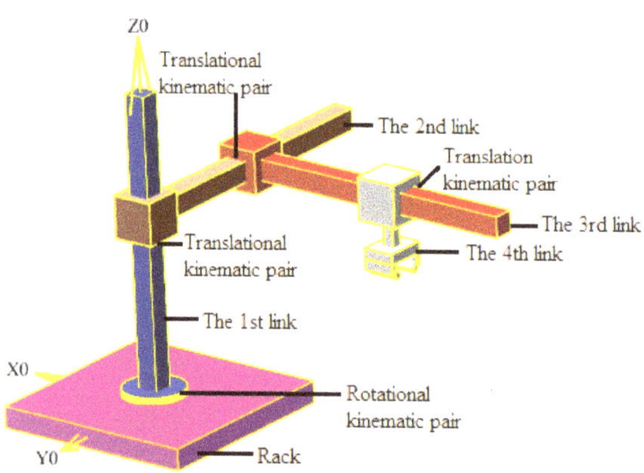

Figure 1. A spatial manipulator with four degrees of freedom (RTTT), consisting of five links connected by one revolute and three prismatic kinematic pairs.

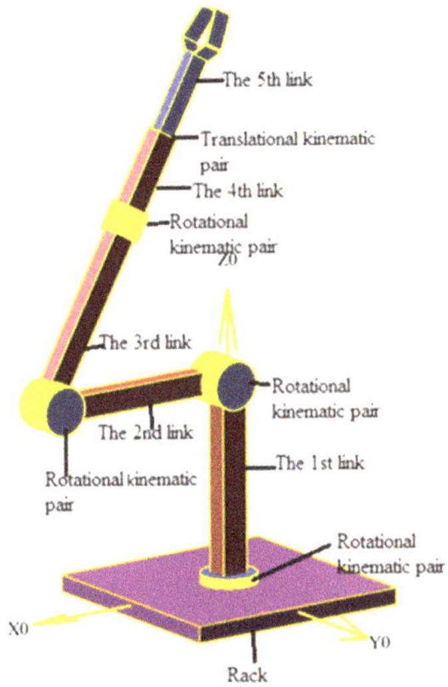

Figure 2. A manipulator with five degrees of freedom (RRRRT) consists of six links connected by four rotational and one translational kinematic pair.

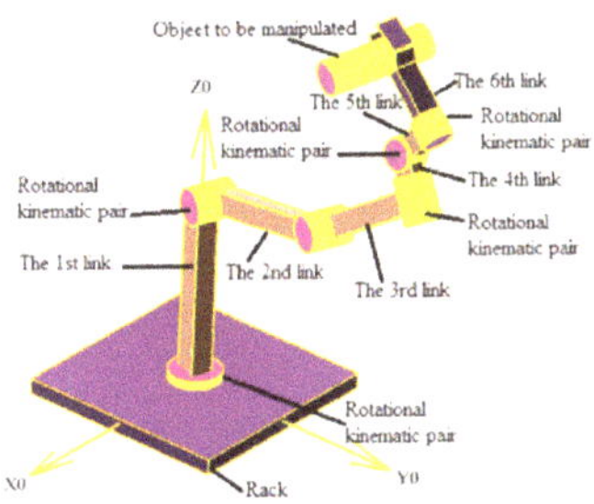

Figure 3. A spatial manipulator with six degrees of freedom (RRRRRR), consisting of seven links connected by six revolute kinematic pairs.

To obtain numerical values of the kinematic, dynamic, and strength characteristics of these manipulators, it is necessary to establish Denavit–Hartenberg coordinate systems on their links. When constructing the 3D model of a manipulator in its initial position, the length of the links should be designed relative to the base coordinate system along the same axis as the link lengths in the Denavit–Hartenberg system.

2.2. Position and Orientation of Manipulator Links in Space

In this study, the Denavit–Hartenberg method is used to determine the position and orientation of the manipulator's links in space. This method was the first to apply 4×4 homogeneous transformation matrices to describe the spatial geometry of a manipulator, providing a universal algorithm for deriving its kinematic equations.

A specific choice of coordinate frames for the manipulator's links allows the transition from one coordinate system to another to be described using only four parameters, rather than six as in the general case. The coordinate system $O_{i-1}X_{i-1}Y_{i-1}Z_{i-1}$ can be transformed into the system $O_iX_iY_iZ_i$ through a sequence of one rotation, two translations, and another rotation. The resulting transformation matrix from the coordinate system $O_{i-1}X_{i-1}Y_{i-1}Z_{i-1}$ to $O_iX_iY_iZ_i$ is given by the following:

$$A_i^{i-1} = \begin{bmatrix} \cos(\theta_i) & -\sin(\theta_i)\cos(\alpha_i) & \sin(\theta_i)\sin(\alpha_i) & a_i\cos(\theta_i) \\ \sin(\theta_i) & \cos(\theta_i)\cos(\alpha_i) & -\cos(\theta_i)\sin(\alpha_i) & a_i\sin(\theta_i) \\ 0 & \sin(\alpha_i) & \cos(\alpha_i) & d_i \\ 0 & 0 & 0 & 1 \end{bmatrix}, \tag{1}$$

where θ_i is the angle by which the axis X_{i-1} must be rotated around the axis Z_{i-1} to align with the axis X_i. If the kinematic pair $(i-1, i)$ is rotational, then θ_i is a generalized coordinate; the parameter d_i represents the distance from the origin $O_{i-1}X_{i-1}Y_{i-1}Z_{i-1}$ of the coordinate system to the intersection of the axis Z_{i-1} with the axis X_i, measured along the axis Z_{i-1}. If the kinematic pair $(i-1, i)$ is translational, then d_i is a generalized coordinate; a_i is the distance between the intersection of the axis Z_{i-1} with the axis X_i and the origin $O_iX_iY_iZ_i$ of the coordinate system, measured along the axis X_i; α_i is the angle of rotation of the axis Z_{i-1} around the axis X_i until it coincides with the axis Z_i.

Let us represent the matrix A_i^{i-1} in the following form:

$$A_i^{i-1} = \begin{bmatrix} R_i^{i-1} & \vec{O}_i^{\,i-1} \\ 0 & 1 \end{bmatrix}, \tag{2}$$

where

$$R_i^{i-1} = \begin{bmatrix} \cos(\theta_i) & -\sin(\theta_i)\cos(\alpha_i) & \sin(\theta_i)\sin(\alpha_i) \\ \sin(\theta_i) & \cos(\theta_i)\cos(\alpha_i) & -\cos(\theta_i)\sin(\alpha_i) \\ 0 & \sin(\alpha_i) & \cos(\alpha_i) \end{bmatrix}$$

The matrix R_i^{i-1} defines the orientation of the axes of the coordinate system $O_iX_iY_iZ_i$ relative to the coordinate system $O_{i-1}X_{i-1}Y_{i-1}Z_{i-1}$:

$$\vec{O}_i^{\,i-1} = \begin{bmatrix} a_i\cos(\theta_i) \\ a_i\sin(\theta_i) \\ d_i \end{bmatrix}, \tag{3}$$

The vector $\vec{O}_i^{\,i-1}$ indicates the position of point O_i, which serves as the origin of the coordinate system $O_iX_iY_iZ_i$, within the coordinate system $O_{i-1}X_{i-1}Y_{i-1}Z_{i-1}$.

The position and orientation of the i-th link of the manipulator relative to the base coordinate system $O_0X_0Y_0Z_0$ are determined by the following expression:

$$A_i^0 = A_1^0 A_2^1 A_3^2 \ldots A_i^{i-1} = \begin{bmatrix} R_i^0 & \vec{O}_i^{\,0} \\ 0 & 1 \end{bmatrix}. \tag{4}$$

Using the matrix A_i^0, the relationship between the position vectors of a given point in the coordinate systems $O_0X_0Y_0Z_0$ and $O_iX_iY_iZ_i$ can be established:

$$\vec{r}_i^{\,0} = A_i^0 \vec{r}_i^{\,i}, \tag{5}$$

where $\vec{r}_i^{\,i} = [x_i \quad y_i \quad z_i \quad 1]^T$ is a column matrix that defines the position of any point on the i-th link relative to the coordinate system $O_iX_iY_iZ_i$, which is rigidly attached to the i-th link; and $\vec{r}_i^{\,0} = [x_0 \quad y_0 \quad z_0 \quad 1]^T$ is a column matrix that defines the position of the same point relative to the base coordinate system $O_0X_0Y_0Z_0$, which is rigidly attached to the base.

2.3. Vectors of Angular Velocities, Linear Velocities, and Accelerations of Manipulator Link Points

The vectors of angular velocities and accelerations of the links, as well as the linear velocities and accelerations of the manipulator link points, are determined relative to the base coordinate system $O_0X_0Y_0Z_0$ and the Denavit–Hartenberg coordinate systems associated with the links $O_iX_iY_iZ_i$ using Newton–Euler recursive equations. The components of these vectors enable the determination of all types of dynamic loads and the identification of their distribution patterns.

In addition to these vector components, determining the dynamic loads acting at the cross-sections of the links requires knowledge of the linear dimensions of the cross-sections and the specific weight of the link material.

In this study, the angular velocity vector $\vec{\omega}_i$ of the *i*-th link relative to the base coordinate system is represented as follows:

$$\vec{\omega}_i = \begin{cases} \vec{\omega}_{i-1} + R^0_{i-1} \times \vec{z}_0 \times \dot{q}_i,\ if\ the\ i\text{-}th\ kinematic\ pair\ is\ rotational, \\ \vec{\omega}_{i-1},\ if\ the\ i\text{-}th\ kinematic\ pair\ is\ translational, \end{cases} \tag{6}$$

where $i = 1, 2, \ldots, n$, $\vec{z}_0 = (0,0,1)^T$, and R^0_{i-1} is defined by Equation (4).

Then, the angular acceleration $\vec{\varepsilon}_i$ of the *i*-th link relative to the base coordinate system is determined by the following expression:

$$\vec{\varepsilon}_i = \begin{cases} \vec{\varepsilon}_i = +R^0_{i-1} \times \vec{z}_0 \times \ddot{q}_i + \vec{\omega}_{i-1} \times \left(R^0_{i-1} * \vec{z}_0 * \dot{q}_i\right), \\ if\ the\ i\text{-}th\ kinematic\ pair\ is\ rotational, \\ \vec{\varepsilon}_{i-1},\ if\ the\ i\text{-}th\ kinematic\ pair\ is\ translational. \end{cases} \tag{7}$$

For the linear velocities and accelerations of the origin of the coordinate system of the *i*-th link of the manipulator relative to the base coordinate system, the following relations hold:

$$\vec{v}_i = \begin{cases} \vec{v}_{i-1} + \vec{\omega}_i \times \vec{\rho}^{\,z}_i,\ if\ the\ i\text{-}th\ kinematic\ pair\ is\ rotational, \\ \vec{v}_{i-1} + \vec{\omega}_i \times \vec{\rho}^{\,z}_i + R^0_{i-1}\vec{z}_0\dot{q}_i, \\ if\ the\ i\text{-}th\ kinematic\ pair\ is\ translational, \end{cases} \tag{8}$$

where $\vec{\rho}^{\,z}_i = R^0_{i-1} {}^* \vec{O}^{\,i-1}_i$.

$$\vec{a}_i = \begin{cases} \vec{a}_{i-1} + \vec{\omega}_i \times \left(\vec{\omega}_i \times \vec{\rho}^{\,z}_i\right) + \vec{\varepsilon}_i \times \vec{\rho}^{\,z}_i, \\ if\ the\ i\text{-}th\ kinematic\ pair\ is\ rotational, \\ \vec{a}_{i-1} + \vec{\omega}_i \times \left(\vec{\omega}_i \times \vec{\rho}^{\,z}_i\right) + 2\vec{\omega}_i \times (R^0_{i-1}\vec{z}_0\dot{q}_i) + \vec{\varepsilon}_i \times \vec{\rho}^{\,z}_i + R^0_{i-1}\vec{z}_0\ddot{q}_i, \\ if\ the\ i\text{-}th\ kinematic\ pair\ is\ translational. \end{cases} \tag{9}$$

The linear velocity and acceleration of point $\vec{P}^{\,i}_i = [x_i \quad y_i \quad z_i]^T$ of the *i*-th link relative to the base coordinate system are determined by the following expressions:

$$\vec{v}_{iP} = \vec{\omega}_i \times (R^0_i \vec{P}^{\,i}_i) + \vec{v}_i, \tag{10}$$

$$\vec{a}_{iP} = \vec{a}_i + \vec{\omega}_i \times \left(\vec{\omega}_i \times (R^0_i \vec{P}^{\,i}_i)\right) + \vec{\varepsilon}_i \times (R^0_i \vec{P}^{\,i}_i). \tag{11}$$

The angular velocity vector $R_0^i\vec{\omega_i}$ and the angular acceleration vector $R_0^i\vec{\varepsilon_i}$ of the i-th link relative to the coordinate system $O_iX_iY_iZ_i$, which is rigidly attached to the manipulator's links, are determined by Equations (12) and (13) given below:

$$R_0^i\vec{\omega_i} = \begin{cases} R_0^i * \left(\vec{\omega}_{i-1} + R_{i-1}^0 * \vec{z}_0 * \dot{q}_i\right), \\ \quad if\ the\ i\text{-}th\ kinematic\ pair\ is\ rotational, \\ R_0^i * \vec{\omega}_{i-1},\ if\ the\ i\text{-}th\ kinematic\ pair\ is\ translational, \end{cases} \tag{12}$$

where $R_0^i = \left(R_i^0\right)^{-1}$, and R_i^0 is defined by Equation (4).

$$R_0^i\vec{\varepsilon_i} = \begin{cases} R_0^i * \left(\vec{\varepsilon}_{i-1} + R_{i-1}^0 * \vec{z}_0 * \ddot{q}_i + \vec{\omega}_{i-1} \times \left(R_{i-1}^0 * \vec{z}_0 * \dot{q}_i\right)\right), \\ \quad if\ the\ i\text{-}th\ kinematic\ pair\ is\ rotational, \\ R_0^i * \vec{\varepsilon}_{i-1},\ if\ the\ i\text{-}th\ kinematic\ pair\ is\ translational. \end{cases} \tag{13}$$

The vectors of linear velocities $R_0^i\vec{v_i}$ and acceleration $R_0^i\vec{a_i}$ of point O_i, which serves as the origin of the coordinate system $O_iX_iY_iZ_i$, relative to the coordinate system $O_iX_iY_iZ_i$, are determined by the following expressions:

$$R_0^i\vec{v_i} = \begin{cases} R_0^i * \left(\vec{v}_{i-1} + \vec{\omega}_i \times \vec{\rho}_i^{\,z}\right),\ if\ the\ i\text{-}th\ kinematic\ pair\ is\ rotational, \\ R_0^i * \left(\vec{v}_{i-1} + \vec{\omega}_i \times \vec{\rho}_i^{\,z} + R_{i-1}^0 \vec{z}_0 \dot{q}_i\right), \\ \quad if\ the\ i\text{-}th\ kinematic\ pair\ is\ translational. \end{cases} \tag{14}$$

$$R_0^i\vec{a_i} = \begin{cases} R_0^i * \left(\vec{a}_{i-1} + \vec{\omega}_i \times \left(\vec{\omega}_i \times \vec{\rho}_i^{\,z}\right) + \vec{\varepsilon}_i \times \vec{\rho}_i^{\,z}\right), \\ \quad if\ the\ i\text{-}th\ kinematic\ pair\ is\ rotational, \\ R_0^i * \left(\vec{a}_{i-1} + \vec{\omega}_i \times \left(\vec{\omega}_i \times \vec{\rho}_i^{\,z}\right) + 2\vec{\omega}_i \times \left(R_{i-1}^0 \vec{z}_0 \dot{q}_i\right) + \vec{\varepsilon}_i \times \vec{\rho}_i^{\,z} + R_{i-1}^0 \vec{z}_0 \ddot{q}_i\right), \\ \quad if\ the\ i\text{-}th\ kinematic\ pair\ is\ translational. \end{cases} \tag{15}$$

During the motion of the manipulator, the points $\vec{P}_i^{\,i} = [x_i \quad y_i \quad z_i]^T$, which are rigidly attached to the link in the coordinate system $O_iX_iY_iZ_i$, acquire linear velocities and accelerations relative to this coordinate system. These quantities can be determined using the following equations:

$$R_0^i \vec{v}_{iP} = R_0^i * \left(\vec{\omega}_i \times \left(R_i^0 \vec{P}_i^{\,i}\right) + \vec{v}_i\right), \tag{16}$$

$$R_0^i \vec{a}_{iP} = R_0^i * \left(\vec{a}_i + \vec{\omega}_i \times \left(\vec{\omega}_i \times \left(R_i^0 \vec{P}_i^{\,i}\right)\right) + \vec{\varepsilon}_i \times \left(R_i^0 \vec{P}_i^{\,i}\right)\right). \tag{17}$$

2.4. Distributed Dynamic Loads Induced by Link Mass During Manipulator Motion

In this study, it is assumed that the cross-sections of the links remain constant and that the mass is distributed along the link's axis. Based on this assumption, the patterns of dynamic load distribution along the link's axis are determined.

For a manipulator with constant link cross-sections, the intensity of gravitational force distribution relative to the base coordinate system at any cross-section of the i-th link is given by the following vector (unit of measurement (N/m)):

$$\vec{f_i} = [0, 0, -\gamma_i s_i]^T, \tag{18}$$

where γ_i is the specific weight of the material of the *i*-th link (N/m^3), and s_i is the cross-sectional area of the *i*-th link (m^2).

In the coordinate system $O_iX_iY_iZ_i$, which is rigidly attached to the *i*-th link, the intensity of the gravitational force distribution is given by the following vector:

$$\overrightarrow{R_0^i f_i} = R_0^i * \overrightarrow{f_i}, \tag{19}$$

where R_0^i is the matrix that defines the orientation of the axes of the coordinate system $O_0X_0Y_0Z_0$ relative to the coordinate system $O_iX_iY_iZ_i$.

Due to the accelerations of points located along the link's axis, inertia forces arise, distributed along the axes of the manipulator's links. If the link's cross-section is assumed to be constant, then the mass per unit length of the link (cross-sectional mass or mass distribution intensity along the link's axis) is given by $\frac{\gamma_i s_i}{g}\left(\frac{N*s^2}{m^2}\right)$. To determine the intensity of inertia forces arising from the linear accelerations of points along the axes of the coordinate system $O_iX_iY_iZ_i$, which is rigidly attached to the *i*-th link, we construct the inertia intensity mass matrix, which is expressed as follows:

$$m_i = \begin{pmatrix} \frac{\gamma_i s_i}{g} & 0 & 0 \\ 0 & \frac{\gamma_i s_i}{g} & 0 \\ 0 & 0 & \frac{\gamma_i s_i}{g} \end{pmatrix}. \tag{20}$$

Then, the intensity of inertia forces in the cross-sections of the *i*-th link is determined by the following expression (unit of measurement (N/m)):

$$\overrightarrow{q}_i = -m_i * \overrightarrow{R_0^i a_{pi}}. \tag{21}$$

The vector of the distribution intensity of dynamic loads, arising from the intensity of the distributed self-weights of the link cross-sections and the intensity of the distributed inertia forces due to the linear accelerations of the cross-sections, is determined by the following equation:

$$\overrightarrow{F}_i = \overrightarrow{R_0^i f_i} + \overrightarrow{q}_i. \tag{22}$$

To determine the intensity of the distribution of dynamic torsional moments of inertia, arising due to the self-mass of the links during their rotation about their own axes under the influence of angular velocities and accelerations, we calculate the intensity of the axial moments of inertia in the cross-sections of the link.

Figure 4 shows an image of the link's cross-section, depicted as a quadrilateral representing a thin plate.

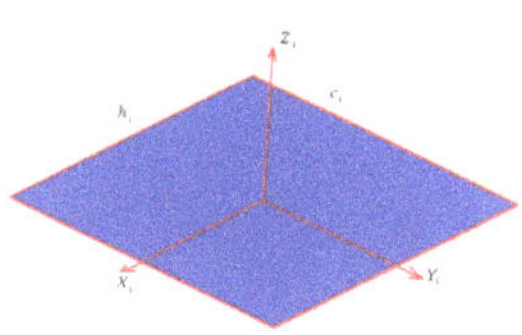

Figure 4. An illustration of the link's cross-section in the form of a quadrilateral representing a thin plate.

If the axes X_i, Y_i, and Z_i of the cross-section of the i-th link are oriented as shown in Figure 4, then the axial moments of inertia relative to these axes are determined by the following expressions (unit of measurement $(N * s^2)$):

$$I_{xi} = \frac{\gamma_i s_i c_i^2}{12g}, \; I_{yi} = \frac{\gamma_i s_i h_i^2}{12g}, \; I_{zi} = \frac{\gamma_i s_i (c_i^2 + h_i^2)}{12g}, \tag{23}$$

where h_i, c_i are the dimensions of the cross-section of the i-th link, and $s_i = h_i * c_i$ is the cross-sectional area of the i-th link.

The matrix of the intensity of axial moments of inertia for the cross-section of the i-th link is given by the following:

$$I_i = \begin{pmatrix} I_{xi} & 0 & 0 \\ 0 & I_{yi} & 0 \\ 0 & 0 & I_{zi} \end{pmatrix}. \tag{24}$$

The intensities of the dynamic torsional inertia moments arising in the cross-sections of the i-th link are determined using the following relation (unit of measurement—N):

$$\vec{M}_i = -\left(I_i * \left(R_0^i \times \vec{\varepsilon}_i\right) + \left(R_0^i * \vec{\omega}_i\right) \times \left(I_i * \left(R_0^i * \vec{\omega}_i\right)\right)\right). \tag{25}$$

Using the presented algorithms, program codes have been developed for constructing visual diagrams of dynamic loads distributed on the manipulator's links.

3. Determination of Dynamic Loads Arising in the Cross Sections of the RRRRT Manipulator Links and Construction of Their Visual Diagrams

In this study, we examine a 3D model of the RRRRT manipulator (Figure 2), controlled by generalized coordinates. To obtain numerical values for the kinematic and dynamic characteristics of the manipulator's links, it is necessary to rigidly associate the links with Denavit–Hartenberg coordinate systems. These coordinate systems must be constructed following the rules of the Denavit–Hartenberg method. Figure 5 illustrates how the links of the considered manipulator are associated with the Denavit–Hartenberg coordinate systems.

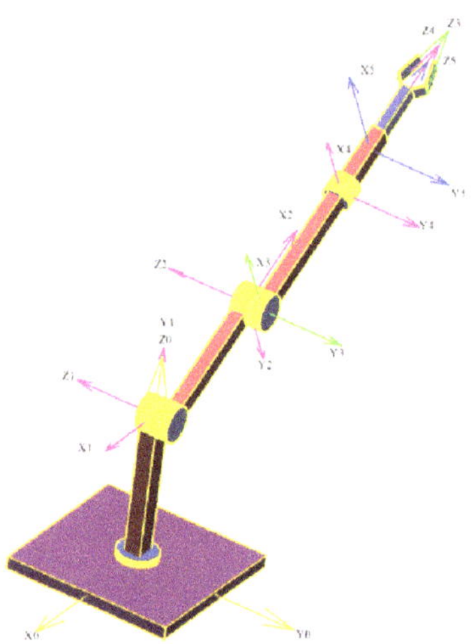

Figure 5. The five-degree-of-freedom (RRRRT) manipulator consists of six links connected by four revolute pairs and one prismatic pair. The links are rigidly associated with coordinate systems constructed according to the rules of the Denavit–Hartenberg method.

After constructing the Denavit–Hartenberg coordinate system for the RRRRT manipulator, the parameters of these coordinate systems can be determined. These parameters are presented in Table 1.

Table 1. Denavit–Hartenberg coordinate system parameters for the RRRRT manipulator.

Kinematic Pairs	Links Forming Kinematic Pairs	Types of Kinematic Pairs	Values of Coordinate System Parameters θ_i	d_i	a_i	α_i
1	0, 1	rotational	θ_1	0.85	0	$\frac{\pi}{2}$
2	1, 2	rotational	θ_2	−0.09	0.8	0
3	2, 3	rotational	θ_3	−0.09	0	$-\frac{\pi}{2}$
4	3, 4	rotational	θ_4	0.8	0	0
5	4, 5	translational	0	d_5	0	0

The generalized coordinate motion laws of the manipulator are defined as follows: $\theta_1 = \dot{\theta}_1 * t, \dot{\theta}_1 = \pi, \ddot{\theta}_1 = 0$, where the unit of time is one second, determined by the relation $t = \frac{i}{k}$, where i = 0...36 represents the discrete positions of the manipulator, and k = 36 is the total number of positions; $\theta_2 = \dot{\theta}_2 * t, \dot{\theta}_2 = \pi, \ddot{\theta}_2 = 0$; $\theta_3 = -\frac{\pi}{2} * \dot{\theta}_3 * t, \dot{\theta}_3 = \frac{2\pi}{3}, \ddot{\theta}_3 = 0$; $\theta_4 = \dot{\theta}_4 * t, \dot{\theta}_4 = \pi, \ddot{\theta}_4 = 0$; $d_5 = \dot{d}_5 * t, \dot{d}_5 = 0.35, \ddot{d}_5 = 0$.

The cross-sectional areas of the manipulator's links (unit of measurement m^2) are as follows: $s_1 = 0.012,\ s_2 = 0.008,\ s_3 = 0.008,\ s_4 = 0.008,\ s_5 = 0.0048$.

The specific weights of the materials of the manipulator's links (unit of measurement N/m^3) are as follows: $\gamma_1 = \gamma_2 = \gamma_3 = \gamma_4 = \gamma_5 = 78 \times 10^3$.

The cross-sectional dimensions of the manipulator's links (unit of measurement m) are as follows: $c_1 = 0.12,\ h_1 = 0.1; c_2 = 0.1,\ h_2 = 0.08; c_3 = 0.1,\ h_3 = 0.08; c_4 = 0.1,\ h_4 = 0.08$; $c_5 = 0.06,\ h_5 = 0.04$.

The implementation of the developed algorithms and software enabled the construction of diagrams representing the distribution of dynamic loads across all manipulator links. These diagrams are positioned in mutually perpendicular planes formed by the principal axes of the cross-sections and the longitudinal axes of the manipulator links. The results are presented in Figures 6–9.

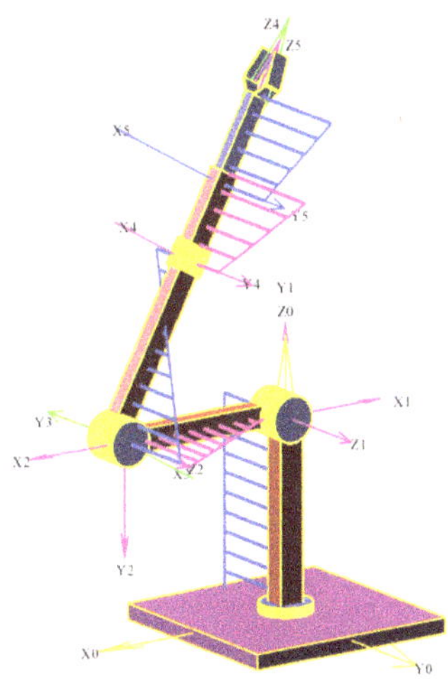

Figure 6. The figure illustrates the patterns of distribution of longitudinally distributed (along the axis of the links) dynamic loads induced by the self-weight of the links in the cross-sections of the links of the RRRRT manipulator.

The diagram of the longitudinal dynamic load distributed along the Y_1 axis of the first link is depicted on the Y_1Z_1 plane and is colored blue. The direction of the distributed longitudinal dynamic loads in the cross-sections, where the ordinates are oriented opposite to the Z_1 axis, is also directed against the Y_1 axis. Conversely, the direction of the distributed longitudinal dynamic forces in the cross-sections, where the ordinates are oriented toward the Z_1 axis, corresponds to the direction of the Y_1 axis.

The diagram of the longitudinal dynamic load distributed along the X_2 axis of the second link is presented on the X_2Z_2 plane in magenta. If the ordinates are directed toward the Z_2 axis, the distributed longitudinal dynamic loads in these cross-sections are aligned with the X_2 axis. Conversely, if the ordinates are directed opposite to the Z_2 axis, the load is oriented in the opposite direction.

The diagram of the longitudinal dynamic load distributed along the Z_3 axis of the third link is drawn in blue on the Y_3Z_3 plane. If the ordinates are directed toward the Y_3 axis, the distributed longitudinal dynamic loads in these cross-sections are aligned with the Z_3 axis. Conversely, if the ordinates are directed in the opposite direction, the loads will be oriented in the opposite direction.

The diagram of the longitudinal dynamic load distributed along the Z_4 axis of the fourth link is presented on the Y_4Z_4 plane in magenta. If the ordinates are directed toward the Y_4 axis, the direction of the distributed longitudinal dynamic loads in these cross-sections will correspond to the Y_4 axis. Conversely, if they are directed in the opposite direction, the loads will be oriented accordingly in the opposite direction.

The diagram of the longitudinal dynamic force distributed along the Z_5 axis of the fifth link is presented on the Y_5Z_5 plane in blue. If the ordinates are directed toward the Y_5 axis, the direction of the distributed longitudinal dynamic loads in these cross-sections will correspond to the Y_5 axis. Conversely, if they are directed in the opposite direction, the loads will be oriented accordingly in the opposite direction.

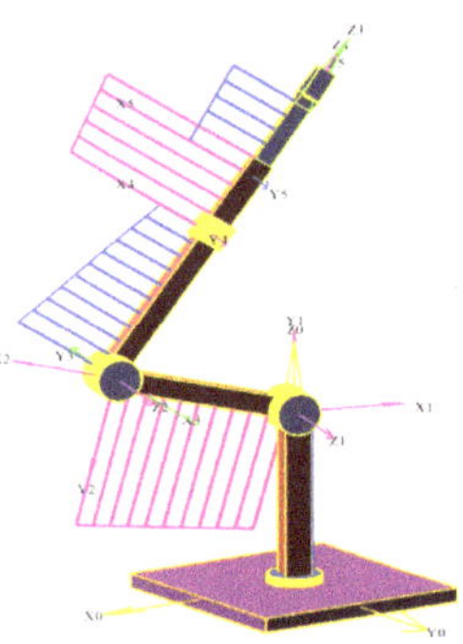

Figure 7. The patterns of distribution of transversely vertically distributed dynamic loads, arising from the self-weight of the links, in the cross-sections of the links of the RRRRT manipulator are presented.

The diagram of the transverse vertical dynamic load distributed along the X_2 axis of the second link is depicted in magenta on the X_2Y_2 plane. If the ordinates are directed toward the Y_2 axis, the distributed transverse vertical dynamic loads in these sections are also oriented toward the Y_2 axis. Conversely, if the ordinates are directed in the opposite direction, the distributed transverse vertical dynamic loads will be oriented in the direction opposite to the Y_2 axis.

The diagram of the transverse vertical dynamic load distributed along the Z_3 axis of the third link is shown in blue on the Y_3Z_3 plane. If the ordinates are directed toward the Y_3 axis, the distributed transverse vertical loads in these sections are aligned with the Y_3

axis. Conversely, if they are directed in the opposite direction, the distributed transverse vertical dynamic loads will be oriented against the Y_3 axis.

The diagram of the transverse vertical dynamic load distributed along the Z_4 axis of the fourth link is presented in magenta on the Y_4Z_4 plane. If the ordinates are directed toward the Y_4 axis, the distributed transverse vertical dynamic loads in these sections are also aligned with the Y_4 axis. Conversely, if the ordinates are directed in the opposite direction, the distributed transverse vertical dynamic loads will be oriented against the Y_4 axis.

The diagram of the transverse vertical dynamic load distributed along the Z_5 axis of the fifth link is depicted in blue on the Y_5Z_5 plane. If the ordinates are directed toward the Y_5 axis, the distributed transverse vertical dynamic loads in these sections are aligned with the Y_5 axis. Conversely, if they are directed in the opposite direction, the distributed vertical dynamic loads will be oriented against the Y_5 axis.

The diagram of the horizontal dynamic load distributed along the X_2 axis of the second link is presented in magenta on the X_2Z_2 plane. If the constructed ordinates are directed toward the Z_2 axis, the intensity of the horizontally distributed dynamic loads in these sections is also directed toward the Z_2 axis. Conversely, if the ordinates are directed opposite to the Z_2 axis, the intensity of the distributed horizontal dynamic loads is directed against the Z_2 axis.

The diagram of the transverse horizontal dynamic load distributed along the Z_3 axis of the third link is depicted in blue on the Y_3Z_3 plane. The ordinates constructed along the Y_3 axis indicate the magnitude and direction of the intensity of the distributed horizontal dynamic load in these sections. If the ordinates are directed toward the Y_3 axis, the distributed horizontal dynamic loads in these sections are also directed toward the Y_3 axis. Conversely, if the ordinates are directed oppositely, the distributed horizontal dynamic loads in these sections are oriented in the opposite direction from the Y_3 axis.

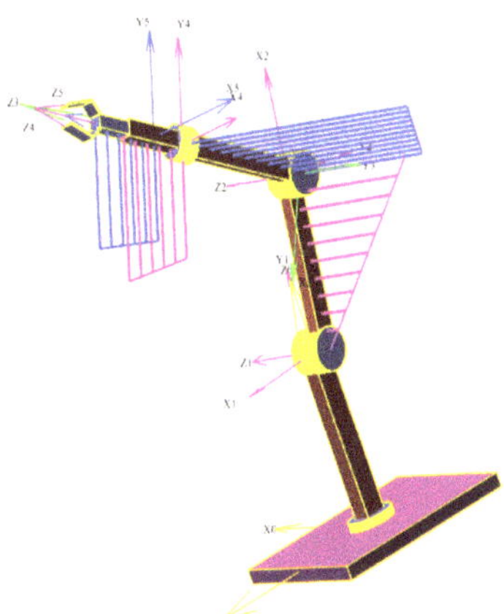

Figure 8. The diagrams of horizontally distributed dynamic loads arising from the self-weight in the cross-sections of the RRRRT manipulator links are presented.

The diagram of the transverse horizontal dynamic load distributed along the Z_4 axis is represented in magenta on the Y_4Z_4 plane. If the obtained ordinates are directed toward the Y_4 axis, the distributed horizontal dynamic loads in these sections are also directed toward the Y_4 axis. Conversely, if they are directed oppositely, the distributed horizontal dynamic loads in these sections are oriented in the opposite direction toward the Y_4 axis.

The diagram of the transverse horizontal dynamic load distributed along the Z_5 axis is shown in blue on the Y_5Z_5 plane. If the obtained ordinates are directed toward the Y_5 axis, then the distributed horizontal dynamic loads in these sections are also directed toward the Y_5 axis. If they are directed oppositely, then the direction of the horizontal dynamic loads in these sections will be opposite to the direction of the Y_5 axis.

The diagram of the transverse horizontal dynamic load distributed along the Z_5 axis is depicted in blue on the Y_5Z_5 plane. If the obtained ordinates are directed toward the Y_5 axis, the distributed horizontal dynamic loads in these sections are also oriented toward the Y_5 axis. Conversely, if they are directed oppositely, the direction of the horizontal dynamic loads in these sections will be opposite to the Y_5 axis direction.

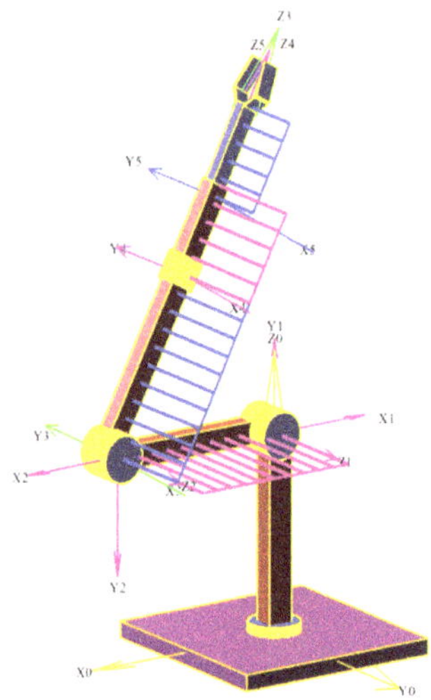

Figure 9. The distributed dynamic torsional moments arising due to the self-mass of the links during their rotation about their own axes are presented in the cross-sections of the links of the RRRRT manipulator.

The distributed torque moments along the second link are represented in the X_2Z_2 plane in magenta. The direction and magnitude of the torque moment in this section are indicated by ordinates aligned along the Z_2 axis. If the ordinate is directed along Z_2, the torque moment in the section induces a counterclockwise rotation about the X_2 axis. Conversely, if the ordinate is oriented in the opposite direction, the section rotates clockwise about the X_2 axis.

The dynamically distributed torque moments on the third link are represented in the Y_3Z_3 plane in blue. If the ordinates are directed toward the Y_3 axis, the distributed moments in these sections induce a clockwise rotation about the Z_3 axis. Conversely, if the ordinates are oriented opposite to the Y_3 axis, the distributed dynamic loads cause a counterclockwise rotation of the section about the Z_3 axis.

The dynamically distributed torque moments on the fourth link are depicted in the Y_4Z_4 plane in magenta. If the ordinates are directed towards the Y_4 axis, the distributed moments in these sections induce a clockwise rotation about the Z_4 axis. Conversely, if the ordinates are oriented in the opposite direction to the Y_4 axis, the distributed dynamic loads cause the section to rotate counterclockwise about the Z_4 axis.

The dynamically distributed torque moments on the fifth link are represented in the Y_5Z_5 plane in blue. If the ordinates are directed towards the Y_5 axis, the distributed moments in these sections induce a clockwise rotation about the Z_5 axis. Conversely, if the ordinates are oriented in the opposite direction to the Y_5 axis, the distributed dynamic loads cause the section to rotate counterclockwise about the Z_5 axis.

4. Verification

The spatial positions and orientations of the links of the RRRRT manipulator can be observed in a visualized 3D computer model created using the developed algorithm and software code in the Maple 2023 environment. Verification is performed by comparing the positions and orientations of these models with extracted numerical values. The kinematic parameters of the manipulator, necessary for determining dynamic loads, were calculated using the recursive Newton–Euler formulas and verified using homogeneous transformation matrices. The diagrams of dynamic loads on the links comply with the distribution laws of the obtained loads.

5. Results and Discussion

Algorithms and software codes were developed and implemented in the Maple 2023 environment to construct three-dimensional computer models of manipulators, providing comprehensive visibility from all perspectives in three-dimensional space. By defining specific patterns of generalized coordinates for the obtained manipulators, their motion in space can be observed.

The Denavit–Hartenberg method was employed to determine the positions and orientations of the links of a five-degree-of-freedom manipulator (RRRRT), which consists of six links connected by one prismatic kinematic and four revolute pairs. The vectors of linear velocities, accelerations, angular velocities, and angular accelerations of the manipulator's link points were determined relative to the Denavit–Hartenberg coordinate systems attached to the links using recursive Newton–Euler equations. The components of these vectors enabled the determination of all four types of dynamic loads and the identification of their distribution patterns.

Algorithms and software codes were developed for determining the distributed dynamic loads occurring in the cross-sections of the RRRRT manipulator links, as well as for visualizing these loads in the form of diagrams illustrating their distribution within the manipulator links throughout the complete operating cycle. The article does not include the codes for computer 3D modeling of the manipulator, dynamic load determination, or the construction of their diagrams.

6. Summary

This study proposes an approach to 3D modeling of spatial manipulators in the Maple 2023 software environment. Algorithms and program codes have been developed to generate computer-based 3D models of manipulators controlled by generalized coordinates. The implementation of these algorithms and codes has enabled the creation of three-dimensional computer models of manipulators with clear visual representations of links, their cross-sections, kinematic pairs, grippers, and loads, varying in structure and degrees of freedom, while ensuring a comprehensive spatial view from all perspectives.

To control the motion of the created computer-based 3D models of manipulators using generalized coordinates, the Denavit–Hartenberg method was applied. Using Newton–Euler recursive equations, the necessary kinematic characteristics of the manipulator links were determined for calculating distributed dynamic loads arising from the self-weight of the links at each cross-section, relative to local coordinate systems rigidly attached to the links.

Algorithms and program codes have been developed for controlling the motion of the 3D manipulator models, as well as for constructing visual diagrams of distributed dynamic loads in mutually perpendicular planes formed by the principal axes of the link cross-sections and axes passing along the longitudinal axes of the links. The implementation of these algorithms and codes enabled the generation of distribution diagrams for all dynamic loads in all links of the moving manipulator. These diagrams visually represent the variations in direction and magnitude of distributed dynamic loads across all link cross-sections throughout the full operational cycle of the manipulator. This approach allows for the consideration of identified dynamic loads in strength and stiffness calculations of manipulator links, which is crucial for designing new innovative manipulators.

Author Contributions: Conceptualization, M.U.; methodology, M.U., T.S., and Y.T.; software, M.U., S.Z., S.P., D.B., and Z.Z.; validation, M.U., S.Z., and S.P.; formal analysis, M.U., Y.T., S.Z., and D.B.; investigation, M.U., Y.T., S.Z., and D.B.; resources, D.B., and Z.Z.; data curation, M.U., and S.Z.; writing—original draft preparation, M.U., and S.Z.; writing—review and editing, M.U., S.Z., and S.P.; visualization, M.U., S.Z., S.P., D.B., and Z.Z.; supervision, M.U., and T.S.; project administration, D.B.; funding acquisition, D.B. All authors have read and agreed to the published version of the manuscript.

Funding: This research was funded by the Science and Higher Education Quality Assurance Committee of the Ministry of Science and Higher Education of the Republic of Kazakhstan, grant number AP22686476.

Data Availability Statement: The data presented in this study are available upon request from the corresponding author.

Conflicts of Interest: The authors declare no conflicts of interest.

References

1. Wu, H.; Deng, H.; Yang, C.; Guan, Y.; Zhang, H.; Li, H. A Robotic Off-line Programming System Based on SolidWorks. In Proceedings of the 2015 IEEE Conference on Robotics and Biomimetics, Zhuhai, China, 6–9 December 2015; pp. 1711–1716.
2. Cekus, D.; Posiadała, B.; Waryś, P. Integration of modeling in solidworks and matlab/simulink environments. *Arch. Mech. Eng.* **2014**, *61*, 57–74. [CrossRef]
3. Kalinichenko, S.V.; Konovalov, K.V.; Sevostyanov, N.E.; Sevostyanov, I.E. Simulation in MATLAB of a vertical walking three-link robot. *AIP Conf. Proc.* **2019**, *2195*, 020008. [CrossRef]
4. Utenov, M.U.; Baigunchekov, Z.Z.; Zhilkibayeva, S.K.; Utenov, N.M. Computational method of determination of internal efforts in links of mechanisms and robot manipulators with statically definable structures considering the distributed dynamically loadings. In Proceedings of the ECCOMAS Congress 2016, Crete Island, Greece, 5–10 June 2016; Volume IV, pp. 8627–8639.
5. Utenov, M.; Zhilkibayeva, S.; Utenov, N. Animation of motion of mechanisms and robot manipulators in the Maple system with the construction of diagrams of internal forces on the links. In Proceedings of the of the 2nd International Conference on Robotics, Control and Automation (ICRCA 2017), Kitakyushu, Japan, 15–18 September 2017; pp. 30–34. [CrossRef]
6. Utenov, M.; Sobh, T.; Baigunchekov, Z.; Zhilkibayeva, S.; Patel, S. Analytical method for determination of internal forces of mechanisms and manipulators. *MDPI Robot.* **2018**, *7*, 53. [CrossRef]
7. Fu, K.; Gonzalez, R.; Lee, K. *Robotics*; Mir Publishers: Moscow, Russia, 1989; 624p.
8. Shahinpur, M. *Course in Robotics*; Publishing House Mir: Moscow, Russia, 1990; 527p.
9. Zhang, X.; Sørensen, R.; Iversen, M.R.; Li, H. Computationally efficient dynamic modeling of robot manipulators with multiple flexible-links using acceleration-based discrete time transfer matrix method. *Robot. Comput.—Integr. Manuf.* **2018**, *49*, 181–193. [CrossRef]
10. Arkouli, Z.; Aivaliotis, P.; Makris, S. Towards accurate robot modelling of flexible robotic manipulators. *Procedia CIRP* **2021**, *97*, 497–501. [CrossRef]
11. Hull, J.; Turner, R.; Asbeck, A.T. Design and preliminary evaluation of two tool support arm exoskeletons with gravity compensation. *Mech. Mach. Theory* **2022**, *172*, 104802.
12. Nguyen, V.L.; Kuo, C.-H.; Lin, P.T. Performance analysis of gravity-balanced serial robotic manipulators under dynamic loads. *Mech. Mach. Theory* **2024**, *191*, 105519.
13. van der Wijk, V. Design and analysis of closed-chain principal vector linkages for dynamic balance with a new method for mass equivalent modeling. *Mech. Mach. Theory* **2020**, *150*, 103815. [CrossRef]
14. Nguyen, V.L. A design approach for gravity compensators using planar four-bar mechanisms and a linear spring. *Mech. Mach. Theory* **2022**, *172*, 104770.
15. Zhou, Y.; Luo, J.; Wang, M. Dynamic coupling analysis of multi-arm space robot. *Acta Astronaut.* **2019**, *160*, 583–593. [CrossRef]
16. Zamora-Ortiz, P.; Carral-Alvaro, J.; Valera, Á.; Pulloquinga, J.L.; Escarabajal, R.J.; Mata, V. Identification of Inertial Parameters for Position and Force Control of Surgical Assistance Robots. *Mathematics* **2021**, *9*, 773. [CrossRef]

17. Wang, L.; Wang, D.; Wu, J. Dynamic performance analysis of parallel manipulators based on two-inertia-system. *Mech. Mach. Theory* **2019**, *137*, 237–253. [CrossRef]
18. Shao, Z.-F.; Tang, X.; Chen, X.; Wang, L.-P. Research on the inertia matching of the Stewart parallel manipulator. *Robot. Comput.-Integr. Manuf.* **2012**, *28*, 649–659.
19. Zhao, Y.-S.; Ren, J.-Y.; Huang, Z. Dynamic loads coordination for multiple cooperating robot manipulators. *Mech. Mach. Theory* **2000**, *35*, 985–995. [CrossRef]
20. Korayem, M.H.; Dynamic, A.B. load carrying capacity of robotic manipulators with joint elasticity imposing accuracy constraints. *Robot. Auton. Syst.* **1994**, *13*, 219–229. [CrossRef]

robotics

MDPI

Article

Dimensional Synthesis of Parallel Robots Using Bilevel Optimization for Design Optimization and Resolution of Functional Redundancy

Moritz Schappler

Institute of Mechatronic Systems, Leibniz University Hannover, 30823 Garbsen, Germany;
moritz.schappler@imes.uni-hannover.de

Abstract: Parallel-kinematic machines or parallel robots have only been established in a few applications where their advantage over serial kinematics due to their high payload capacity, stiffness, or dynamics with their limited workspace-to-installation-space ratio pays out. However, some applications still have not yet been sufficiently or satisfactorily automated in which parallel robots could be advantageous. As their performance is much more dependent on their complex dimensioning, an automated design tool—not existing yet—is required to optimize the parameterization of parallel robots for applications. Combined structural and dimensional synthesis considers all principally possible kinematic structures and performs a separate dimensioning for each to obtain the best task-specific structure. However, this makes the method computationally demanding. The proposed computationally efficient approach for dimensional synthesis extends multi-objective particle swarm optimization with hierarchical constraints. A cascaded (bilevel) optimization includes the design optimization of components and the redundancy resolution for tasks with rotational symmetry, like milling. Two case studies for different end-effector degrees of freedom demonstrate the broad applicability of the combined structural and dimensional synthesis for symmetric parallel robots with rigid links and serial-kinematic leg chains. The framework produces many possible task-optimal structures despite numerous constraints and can be applied to other problems as an open-source MATLAB toolbox.

Keywords: parallel robot; parallel-kinematic machine; dimensional synthesis; bilevel optimization; constraint handling; design optimization; redundancy resolution

1. Introduction

Despite a research history of more than three decades and potential theoretical advantages [1], *parallel robots* have not been successful in the market yet except for single examples [2]. Parallel robots can, in principle, encounter some disadvantages of serial robots, especially regarding dynamics, accuracy, moving mass, and related energy consumption. Since many problems of fundamental research have already been solved [2–4], the still high and increasing publication activity on rigid parallel robots can be attributed to reaching broader (yet non-commercial) applications. *Further research on their optimal design is required*, as optimal and task-specific design is essential for the deployment of parallel robots to provide new and improved automation of tasks. This paper *contributes* to that field by *combining methods* for structural and dimensional synthesis (Section 1.1), resolution of functional redundancy (Section 1.2), and design optimization (Section 1.3) in one *framework* based on the concept of *bilevel optimization*. There, two optimization problems are nested into each other, which holds for engineering design problems [5], such as robot synthesis.

1.1. Structural and Dimensional Synthesis of Parallel Robots

The systematic *structural synthesis* of parallel robots (for finding the number and alignments of joints and kinematic chains called legs) was driven by mathematical formulations such as screw theory [6], the theory of linear transformations [7], and other concepts to take into account the higher complexity of parallel over serial robots. The importance of the *dimensional synthesis* (for determining numerical values for the parameters) was highlighted by several authors, e.g., [8] (p. 124), [9] (p. 175), and [1] (p. 25), as being at least as important for the robot performance as the structural synthesis. This distinguishes parallel from serial robots, where the dimensioning is much more intuitive. Specific further studies on dimensional synthesis have been performed for single parallel robots, e.g., [9–19]. Many more references could extend the non-exhaustive list since proper dimensioning is necessary for all parallel-robot prototypes. The concept of *combined structural and dimensional synthesis* optimizes multiple structures simultaneously and only then selects the best structure. This was first realized by [9,10] in a basic example of a few non-redundant parallel robots. The concept was only pursued systematically by a few others, such as [20] for three-DoF parallel robots, *leaving aspects like functional redundancy uncovered.*

1.2. Functional Redundancy

The operational space is the physical motion space of the robot end effector, and the task space is the sub-set coordinate space to define a robot task. These spaces are congruent for most but not all tasks [21] (p. 88). Many *industrially relevant tasks* like welding, milling, or drilling only require five degrees of freedom (DoF) in their task space, while one coordinate of the six-DoF operational space can be set arbitrarily, resulting in *functional redundancy*. The aspect was mainly investigated for serial robots. For parallel robots, kinematic models are more complex and the minimal coordinates are already the operational-space coordinates; therefore, fewer publications use dedicated models [22] but rather incorporate the aspect into existing models by gradient-descent [23] or within an overlying optimization [24–29], which can become computationally expensive.

Functional redundancy has not yet been considered explicitly *as part of the dimensional synthesis* of parallel robots. In the structural synthesis, no dedicated consideration is necessary since the operational space has to be regarded there. Tasks or manipulators with five degrees of freedom are missing in the relevant works on *combined synthesis*, e.g., in [9,20,30]. Functional (or task) redundancy may look counter-intuitive for task-specific robots at first. Including this in the synthesis is motivated by many parallel robots having six end-effector degrees of freedom. In contrast, five DoFs are difficult to realize. Excluding functionally redundant robots from the synthesis *would strongly limit the combined-synthesis approach for these most relevant tasks.* As shown in the literature, *exploiting functional redundancy is critical for the performance and feasibility of robots in these tasks.*

1.3. Design Optimization

Lifting the structural assumptions from the literature requires a more detailed look at those non-kinematic parameters of parallel robots subject to *design optimization*. Higher internal bending moments of multi-joint leg chains may require stronger dimensioning of limbs, which increases self-collisions. Required high actuator torques may lead to bulky motors that can not be integrated into the construction anymore. Static compensation springs may reduce the latter effect if they are well designed, eventually favored by a specific kinematic structure. Including these aspects is *necessary for a robust comparison of one kinematic structure to another* and is investigated together with the other facets.

1.4. Motivation and Structure of the Paper

Methods for the *combined structural and dimensional synthesis* of serial and parallel robots have been proposed and provide *well-argued advantages over a classical two-stage synthesis*. Furthermore, many methods exist for the resolution of functional redundancy. However, the two concepts have yet to be combined even though *tasks enabling functional redundancy are the most industrially relevant type of task*. Moreover, design optimization is mainly covered for serial rather than parallel robots. By missing tools for systematic optimization, the performance of parallel robots cannot be fully exploited, and *industrial tasks may be realized in a sub-optimal way or not at all if no robot exists for them yet*.

The overall workflow of the combined robot synthesis is summarized in Figure 1. Next; the single aspects are elaborated in more detail in comparison to existing works in Section 2. The materials and methods to encounter the gaps in the literature are the subject of Section 3, followed by results in the form of two case studies in Sections 4 and 5 and a discussion and conclusions in Sections 6 and 7.

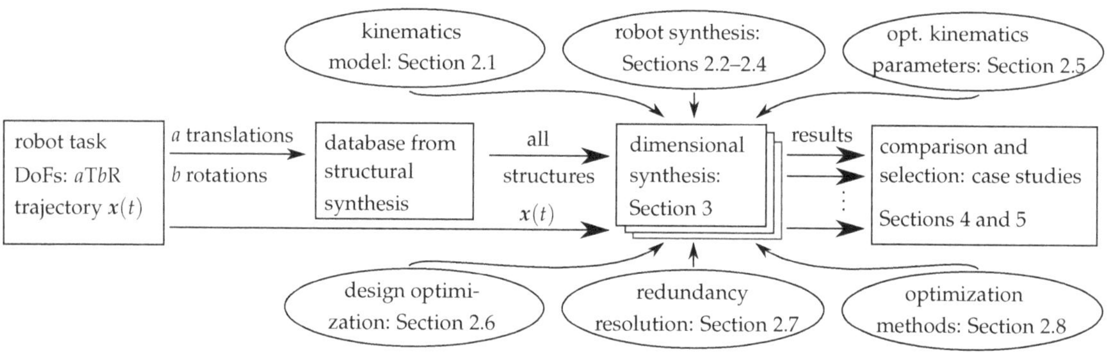

Figure 1. Overview of the procedure for combined structural and dimensional synthesis, which structures the paper. Abbreviations: degree of freedom (DoF); optimal (opt.)

2. Related Work

The literature relevant to the dimensional robot synthesis is discussed in the following subsections. First, preliminaries regarding the kinematics modeling of parallel robots are introduced in Section 2.1. The proposed framework of hierarchical optimization from Section 3 structures the remaining text. It is separated into an overview of approaches for the robot-synthesis problem in general in Sections 2.2–2.4, followed by the optimization of kinematic parameters (Section 2.5), other design parameters (Section 2.6), and robot motion (Section 2.7). Optimization methods are discussed in Section 2.8 regarding their usability within the different optimization problems of the robot synthesis.

2.1. Preliminaries on Parallel Robots: Functional Redundancy and Kinematic Constraints

For a robot manipulator, *intrinsic redundancy* exists if the joint space has a higher dimension than the operational space. *Functional redundancy* is defined as an operational-space dimension higher than the task-space and the joint-space dimension [21,31–33]. Both forms of redundancy are combined under the term "kinematic redundancy" [21] (p. 88).

The occurrence of functional redundancy is illustrated in Table 1, where the most relevant spatial robot operational-space degrees of freedom (in rows; "robot DoF") and task-space DoFs (in columns; "task DoF") are combined in a matrix, and the cells show the effect on the degree of functional redundancy n_{R}. In some tasks, n_{C} additional constraints have to be defined, which implicitly corresponds to the definition of coordinates.

Table 1. Overview of combinations of robot end-effector DoFs (rows) and task DoFs (columns) with the degree of redundancy n_R (highlighted in blue) and number of resulting constraints n_C. If not given, $n_R = 0$ or $n_C = 0$ holds. Naming of tasks f.l.t.r.: rotation fixed (3T0R); only planar rotation that is arbitrary (3T0*R); only planar rotation that is defined (3T1R), pointing (3T2R); free motion in space (3T3R). The symbol "×" marks impossibility and row and column headings use bold font.

Task DoFs →	**3T0R**	**3T0*R**	**3T1R**	**3T2R**	**3T3R**
Robot DoFs ↓	**(Rot. Fix.)**	**(Rot. Arbitr.)**	**(Rot. Def.)**	**(Pointing)**	**(Free)**
3T0R (3)	$n_R = 0$	$n_R = 0$	×	×	×
3T0*R (3)	×	$n_R = 0$	×	×	×
3T1R (4)	$n_C = 1$	$n_R = 1$	$n_R = 0$	×	×
3T2R (5)	$n_C = 2$	$n_C = 2$	$n_C = 1$	$n_R = 0$	×
3T3R (6)	$n_C = 3$	$n_C = 2$; $n_R = 1$	$n_C = 2$	$n_R = 1$	$n_R = 0$

Redundancy for *parallel robots* has been extensively studied for the aspects of *kinematic redundancy* (additional joints in a leg chain) or *actuation redundancy* (additional actuators or actuated leg chains) [34,35]. *Functional redundancy* for parallel robots (more platform DoFs than required by the task) is distinguished since parallel robots are characterized by the platform DoFs as minimal coordinates, unlike serial robots, characterized by the joint DoFs. Functional redundancy is addressed less in the literature and is not mentioned in the otherwise extensive review articles [34,35] on redundancy for parallel robots. In the following, one degree of functional redundancy is assumed, corresponding to the tool rotation in machining tasks (3T2R). Some references show the possibility of improving the performance and avoiding limits by exploiting functional redundancy, such as for a planar 3-RRR robot [23,28,36,37], for a 6-PUS [25] and 6-UPS robot [24,25,29], and a combination of two three-DoF parallel robots [26]. The notation for parallel robots includes the number of identical leg chains and the type of their joints counted from the base: revolute (R), prismatic (P), universal (U), and spherical (S). Underlining denotes an actuated joint. In total, the possible improvement can be assumed to extend to all parallel robots since they are susceptible to singularities, joint limitations, and/or self-collisions.

A general approach for a kinematics model of parallel robots is reported in introductory textbooks [1,7,38]. The considered parallel robot consists of m legs with n_i joint coordinates $\boldsymbol{q}_i$ each, connected at a moving platform. Additionally to the active joints $q_{\mathrm{a},i}$, explicitly all passive joints $q_{\mathrm{p},i}$, including the platform-coupling joints, are part of the coordinates $\boldsymbol{q}_i$ of leg i. All n_q joint coordinates of the parallel robot are stacked in the vector $\boldsymbol{q} = [\boldsymbol{q}_1^{\mathrm{T}}, \ldots, \boldsymbol{q}_m^{\mathrm{T}}]^{\mathrm{T}}$, and all n_{q_a} active joints are assembled in the vector $\boldsymbol{q}_\mathrm{a} = [q_{\mathrm{a},1}, \ldots, q_{\mathrm{a},m}]^{\mathrm{T}}$.

A sketch of the assembly of the robot's leg chains (corresponding to solving the inverse-kinematics problem) is depicted in Figure 2a with the necessary coordinate systems for kinematic modeling adapted from [38], including a robot base frame $(\mathrm{CS})_0$, leg-chain base frames $(\mathrm{CS})_{\mathrm{A}_i}$, and cut coupling-joint frames $(\mathrm{CS})_{\mathrm{C}_i}$ at the end of the leg chains. The desired pose of the moving-platform frame $(\mathrm{CS})_\mathrm{D}$ is expressed with the minimal coordinates $\boldsymbol{x}$.

The relation between joint coordinates $\boldsymbol{q}$ and platform coordinates $\boldsymbol{x}$ is established with the kinematic-constraint equations, for which, most commonly, the vector loop

$$\boldsymbol{\delta}_{\mathrm{t},i}(\boldsymbol{q}_i, \boldsymbol{x}) = -{}_{(0)}\boldsymbol{r}_{\mathrm{A}_i\mathrm{B}_i}(\boldsymbol{x}) + {}_{(0)}\boldsymbol{r}_{\mathrm{A}_i\mathrm{C}_i}(\boldsymbol{q}_i) \overset{!}{=} \boldsymbol{0} \quad (1)$$

between the position of the platform-coupling point B_i relative to the base-coupling point A_i is used for each leg chain i [1]. If the inverse-kinematics problem is solved with $\boldsymbol{\delta} = \boldsymbol{0}$, the differential kinematics of the parallel robot are calculated with the time derivative

$$\frac{\mathrm{d}}{\mathrm{d}t}\boldsymbol{\delta}(\boldsymbol{q}_\mathrm{a}, \boldsymbol{x}) = \boldsymbol{\delta}_{\partial \boldsymbol{q}_\mathrm{a}}\dot{\boldsymbol{q}}_\mathrm{a} + \boldsymbol{\delta}_{\partial \boldsymbol{x}}\dot{\boldsymbol{x}} \overset{!}{=} \boldsymbol{0}, \qquad \dot{\boldsymbol{q}}_\mathrm{a} = -\boldsymbol{\delta}_{\partial \boldsymbol{q}_\mathrm{a}}^{-1}\boldsymbol{\delta}_{\partial \boldsymbol{x}}\dot{\boldsymbol{x}} = \boldsymbol{J}^{-1}\dot{\boldsymbol{x}} \quad \text{with} \quad \boldsymbol{\delta}_{\partial \boldsymbol{x}} = \frac{\partial \boldsymbol{\delta}}{\partial \boldsymbol{x}}. \quad (2)$$

The passive-joint coordinates q_P do not occur since they were eliminated in a previous step and a minimal set of constraints δ is defined out of the $\delta_{\mathrm{t},i}$ of (1).

As is discussed in more detail in the author's previous work [22], the constraints (1) cannot be used in the case of functional redundancy and have to be replaced by the kinematic constraints $\boldsymbol{\psi}^\mathrm{T} = \begin{bmatrix} \boldsymbol{\psi}_1^\mathrm{T} & \boldsymbol{\delta}_2^\mathrm{T} & \cdots & \boldsymbol{\delta}_m^\mathrm{T} \end{bmatrix}$ sketched in Figure 2b. The first leg chain is defined as the leading leg chain, and its constraints $\boldsymbol{\psi}_1$ are defined by the position and z-axis of $(\mathrm{CS})_\mathrm{E}$, corresponding to the tool center point and the tool axis, not to the full end-effector rotation. The second and further (following) leg chains with $\boldsymbol{\delta}_{2\cdots m}$ are defined relative to the first chain. Furthermore, all the legs' constraints include a rotational component expressed with Euler angles.

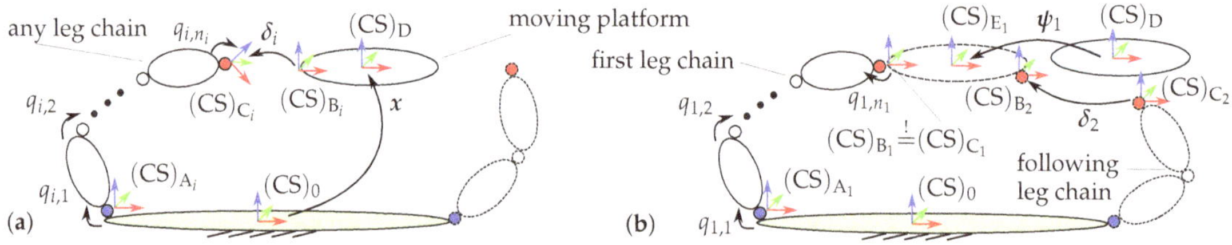

Figure 2. Sketch of the general parallel-robot kinematics for the conventional approach (**a**) and the modified model for functional redundancy with different definitions for first and following leg chains (**b**). Constraints are denoted by δ for the full set or by $\boldsymbol{\psi}$ if a component related to the redundant coordinate was removed.

2.2. *Requirements-Oriented Development Process*

The methods for designing a parallel-robot manipulator to a prototype degree can be obtained from the general design process for mechatronic systems, e.g., visualized by the *V-model* within the updated German standard VDI 2206:2021 [39]. The central aspect of this model is collecting lists of requirements, which can be structured in different ways, e.g., hierarchically, as summarized in [40]. This *requirements-oriented development process* is explained in [41] in the example of parallel robots. The scheme includes many experts and departments working together within the iterations in the design process. For providing an optimized kinematic structure of the parallel robot, the work of [9] is referenced. Dynamics and control are named as separate fields to incorporate a functional robot design.

The use of catalogs to structure expert knowledge is further detailed in [42]. The module-based approach within [8] presents a basis for the overall structural synthesis of parallel robots. The evaluation of constructional elements of parallel robots by [43] (chapter 6) is an example of the view of *design and construction* [41], which is necessary to realize a working prototype.

2.3. *Co-Design Formalism*

Another more mathematical approach for including different modular design problems into a systematic design process of any system is formalized by Censi in [44] within the *theory of co-design*. The design constraints imposed by the selection and dimensioning of subsystems are included in the overall design problem by a tuple of *implementation space* (e.g., selection of the motor for a robot), *functionality space* (torque and speed provided by the motor), and *resource space* (mass, cost, input voltage/current of the motor). The co-design problem can be expressed by a graph. Its solution is given as a Pareto front and represents the proposed implementations by the optimization algorithm. The design of a kinematic structure of a parallel robot within this framework can be seen as a design problem where the implementation consists of structural and dimensional synthesis. It is connected with other design problems, such as the link-dimensioning problem, the joint-construction/selection problem, and the drive-train-selection problem within the parallel-robot co-design problem. The single design problems are summarized in Figure 3.

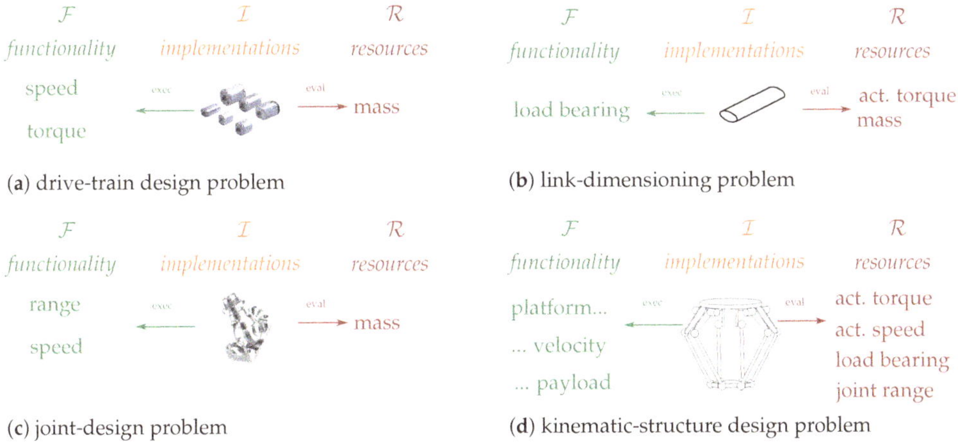

Figure 3. Single design problems within the parallel-robot synthesis in the notation of [44].

Without *recursive co-design constraints*, the parallel-robot design problem can be expressed hierarchically, where implementing the kinematic-structure design problem defines the actuator torque, actuator speed, passive joint range, and load bearing as required resources. They correspond to the functionalities provided by the drive train, the passive joints, and the link dimensioning. The single design problems can be connected similarly to the example in [44], as shown in Figure 4, using Censi's graphical notation, which should not be confused with a signal flow diagram or flow chart. The $\preccurlyeq$ sign for *partially ordered sets* means that, e.g., a kinematic structure requires *at least* a given actuator torque to provide the functionality of the platform payload. The recursive constraint is included by considering that the implementation of the robot's components influences the moving mass.

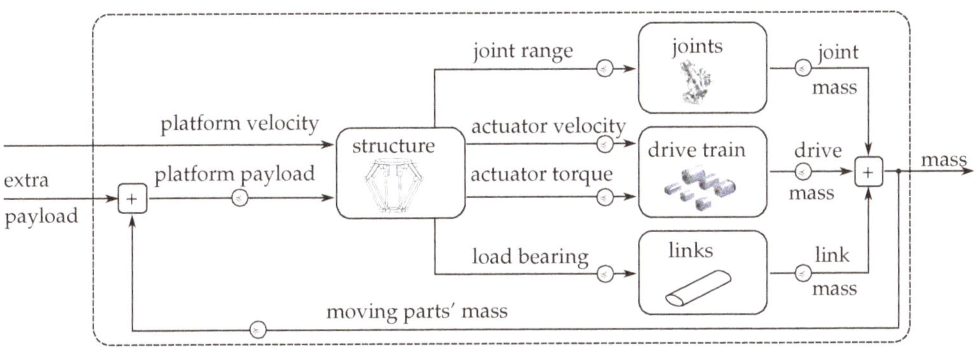

Figure 4. Parallel-robot synthesis as a simplified co-design problem in the notation of [44].

Other quantities could also be added to the diagram, such as the functionality of platform stiffness, which would also be affected by joint and link dimensioning. If considered, the functionality of position accuracy would be influenced mainly by the structure implementation. In a subsequent optimization, the problem graph is solved by a specific solver, and a Pareto diagram of the overall design problem is created from its functionalities (e.g., platform velocity and extra payload) and resources (e.g., mass).

The simplified visualization gives a good overview of the problem structure but obscures some dependencies within the problem, e.g., by summing up the payload and various masses. The kinematic structure influences the computation of the full inverse kinematics, inverse dynamics, and cut forces, which are all required to solve the single design problems. Due to the further *high number of optimization variables* within the kinematic-structure design problem, the concrete applicability of Censi's approach to the parallel-robot design problem has not been investigated further. The monotony of the relation between function and resource space would need to be shown to solve the recursive co-design problem with the solvers proposed by Censi. Instead, a more hierarchical approach is pursued, where the structure-design problem is the starting point, as discussed next.

2.4. Combined Structural and Dimensional Synthesis

The kinematic-structure design problem can be separated into the discrete problem of selecting the structure (*structural synthesis*) and the continuous problem of dimensioning the structure (*dimensional synthesis*). As elaborated above, other design problems can only be addressed if an implementation for the structure-design problem is chosen. In the *combined synthesis* proposed by [9], a kinematic structure is only selected after all principally suited structures have undergone a dimensional synthesis to evaluate their performance in the specific task requirements. For parallel robots, one difficulty lies in a useful pre-selection of structures with general design assumptions, e.g., discussed by [8,9,20]. The restrictions must be broad enough to ensure that structures with good properties are not excluded too early within the process and narrow enough to reduce the computational effort.

In [9], only the linear and classical Delta robots are compared for illustration in the combined synthesis, and the focus is on methods for the comparison of structures with different actuation, especially the application of Pareto optimization. In [20], the combined synthesis of parallel robots with three translational and no rotational degrees of freedom (3T0R DoFs) uses a filtered list of possible structures focusing on the number of joints and feasibility for construction. The optimization uses a single criterion with a weighted sum for multiple objectives and additional kinematics-oriented penalties for violating boundary conditions. Therefore, only one solution emerges for each robot in three tasks: handling, positioning, and guidance. The combined synthesis of serial robots in [30] uses a hierarchical approach for including constraints such as self-collisions within the objective function of the single-objective optimization. Similarly, ref. [45] define their "kinematic-structure design problem" as a loop of the "continuous design problem" over all (serial) structures.

To summarize, the general idea of a combined synthesis is straightforward but may not be pursued by many robot designers due to the *lack of suitable tools*, as stated by [20] (p. 14) in 2016. The lack of generalized optimization strategies within companies specialized in parallel robots, even for the dimensional synthesis of single architectures, such as the hexapod, was remarked on in 2012 by [18] (p. 69). To the author's best knowledge, this has not changed up to 2024—partly due to the lack of open-source culture within the academic community of robot modeling and design, partly due to the lack of commercial and industrial interest in such a software framework. Some publications refer to a specific self-developed software and document its results, but the software remains proprietary. Examples are the parallel-robot database in [46], the parallel-robot design catalog in [42], or the MATLAB kinematics analysis tool MAPS from [8] (p. 105), referenced in [42]. Some activities were recently started, e.g., a benchmark database for (serial-) robot synthesis [47] and published code for optimizing a given parallel-kinematic mechanism [48].

Many references in the literature handle the optimization of single robot structures, as shown in the next section. To perform the combined synthesis based on the underlying ideas, the computational efficiency of the respective tools has to be improved, and steps performed manually have to be carried out automatically since the computational and human capacities are otherwise exceeded. Therefore, a fast evaluation of a parameter set is necessary to solve the high-dimensional optimization problems in adequate time.

The *combined structural and dimensional synthesis* proposed in the conclusions of the theses of Frindt [8] (p. 123 ff.) and Krefft [9] (p. 175 ff.) included several limiting assumptions regarding the robot structure to reduce the complexity. Some of these assumptions can now be lifted, facing the increased computational capabilities twenty years after their research. Thereby, more unexpected solutions can be found within the synthesis. Furthermore, the previous (implicit) limitation of the synthesis to non-redundant robots is a restriction that can be abolished; therefore, functionally redundant structures are considered together with non-redundant structures.

2.5. Dimensional Synthesis of Kinematic Parameters

As stated by [1] and other authors, the performance of a *parallel robot* is strongly influenced by its dimensional parameters, which underlines the importance of *dimensional synthesis*. It can be interpreted in a stricter sense as the optimization of parameters influencing the robot's kinematics and, thereby, only indirectly the robot's dynamics. For a parallel robot, the dimensional parameters are the link lengths, base and platform diameters, and tilting angles of the joints [9,13,14,18,20,49–51]. Parameters solely influencing the dynamics (like link strength) are excluded from this strict definition. Within the leg chains, a perpendicular connection between two joints without skew alignments is assumed in the literature (e.g., [9,20]), which reduces the number of optimization parameters compared to the general serial leg chains, based on [7,30], used in this work.

Parametric CAD models were used to optimize the link dimensions of *serial robots* [52–54]. The approach mixes the optimization of kinematic parameters like link lengths and other design parameters such as material strength and link geometry. Furthermore, using CAD models and co-simulations or multidisciplinary design optimization [52] is too slow and too detailed for early design steps, which should be rather conceptual, cf. [55]. The design problem for open-loop structures does not include kinematic constraints that have to be met, and most of the tested parameters lead to valid solutions, which improves the convergence of the optimization.

More efficient implementations are possible with analytic models used in [13] to design serial and *parallel robots*. In [9], the kinematic parameters of a Hexa (6-R̲US) parallel robot and other structures are optimized based on analytic models and already existing drives. Similarly, ref. [10] optimizes and compares the hexapod (6-UP̲S), HexaGlide (parallel-rail 6-P̲US), and Linapod (vertical 6-P̲US) by Jacobian-based criteria. The results show improved workspace performance characteristics relative to the original design. In a combined "task-based" synthesis, the link lengths and other geometric parameters are optimized for several 3T0R parallel robots in [20] together with the link strength. In [51], four different 3T0R robots are optimized and compared. Several parallel robots were able to fulfill the reference tasks with no clear dominating solution for all tasks, which leads to the conclusion that optimization has to be performed for each reference task to obtain an optimal robot. Other authors perform their dimensional synthesis, e.g., on the linear Delta [11,14], the revolute (conventional) Delta [15,16], a 4-SP̲S [19], a Par4 structure (parallel-hybrid 3T1R Delta) [17], the hexapod (6-UP̲S) [18,49], or the HexaSlide (6-P̲US) [12].

The *robot's positioning relative to the task* primarily influences the performance of serial robots, which are usually placed beside it. Therefore, their dimensional synthesis often includes optimizing the base position [30,45,56–58]. The continuous base-position variables may be discretized for the synthesis of modular robots to adapt to the otherwise purely discrete optimization problem [58]. Depending on the problem formulation, the base position may be modeled and optimized as a "design joint" with zero motion [45]. For parallel robots, the base positioning is of less relevance due to the centered alignment of the robot above (or sometimes below) the task. If a synthesis for a given task is performed, only optimization of the vertical base position is performed [20]. If the performance is evaluated for the robot's workspace [9,10,18], the base position does not have to be considered.

2.6. Design Optimization: Drive Trains, Link Geometry, and Static Balance

The optimization of design parameters that only influence the dynamics and not the kinematics can be summarized within the *design optimization*. The single design problems are referenced in Figure 3.

For the *optimal selection of drive-train components* (see Figure 3a) separate from the kinematics parameters, these can be regarded as given. Optimizing the drive train first and then the link lengths like in [59] is only feasible if the link mass is ignored. Common objective functions are the drive-train mass [60–62], the trajectory cycle time [52], the predicted gear lifetime [63], or monetary costs [63]. Constraints are at least the speed and torque limits of the components [60–62]. It is usually sufficient to regard the inverse rigid-body dynamics without considering electric, thermal, or control effects. The extension of the robot simulation within the optimization has been carried out for thermal dynamics in [60] and electrical and controller dynamics in [64]. In contrast to the other design problems, the selection of motors and gears is a discrete optimization problem since a selection from a drive-train catalog is made [63].

The *optimization of the parameters of the robot links* (see Figure 3b), like the engineering design, can be achieved with parameterized CAD models prepared by a designer for the specific robot [52,53] or by using geometric primitives like hollow cylinders [20,30,65] for approximations. In the latter case, the parameters are, e.g., the diameter and strength of the hollow cylinders. Rounded links or links based on Hermite splines require additional parameters, which is disadvantageous for convergence of the optimization compared to straight links [66]. For CAD optimization, further parameters can, e.g., be the dimensioning and placement of slot holes [53] and the material strength [52].

The *selection and design of passive joints* for parallel robots (see Figure 3c) are driven mainly by the kinematics requirement of the kinematic structure. Selection from catalogs, as described in [43] (p. 162 ff.), has no substantial effect on the overall design problem since the joint mass is low relative to the rest of the structure. The individual development of joints for specific applications is possible [67,68], but general limits of the design alternatives can already be included as a filter into the joint-design problem. Then, a manual joint design can be made after the dimensional synthesis.

Some authors optimize kinematic and design parameters within the same optimization, e.g., [20,53,65]. Other references only cover design optimization with given kinematic parameters [52,61]. Design parameters are often chosen after the dimensional synthesis is performed or within an iterative but non-automated process, e.g., in [69]. A comparison of a single-level and a superior multi-level strategy is investigated, e.g., in [52], where the actuator type is optimized in an outer (global) optimization, and the link strength is optimized in an inner (local) optimization. However, the study only covers design and not dimensional variables. Hierarchical structuring of a dimensional and then a design optimization in an automated framework is mostly beneficial within a combined synthesis and has not been observed within the references listed here.

2.7. Optimizing Robot Motion Within the Dimensional Synthesis

If the dimensional synthesis is performed based on a simulated robot motion, optimizing this motion influences the performance evaluation and, thereby, the synthesis result. Motion optimization is possible if a redundant DoF is available (i.e., intrinsic or functional redundancy), if multiple solutions to the inverse-kinematics problem exist (e.g., to discard those leading to collisions), or if the motion is subject to parameterization (e.g., for point-to-point motion with freedom for the intermediate motion), cf. [70] (p. 54 ff.). Many of the references for parallel robots above perform a sampling-based performance evaluation for the complete workspace without a dedicated motion, which is more precise regarding some criteria but time-consuming. Therefore, handling robot motion is explicitly addressed in the following.

Some works exploiting *intrinsic redundancy* within the synthesis exist for *serial robots*. In [56], intrinsically redundant serial robots with up to ten joints are optimized for cluttered environments using a conjugate-gradient method with a problem formulation based on a Lagrangian function. As only the end-effector position with arbitrary orientation is considered, the cases are also functionally redundant. The focus is only on feasible, especially collision-free, solutions, which is ensured by a *point-to-point path planning* based on the probabilistic-roadmap method for a few reference positions.

A *concurrent optimization* of the inverse kinematics for several reference points and the design parameters of a kinematically redundant seven-DoF serial manipulator was investigated in [57], methodologically similar to kinematic calibration. Constraints from joint or workspace limitations were included by nullspace projection of one redundant DoF, not considering the additional functional redundancy of the investigated drilling task. A five-DoF arm was optimized similarly for a 3T2R task by [71], where collisions and pose errors were used as optimization constraints. In [45], the polynomial parameters of a collocated trajectory are optimized together with the design variables for a serial milling robot with three to six DoFs. The seven design variables modeled as virtual joints and the joint configuration of a 21-DoF hyper-redundant robot are optimized by sampling the design variables along the reference trajectories by a task-priority IK algorithm in [72], extending the similar approach without redundancy from [73]. The second step is to validate the candidate designs for the reference tasks. The deduction of design optimality by averaging locally optimal parameters seems questionable, at least for a transfer to parallel robots. The position and orientation error of the 3T0R task with arbitrary rotation is used as a scalar weighted-sum objective. The inability of the concurrent approach, discussed in this paragraph, to incorporate multiple IK solutions is admitted in the summary of [71].

The concurrent optimization is explicitly rejected due to the high number of variables in [74] for their *two-level optimization* of the design parameters and the poses of a four-DoF robot. The pose error for 16 reference points is taken as a common objective in both levels, which is necessary due to the 3T2R task definition with a scalar inequality constraint on the tilting orientation. The IK with joint constraints in a non-redundant case is solved in [75] by a lower-level particle swarm optimization to find all IK solutions. For the synthesis of modular serial robots with five and six DoFs, ref. [58] use a *two-stage approach* with a lower-level optimization of a quasi-static objective for independent robot postures for each sampled reference point of a 3T2R task. In [76], multiple circle trajectories were used within the dimensional synthesis of a seven-DoF serial manipulator, taking a closed-loop clamping weighted least-norm approach for the inverse kinematics as a variant of CLIK. The redundant DoF is only utilized for avoiding joint limits, and due to the pseudoinverse damping for stability near singularities (without avoiding them), the resulting trajectory tracking error has to be considered as an objective for the upper-level particle swarm optimization for the synthesis. Allowing tracking errors in this and other references complicates comparing multiple structures as fewer performance-oriented objectives can be optimized.

Some references cover a more general case of optimizing robot *trajectory motion* within a dimensional synthesis. In [77], this is regarded as a co-design problem using the mathematical description from the field of *bilevel optimization* [5]. The co-design formulation in [77] relates to the robot design and its motion, which are solved by a gradient-based optimization approach applied to an optimal-control dynamic motion planner. The problem strongly arises in the synthesis of legged robots handled and referenced there since gait trajectories are dependent on the leg kinematics, whereas, for the design of fixed-base manipulators, a reference task-space trajectory or pose can be assumed as given. Similarly, in [78], the states (joint positions, velocities, contact forces) of different legged robots are optimized at discretized knot points for dynamics integration together with the design parameters using analytic derivatives of objective and constraints.

To summarize, although some of the synthesis examples from the literature are for functionally redundant parallel robots (e.g., parallel-robot machine tools [50] (p. 147 ff.)), the author is not aware of any explicit exploitation of *functional redundancy* within the *parallel*-robot synthesis.

2.8. Methods for Optimization

The comparison of multiple parallel robots (whether two different structures or two sets of parameters for one structure) should be performed using *multiple criteria* [1], representing all robot requirements. Using a single objective (SO) poses the problem of selecting a compromise solution [9] (p. 112) and [19]. Combining multiple criteria within a weighted sum suitable for SO optimization raises the question of defining the weights [69]. A multi-objective (MO) border solution may be obtained by SO optimization when limit values for secondary objectives are known and are taken as constraints. SO optimization leads to feasible results for the dimensional synthesis of parallel robots in some cases, as shown by [49], [50] (p. 139 ff.) and [20] with a genetic algorithm (GA), and by [65,79,80] with the particle swarm optimization (PSO). In [19], SO results were found to be infeasible, and an MO optimization was performed for the design problem of a parallel-robot platform for ankle rehabilitation. Mainly, genetic algorithms were employed for *multi-objective optimization*, such as the strength Pareto evolutionary algorithm [9,14] or the nondominated sorting genetic algorithm [19,81]. Some authors do not specify their MO-GA variant, such as [69], only describe it [10] (p. 85 ff.) or refer to the MATLAB implementation [82], like [51]. *MO-PSO* has only been used more recently [83–86].

If (one of) the initial guess(es) for the kinematic parameters already leads to a valid solution, *gradient-based optimization methods* like SQP can also lead to good solutions [87], as reported by [88] for seven parameters of a serial robot, by [13] for optimization of a parallel robot with six parameters, or by [12] for seven parameters of a HexaSlide. In [18] (p. 78 ff.), the optimization of a hexapod with a steepest-ascent hill-climbing algorithm was performed based on a single workspace criterion. For *problems of higher dimension and without manually provided initial solutions*, like in the combined synthesis of parallel robots, these approaches are not likely to be functional in generality, especially for finding optima far away from the initial value [87]. Therefore, *evolutionary algorithms* such as the GA *are more feasible for the dimensional-synthesis problem* [9] (pp. 112 f., 141). Another argument for population-based methods like the GA is their ability to support multiple objectives [57], in contrast to gradient-based methods, designed mainly for SO problems. Other methods used in the context of parallel robots are ant colony optimization, the artificial bee colony algorithm, and differential evolution [89], or a combination of local (gradient-free Nelder–Mead simplex) and global search [48].

Particle swarm optimization is reported to have better results than genetic algorithms for constrained nonlinear optimization problems [90], such as dimensional synthesis. A general introduction and comparison of PSO and GA is given by [90]—and by [89] in the context of parallel-robot dimensional synthesis. Both algorithms are population-based and perform an optimization of a variable by evaluating a fitness function, which may be subject to constraints. The optimization variable is called particle (PSO) or gene (GA) and is usually multi-dimensional. So is the fitness function if an MO (Pareto) optimization is performed. An improvement of PSO over GA for parallel-robot dimensional synthesis was, e.g., recognized by [80,87]. One reason for better results of PSO is that not only do the parameters of the current iteration carry information but past iterations are also taken into account to generate a new set of parameters [91]. Furthermore, PSO is easier to implement and parameterize [84]. Avoiding the GA's selection step allows for gradually improving all particles [65].

Constraint handling [92,93] is central to the validity of the robot synthesis and the convergence of the PSO. From the existing constraint-handling strategies, the static-penalty approach is the simplest one. It adds a *constraint-violation penalty* term to the *objective function*, producing the *fitness function*. The penalty is static w.r.t. iteration number but can vary by the degree of violation or kind of constraint violated. This approach is already feasible for the dimensional synthesis. Other strategies not pursued in this work are, e.g., the exterior penalty, which continues the computation of constraint-violating particles, or the dynamic penalty, which varies with the iteration number.

Only for parallel robots of low complexity does a (partly) *manual optimization* of the robot parameters by a user seem promising to provide feasible results. The general requirement for an assistive design tool is an "interactive speed" with a reaction time of at most a minute, better seconds [77]. For instance, the graphical mechanism-analysis tool of [94] supports manual iterations of parallel-robot parameters by a pre- and postprocessor, which is sufficient for educational purposes of creating academic examples. The complexity of the optimization problem requires, in most cases, a more time-consuming, non-interactive optimization, as discussed above. A gradient-based optimization has the potential to be sufficiently fast as a basis for an interactive tool, as shown by [77] for mobile robots. Applying an approach with *symbolic* gradients, e.g., via CasADi [95], is hindered by the many discontinuous or complex objectives, cf. [48], due to the maximum function that needs numeric evaluation. This leaves only numeric derivatives, which can be expensive and inaccurate [78]. Moreover, due to the much stronger influence of constraints on the parallel-robot synthesis and the associated tendency of local optima [10] (p. 83), the gradient-based approach is not further pursued in favor of particle swarm optimization.

2.9. Summary of Existing Works and Contributions

The examples from the state of the art on parallel-robot dimensional synthesis (Section 2.5) show that good solutions can be found despite the optimization problem's high complexity. However, all authors except [20,51] perform an optimization of only one pre-selected structure (Section 2.4). For single-robot optimization, manual preparation is performed by carefully selecting initial values, parameter limits, and kinematic, dynamic, and stiffness models. A combination of dimensional synthesis and design optimization (Sections 2.3 and 2.6) for parallel robots is performed by [20,65] for one design sub-problem. Otherwise, only a few authors followed this approach and only for serial robots. The optimization of the robot's motion (Section 2.7) is only occasionally performed, not for parallel robots, and mainly for avoiding constraints or without systematic evaluation of the interplay of objectives. Paramount for the success of the optimization is a feasible structuring of the problem and a selection and configuration of the optimization algorithm (Section 2.8).

The research gap is addressed by the *contributions* of this paper: a framework for the combined structural and dimensional synthesis of parallel robots.

1. It includes design optimization—for the first time in parallel-robot synthesis in a bilevel (cascaded) optimization of two nested evolutionary algorithms.
2. For the first time, the resolution of functional redundancy is included in the synthesis of parallel robots—likewise by bilevel optimization but also using a classical algorithm.
3. The complete implementation using MATLAB is provided as open source [96].
4. Two case studies show the applicability to specific robot tasks with different degrees of freedom, next to three other case studies that were already published: [68,97,98].
5. The framework is validated against simulative results from the literature.

3. Materials and Methods

The algorithm structure developed in the following considers the high computational requirements within the combined robot synthesis, which arise from the repetition of the synthesis for many different architectures. It is necessary to perform a design optimization to compare different implementations of different architectures. As a general assumption, in a cascaded (bilevel) optimization, the inner optimization loop needs to converge to its optimum before evaluating the outer loop. The single cascaded optimization loops are the combined synthesis for *all* robots from a previously performed structural synthesis, the dimensional synthesis of *one* robot detailed next, the design optimization (Section 3.3), and the nullspace optimization of redundant degrees of freedom (Sections 3.4 and 3.5). A decoupling of the single optimization problems is attempted to reduce the complexity and computational effort of the co-design problem introduced in Section 2, Figure 4. In the following sections, the dimensional-synthesis optimization problem is elaborated in detail regarding the overall scheme in Section 3.1 and the definition of constraints, objectives, and optimization variables in Section 3.2, summarized in Figure 5.

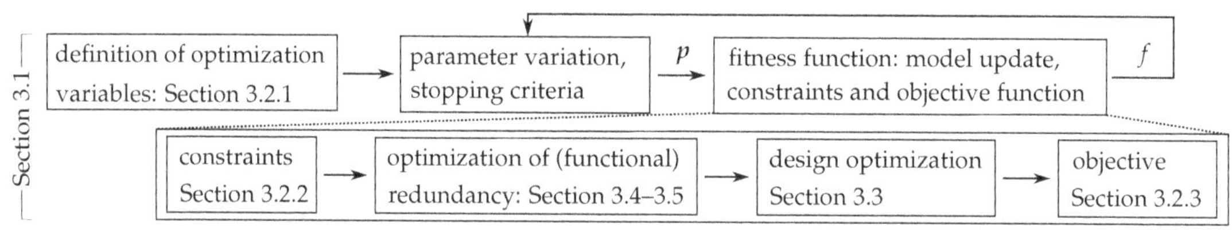

Figure 5. Flowchart diagram summarizing the dimensional-synthesis optimization problem.

Parts of the concept have been previously published in preliminary form in the conference articles [99] (Sections 3.1 and 3.2), [97] (Sections 3.3 and 3.4), [100] (Section 3.3.2), and [101] (Section 3.5). Concepts regarding the kinematics model and methods for redundancy resolution from [22] (summarized in Section 2.1) and [29] are used within the algorithm on a lower level and referred to but are not necessary for comprehension.

3.1. Optimization Problem, Hierarchical Constraints, and Optimization Scheme

Within the dimensional synthesis, a set of robot kinematic parameters $\boldsymbol{p}$ is optimized to obtain an optimal solution $\boldsymbol{p}_{\mathrm{opt}}$ in a single-objective optimization (SOO) or a set of Pareto-optimal solutions $\{\boldsymbol{p}_{\mathrm{opt},1}, \ldots, \boldsymbol{p}_{\mathrm{opt},n_{\mathrm{Pareto}}}\}$ in a multi-objective optimization (MOO).

Both SO and MO optimizations were implemented and tested using either the MATLAB global optimization toolbox [82] or the MO particle swarm optimization (PSO) algorithm from [102]. The term "particle" is used synonymously for the parameter vector $\boldsymbol{p}$ and "fitness" for the optimization function f since the PSO proved better results than the genetic algorithm (see Figure A5 in Appendix A) and is used throughout this paper.

3.1.1. Optimization Problem

The optimization is written as a minimization, following the standard convention for the SO problem

$$p_{\text{opt}} = \underset{p \in P}{\operatorname{argmin}} f(p), \tag{3}$$

where P is the set of permissible parameter values. The set of MO solutions P_{opt} for n_{obj} objectives is defined by

$$f_i(p) \geq f_i(p_{\text{opt},j}) \quad \forall\ i \in \{1, \ldots, n_{\text{obj}}\} \text{ and } \forall\ p \in P \setminus P_{\text{opt}} \quad \text{with} \quad p_{\text{opt},j} \in P_{\text{opt}} \subset P. \tag{4}$$

3.1.2. Hierarchical Constraints

Many parameter combinations do not present a valid solution since they violate some of the several constraints that have to be met to obtain a feasible parallel robot. As the computation time is the bottleneck in the optimization process, the chosen way of *handling the constraints* is to abort the processing of a particle once one of the constraints is not met. This procedure is already sketched in [52] in the example of one constraint and in [30] regarding several constraints with constant penalties regardless of their degree of violation. Following [30], the penalty terms v_i are defined hierarchically according to the infeasibility of the solution with

$$v_1(p) > v_2(p) > \cdots > z(p), \tag{5}$$

where z represents the objective function, and v_i is the penalty term for the constraint i, checked in the order of ascending index i. For example, geometric infeasibility (leg chains are too short to connect to the moving platform) is more severe than a joint-limit violation, which leads to $v_{\text{leglength}} > v_{\text{jointrange}}$. The SO *fitness function* implements the abortion upon constraint violation by piece-wise definition in an if-else manner, extending the concept of "expanded objective function" of [93] as

$$f(p) = \begin{cases} v_1(p) & \text{if constraint 1 is violated} \\ v_2(p) & \text{(else) if constraint 2 is violated and constraint 1 is not} \\ \vdots & \vdots \\ z(p) & \text{(else) if no constraint is violated.} \end{cases} \tag{6}$$

The approach adapts the *static-penalty approach* [93], which is termed "*hierarchical constraints*" here due to the constraints' ordered values. The terms "fitness function" f and "objective function" z are distinguished since the former also includes the constraints' penalties v_i in the chosen approach.

As an extension to [30], the *degree of violation of the constraint* increases the *penalty*, which requires each constraint to have a dedicated interval $v_{i,\min} \leq v_i < v_{i,\max}$ without intersection with the penalties of other constraints. For example, if for one particle p_1, the legs are shorter than for another particle p_2 and, in both cases, the platform cannot be reached, $v_{\text{leglength}}(p_1) > v_{\text{leglength}}(p_2)$ holds to have a higher penalty for the first case. The degree of violation $\rho > 0$ in this case is $\rho(p_1) > \rho(p_2)$. It is normalized to obtain the constraint penalties v_i with the function

$$v_i(\rho) = v_{i,\min} + f_{\text{norm}}(\rho) \cdot (v_{i,\max} - v_{i,\min}) \quad \text{with} \quad f_{\text{norm}}(\rho) = \frac{2}{\pi} \arctan\left(\frac{\rho}{\rho_{\text{scale}}}\right). \tag{7}$$

Since it is not always possible to determine the upper limit of ρ for any parameter value $\boldsymbol{p}$, the saturation of (7) ensures that each constraint penalty v_i stays within the dedicated range. The scaling ρ_{scale} adjusts the range of the values of ρ, which lie in the saturation of $f_{\mathrm{norm}} \approx 1$. If the upper limit of ρ is known, e.g., $\rho_{\mathrm{max}} = 1$ for 100% of points failed, linear scaling is used instead in (7) with $f_{\mathrm{norm,lin}}(\rho) = \rho/\rho_{\mathrm{max}}$. The performance criteria c within the objective function z undergo a similar normalization $z = f_{\mathrm{norm}}(c)$ to ensure they stay within the interval $0 \leq z < z_{\mathrm{max}}$.

3.1.3. Optimization Scheme

The overall flow chart of the optimization is depicted in Figure 6, with examples of some constraints for parallel robots.

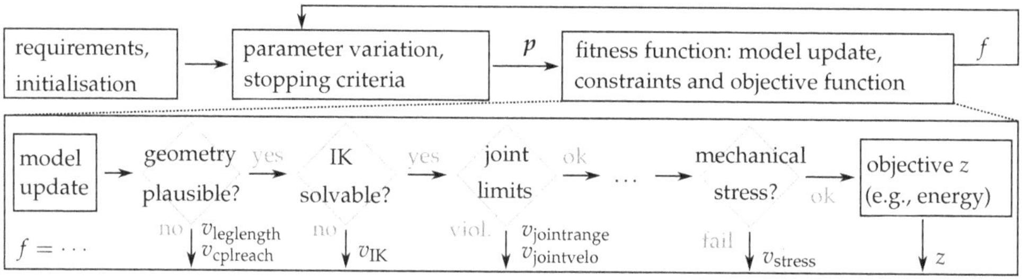

Figure 6. Flowchart diagram of the particle swarm optimization and the structure of the fitness function with hierarchical constraints. Mod. from [99].

The constraints are computed for reference points and a reference trajectory in the robot workspace, and the objectives are also based on the reference trajectory. Therefore, the inverse kinematics (IK) need to be solved after the geometric plausibility check. An investigation of the complete robot workspace by geometric derivation (like in [18]) or discretization is out of scope due to the high computational effort in the general case. This aligns with the argumentation of [10] (p. 151) that disadvantageous positions (regarding achievable tilting angles) are located at the border of the workspace, and assessing them is sufficient for synthesis. The task, including reference points and trajectory, is expressed in the world frame $(\mathrm{CS})_{\mathrm{W}}$. Since the robot base position can be optimized, the task has to be transformed to the robot base frame $(\mathrm{CS})_0$ for every particle as part of the block "model update". Additionally, dynamics parameters (e.g., link mass) are set based on the updated kinematic parameters (e.g., link length) and a simplified geometric model using hollow cylinders for the links. The constraints and the objective function are explained in more detail in the following sections.

In the case of an MOO with n_{obj} objectives, the procedure stays the same, and since only one violated constraint i is computed, $\boldsymbol{f} = [v_i, \ldots, v_i]^{\mathrm{T}}$ is set for constraint violation, and $\boldsymbol{f} = [z_1, z_2, \ldots, z_{n_{\mathrm{obj}}}]^{\mathrm{T}}$ in case of no violation. In most cases, $n_{\mathrm{obj}} = 2$ or $n_{\mathrm{obj}} = 3$ is used to obtain two- or three-dimensional Pareto fronts, which can still be inspected manually and visualized. To be able to change the criteria after the optimization, all possible criteria have to be used in the optimization (i.e., $n_{\mathrm{obj}} > 3$), and then multiple 2D or 3D diagrams are created from the Pareto-dominant particles regarding the selected criteria.

3.2. *Constraints, Objectives, and Optimization Variables*

Constraints are checked in a hierarchical order from severe violations of plausibility to violations of more detailed limits regarding a possible practical implementation. The enumeration of constraints is given in Section 3.2.2 for later reference. Once all constraints are met, the objective function is computed. Several performance criteria are available from the state of the art on parallel robots [1] and parallel-robot synthesis [9]. The objective function should be chosen based on physical criteria independent of the specific robot structure. Otherwise, comparing different structures is not feasible, cf. [9] (p. 175 ff). A list of possible objective functions is given in Section 3.2.3. To reduce the nonlinearity of the fitness function f with respect to the optimization parameters $\boldsymbol{p}$, a conversion between optimization parameters and physical kinematic parameters $\boldsymbol{p}_{\mathrm{kin}}$ of the robot is performed. The concept primarily aims to reduce the coupling between parameters, which allows changing one parameter without losing the validity of the structure. The kinematic parameters subject to optimization are discussed in detail in the following Section 3.2.1.

3.2.1. Optimization Variables

The optimization variables in the dimensional synthesis affect the geometry of the parallel robot, shown exemplarily in Figure 7. Only parallel robots with identical leg chains are considered with coupling joints on the fixed base and moving platform that are set symmetrically on a circle circumference with radius r_{b}. The parameters for the base-coupling joints are summarized in Table 2. The platform-coupling joints follow the same order with parameters using the index "p" instead of "b." The direction of the coupling joint's axis of rotation or translation is selected according to geometric principles, which are also commonly found in existing robot structures in the literature. The main principles are shown in Figure 8 as vertical (v), tangential (t), radial (r), or conical (c) with respect to the base plane.

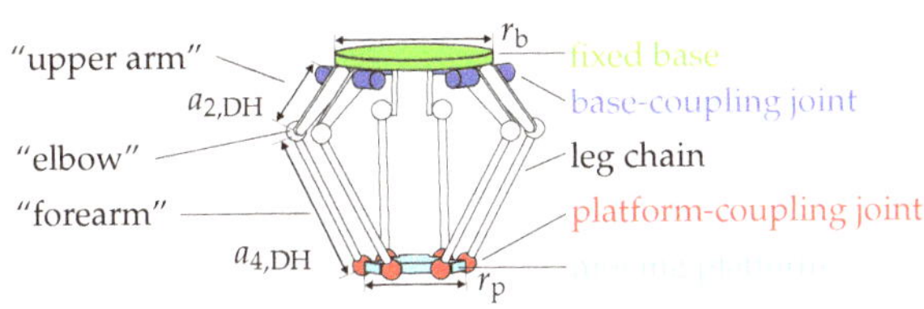

Figure 7. Parallel robot with annotation of geometric structural entities. Mod. from [103].

Table 2. Dimensional parameters for the base-joint alignments in this paper.

Position	**Symmetric on Circle Circumference**				**Pairwise**			
Direction	v	t	r	c	V	T	R	C
Parameters	r_{b}	r_{b}	r_{b}	$r_{\mathrm{b}}, \gamma_{\mathrm{b}}$	$r_{\mathrm{b}}, d_{\mathrm{b}}$	$r_{\mathrm{b}}, d_{\mathrm{b}}$	$r_{\mathrm{b}}, d_{\mathrm{b}}$	$r_{\mathrm{b}}, d_{\mathrm{b}}, \gamma_{\mathrm{b}}$
Figure	8a	8b	8c	8d	9a	9b	9c	9d

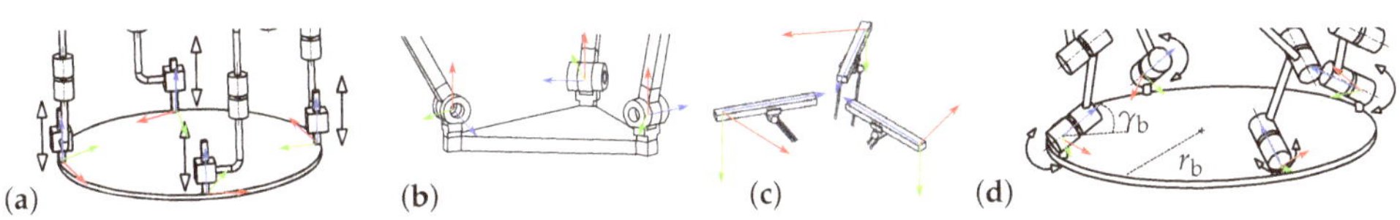

Figure 8. Geometric principles for circular alignments of the base-coupling joint with given coupling joint frame (the blue z-axis corresponds to first joint axis): (**a**) vertical (mod. from [Figure 9.10g] in [6]), (**b**) tangential (mod. from [Figure 9.14] in [104]), (**c**) radial (mod. from [105]), (**d**) conical ([Figure 9.10f] in [6]).

Six-DoF robots are usually built with a *pairwise alignment* of the base-coupling joint, as shown in Figure 9. In this case, the coupling-joint pair centers are aligned on a regular triangle, and the joint pairs are set around the angle bisector with a given distance d_{b}. Likewise, a circle with a radius r_{b} can be used for the joint center definition [106], cf.

Figure 9c. Alternatively, a geometric description with two parameters by the joint positions' radius and an angle offset is possible, cf. [49] and [18] (p. 70). A distinction of a pairwise alignment to the previous modes is noted by capital letters, i.e., V for vertical, R for radial, T for tangential, and C for conical. The axes of the joint pairs are parallel (oriented at the regular triangle). Therefore, the conical alignment "C" geometrically corresponds to a pyramid.

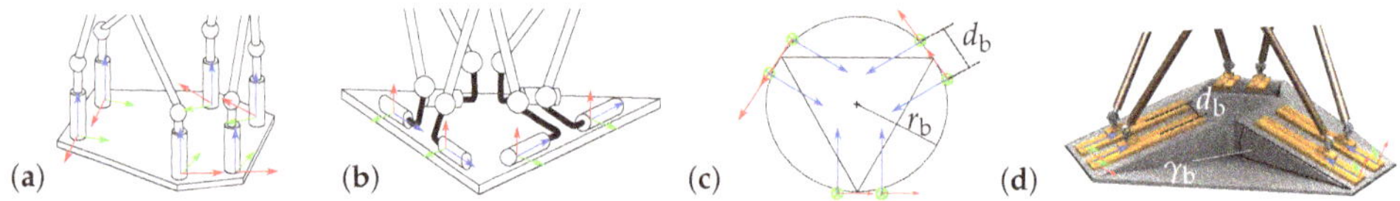

Figure 9. Geometric principles for pairwise circular alignments of the base-coupling joint with given coupling joint frames: (**a**) Vertical (mod. from [Figure 2f] in [103]), (**b**) Tangential (mod. from [Figure 2a] in [103]), (**c**) Radial (top view on base), (**d**) Conical/pyramidal (source: Daniel Ramirez, LUH; mod).

The parameter in the optimization parameters $\boldsymbol{p}$ without direct correspondence in the physical kinematic parameters $\boldsymbol{p}_{\text{kin}}$ is

1. the scaling parameter p_{scale}, which increases the size of the whole robot (platform and legs) without changing its overall kinematics characteristics.

The optimization parameters corresponding to the *base and platform* are

2. the base scaling parameter p_{b}, giving the base radius by $r_{\text{b}} = p_{\text{scale}} p_{\text{b}}$,
3. the platform scaling parameter p_{plfscale} that sets the moving-platform radius relative to the base by $r_{\text{p}} = p_{\text{plfscale}} r_{\text{b}}$,
4. the base-coupling-joint pair-distance scaling parameter p_{bpdscale}, giving the distance $d_{\text{b}} = p_{\text{bpdscale}} r_{\text{b}}$ of the joint pairs, in case of the alignments V, T, R, or C,
5. the similar pair-distance scaling parameter p_{ppdscale} with $d_{\text{p}} = r_{\text{p}} p_{\text{ppdscale}}$ for the moving platform for platform-coupling-joint alignments V, T, and R,
6. the elevation γ_{b} of the base-coupling joint, in case of a conical or pyramidal alignment of the first joint axis of the leg chains (alignments c and C), and
7. the angle inclination γ_{p} for platform-coupling-joint alignment c.

The angular parameters 6 and 7 are optimized without scaling since their range of values is limited within $\pm 90^\circ$. If limits are given for the base or platform radius, the scaling of parameters 2 or 3 is omitted, and the physical values are optimized directly as $r_{\text{b}} = p_{\text{b}}$ or $r_{\text{p}} = p_{\text{p}}$ with a new parameter p_{p} instead of p_{plfscale}.

For the *leg chains*, the *kinematic parameters* to optimize are $a_{i,\text{DH}}$, $d_{i,\text{DH}}$, $\alpha_{i,\text{DH}}$, and $\theta_{i,\text{DH}}$, in the modified Denavit–Hartenberg (DH) notation from Khalil [107]. Regarding optimization holds:

8. The leg-length parameters are scaled by $a_{i,\text{DH}} = p_{\text{scale}} p_{a_i}$ and $d_{i,\text{DH}} = p_{\text{scale}} p_{d_i}$, and
9. the leg-angle parameters are optimized directly by $\alpha_{i,\text{DH}} = p_{\alpha_i}$ and $\theta_{i,\text{DH}} = p_{\theta_i}$.

The number of kinematic parameters $\boldsymbol{p}_{\text{DH}} = [p_{a_1}, p_{d_1}, p_{\alpha_1}, p_{\theta_1}, \ldots, p_{a_n}, p_{d_n}, p_{\alpha_n}, p_{\theta_n}]^{\text{T}}$ varies depending on the leg chain (with n joints). The parameters $a_{1,\text{DH}}$ and $d_{1,\text{DH}}$ are removed from the optimization and set to zero since they are redundant with the base size or base z position $r_{\text{b},z}$, depending on the joint alignment. This also holds for the $a_{2,\text{DH}}$ and $d_{2,\text{DH}}$ parameters if the first joint is prismatic. If a prismatic joint i is within the leg chain, a lift cylinder is assumed for technical realization (see constraint 13) with a direct connection between the previous and the next joint, without a lever arm, resulting in $a_{i-1,\text{DH}} = a_{i+1,\text{DH}} = d_{i+1,\text{DH}} = 0$. For some parameters $\alpha_{i,\text{DH}}$ or $\theta_{i,\text{DH}}$, only constant values of 0° or 90° can be set as a result of the structural synthesis. Otherwise, angles are optimized continuously from -90° to 90°. To be able to reproduce results with the parameter models from the literature, perpendicular connections between the joints (by $d_{i,\mathrm{DH}} = 0$) and without skew joint-axes alignments (by rounding p_{α_i} to $\alpha_{i,\mathrm{DH}} \in \{0, \pm 90^\circ\}$) can be enforced.

The *relative position and orientation of robot* (base frame $(\mathrm{CS})_0$) and task within the world frame $(\mathrm{CS})_\mathrm{W}$ are optimized by

10. the base position $_{(\mathrm{W})}\boldsymbol{r}_0 = [r_{\mathrm{b},x}, r_{\mathrm{b},y}, r_{\mathrm{b},z}]^\mathrm{T}$ and
11. the base orientation, expressed as ${}^\mathrm{W}\boldsymbol{R}_0 = \boldsymbol{R}([\varphi_{\mathrm{b},x}, \varphi_{\mathrm{b},y}, \varphi_{\mathrm{b},z}]^\mathrm{T})$ using a rotation matrix $\boldsymbol{R}$ obtained from intrinsic X-Y'-Z'' Euler angles.

Depending on the robot's DoFs and the settings, some parameters remain constant, e.g., the base tilting angles $\varphi_{\mathrm{b},y}$ and $\varphi_{\mathrm{b},z}$ in the case of a translational 3T0R parallel robot. Similarly, an *additional end-effector transformation* between the platform frame $(\mathrm{CS})_\mathrm{P}$ (located in the middle of the moving platform) and the end-effector frame $(\mathrm{CS})_\mathrm{E}$ with the tool center point as origin is optimized by

12. the end-effector position $_{(\mathrm{P})}\boldsymbol{r}_\mathrm{E} = [r_{\mathrm{p},x}, r_{\mathrm{p},y}, r_{\mathrm{p},z}]^\mathrm{T}$ relative to the platform frame and
13. the end-effector orientation expressed as ${}^\mathrm{P}\boldsymbol{R}_\mathrm{E} = \boldsymbol{R}([\varphi_{\mathrm{p},x}, \varphi_{\mathrm{p},y}, \varphi_{\mathrm{p},z}]^\mathrm{T})$.

Again, some of these parameters can remain constant or can be omitted. If an existing commercially available robot is optimized for a given task, only parameters 10–13 have to be optimized. The optimization of a complex kinematic structure leads to a high number of more than ten parameters. The total vector of (possible) optimization parameters is

$$\boldsymbol{p} = [p_\mathrm{scale}, p_\mathrm{b}, p_\mathrm{plfscale}, p_\mathrm{bpdscale}, p_\mathrm{ppdscale}, \gamma_\mathrm{b}, \gamma_\mathrm{p}, \boldsymbol{p}_\mathrm{DH}^\mathrm{T}, r_{\mathrm{b},x}, \ldots, \varphi_{\mathrm{b},z}, r_{\mathrm{p},x}, \ldots, \varphi_{\mathrm{p},z}]^\mathrm{T}. \quad (8)$$

3.2.2. Constraints

The constraints are implemented based on the degree of violation ρ of physical values and the saturation and scaling of (7). The penalties v_i are ordered from severe to minor violations of the robots' plausibility by spanning a wide (arbitrary) range of orders of magnitudes from 0 (perfect) over 10^3 (worst still feasible value) to 10^{14} (already the parameters are infeasible).

In implementing the optimization algorithm, limits on the optimization variables $\boldsymbol{p}$ can only be set directly on the variable and not on the values derived from them. Therefore, more complex parameter limits are manually included as a first stage of constraints:

1. The inclination angle γ_b and γ_p in a conical (c) alignment of base or platform joints should differ about a minimal value of, e.g., 5° from the radial (r, 0°) or vertical alignment (v, $\pm 90^\circ$).
2. Since the joint position in the pairwise alignment of coupling joints is dependent on two parameters r_b and d_b, the effective radius (as the distance of the coupling joints A_i to the center) of the base or platform has to be computed from both.
3. The Denavit–Hartenberg parameters a_i and d_i express the distance between two joints. A minimal joint distance can be required to take the mechanical designs into account.
4. The length of the leg chain is determined from the $a_{i,\mathrm{DH}}$ and $d_{i,\mathrm{DH}}$ parameters and the elongation of prismatic joints within the chain. A maximum length can be demanded to ensure the feasibility of the design.

After handling the above indirect parameter limits, the geometric plausibility is checked

5. by leg lengths l_leg matching the base and platform radii and
6. by checking if all platform-coupling points $\boldsymbol{r}_{\mathrm{B}_i}$ can be reached by all the leg chains for all reference points $\boldsymbol{x}_{\mathrm{ref},j}$.

The required leg length for constraint 5 is $l_{\mathrm{req}} = |r_{\mathrm{P}} - r_{\mathrm{b}}|$. The missing leg length results in $l_{\mathrm{tooshort}} = l_{\mathrm{req}} - l_{\mathrm{leg}}$. If positive, the constraint is violated and normalized to $\rho = l_{\mathrm{tooshort}}/l_{\mathrm{leg}}$, which is then scaled and saturated via f_{norm} to obtain the penalty $v_{\mathrm{leglength}}$. For constraint 6, the required minimum length of the leg chains is $l_{\mathrm{req}} = \max_{i,j} ||\boldsymbol{r}_{\mathrm{B}_i}(\boldsymbol{x}_{\mathrm{ref},j}) - \boldsymbol{r}_{\mathrm{A}_i}||$ for connecting base and platform (see Figure 2) with the same way of obtaining the penalty $v_{\mathrm{leglength2}}$. The example shows that computing constraint 5 requires less computational effort than constraint 6, as the latter includes coordinate transformation from the rotational component of the mobile platform's pose variable $\boldsymbol{x}_{\mathrm{ref},j}$.

Then, the reference points are checked to ensure the prescribed workspace can be reached. This avoids completely discretizing the workspace volume, which is too time-consuming.

7. The inverse-kinematics problem is solved for all reference points $\boldsymbol{x}_{\mathrm{ref},j}$ using the methods from [22]. Failure leads to a constraint violation according to the index of the failing point. This is carried out by assigning $\delta = 10^6$ to the kinematic constraints of failed or not computed points, which is far above the threshold of 10^{-8} for the success of the inverse kinematics implemented via Newton Raphson. In this way, penalizing IK results depending on their violation of the kinematic constraints is possible. The violation term then is $\rho = \mathrm{avg}_j |\delta_j|$, applied component-wise.
8. If the resulting IK solution is a singular configuration, this is counted as a constraint violation. First, leg-chain singularities (type I) and then parallel singularities (type II) are considered.
9. Next, a constraint on the Jacobian condition number is checked for all points.
10. The range of revolute and prismatic joint coordinates is tested against technical limits, which can be obtained as upper limits for plausible values from data sheets, e.g., for universal or spherical joints. For rotatory joint DoFs, the 2π-periodicity has to be regarded using tools from directional statistics (of non-Euclidean spaces) and circular distributions for calculating the range of angles.
11. If absolute values for joint coordinates are already known, e.g., when optimizing the base position of an existing robot, the joint coordinates are checked against these limits.
12. If base-mounted prismatic joints are aligned within the base plane, the robot's allowed base diameter is checked against the necessary length and position of the guide rails of these joints, obtained from the range of prismatic-joint coordinates and assuming leg-chain symmetry.
13. If a prismatic joint exists within the structure, a lift cylinder consisting of an outer cylinder and a push rod is assumed for technical realization. The joint-coordinate limits define the length of the rod and outer cylinder. If the outer cylinder needs to start beyond the previous joint, this is infeasible for technical realization and penalized. The check is performed for each leg separately and for a symmetric design based on all legs' prismatic-joint coordinates.
14. A self-collision of the robot structure leads to a constraint violation with a penalty obtained from the collision bodies' penetration depth, using a simplified geometric model with spheres and capsules.
15. A violation of the allowed installation space is counted as a less severe constraint violation since the robot works if regarded solely. The maximal distance of any part of the robot to the allowed volume is taken for the penalty.
16. A collision with an obstacle object within the workspace is counted as another constraint since this is less severe than the self-collision by the same argumentation.

Not all of these constraints must be checked for all tasks, such as 11, 15, or 16. The limits for constraint 10 or thresholds for constraints 9 and 12 can be set relatively high to obtain a more extensive set of results in the first iterations of the synthesis and can then be tightened.

The reference trajectory is evaluated in the next step using the differential inverse kinematics and the initial joint configuration(s) from the previous step.

17. The failure of the differential IK creates a penalty resulting from the progress achieved.
18. Singularities of types I and II are checked, similar to constraint 8.
19. An inconsistency within the acceleration, velocity, or position between the leg chains can result from errors in the robot models or the check of infeasible kinematic structures within the structural synthesis (performed with the dimensional-synthesis framework) and produces a penalty.
20. A parasitic motion is an end-effector velocity component in an undesired degree of freedom for a robot with reduced mobility, which leads to a penalty.
21. The range of joint coordinates from constraint 10 is checked for each leg chain and for symmetric robots for all leg chains together, assuming a symmetric construction.
22. The feasibility of the lift cylinders for the prismatic joints from constraint 13 is rechecked for the joint configurations within the trajectory.
23. Limits on the velocity and the accelerations of active joints are checked. The limits result from plausibility considerations or may be obtained from data sheets.
24. If the differential IK algorithm leads to a flipping configuration, this is detected by using correlations—if it is not already discovered by the consistency check of constraint 19.
25. Finally, self-collisions, installation space violations, and workspace collisions are checked, similar to constraints 14–16.

Some constraints above, like 19 or 20, are only relevant within a *structural synthesis*, which can also be performed numerically with the dimensional synthesis. In the structural synthesis, many of the constraints that are designed to lead to more feasible solutions (such as joint limits and self-collisions) can be omitted, and only constraints regarding the mathematical solution of the inverse kinematics have to be regarded, such as 7, 8, 17, and 18.

If all the constraints above for the trajectory are met, further constraints are investigated:

26. The threshold for the condition number is checked, similar to constraint 9.
27. A design optimization is performed after success in the constraints above, elaborated in Section 3.3. A penalty is given if no valid solution can be found within the design optimization.
28. After computation of the inverse dynamics, the internal forces (see [99]), and the eventual design optimization, exceeding the material's yield strength with a given safety factor results in a penalty depending on the material-stress violation.
29. Limits on the physical values of the performance criteria, such as stiffness (objective 6 of the next section), precision (objective 1), or actuator force (objective 2), are checked and lead to a penalty corresponding to the relative degree of limit violation.

Further details on the specific implementation of the constraints are omitted for brevity. They can be obtained from the open-source implementation [96] within the files `cds_constraints.m`, `cds_constraints_traj.m` and `cds_fitness.m`. A validation of the concept of hierarchical constraints by means of computation time is given in Figure A6 in Appendix A for the case study from Section 4.

3.2.3. Objective Function

The objectives investigated primarily are

1. maximum position error (precision) of the tool center point, cf. [108],
2. maximum actuator force (in the case of the same type of actuators),
3. maximum actuator velocity (similar to number 2),

4. maximum actuators' rated power as a product of maximum motor torque and speed,
5. energy consumption during the reference trajectory (cf. [99]),
6. stiffness of the end-effector platform,
7. mass of the structure assuming hollow cylinders and a solid platform plate.

Criteria 2 and 3 can be used directly for dimensioning drives since, in combination, they present a speed-torque diagram. If different structures with prismatic and revolute actuation should be compared, the power criterion (no. 4) may support finding motors with small dimensioning. These criteria are advantageous if no automated drive-train optimization is performed. Criteria 1, 5, 6, or 7 should be used according to the task requirements. The maximum trajectory value for criteria 1, 2, 3, or 4 is minimized in the optimization. To maximize the mechanical stiffness with criterion 6, the largest compliance eigenvalue (from the inverse translational stiffness matrix over the trajectory, by the model from [9] (p. 49 ff) is minimized.

Some criteria are designed as indirect measures for the ability to build up a working prototype, such as

8. material stress (internal forces in relation to the material's yield strength; see [99]),
9. length of the links (summed over all leg chains),
10. maximization of the smallest collision distance during reference motion (cf. constraint 14),
11. installation-space volume from a convex volume containing all joint positions throughout the trajectory, obtained by the alpha-shape algorithm [109],
12. footprint, i.e., the floor area taken by the robot projected from the points of the installation space, using the alpha-shape algorithm [109], and
13. used joint-coordinate range in relation to the maximum allowed joint range. If absolute position limits are given, they can also be used for the criterion.

Most of these criteria, like 8, 10, 11, and 13, are also suited as constraints, but using them as objectives may shift the solution towards a desired direction in the design space. When, e.g., minimizing the material stress (no. 8), the link dimensioning will likely not pose a problem, and relatively low internal forces can be expected. An optimization for short links (no. 9) is likely also to produce a compact solution, similar to the optimization for small installation-space volume (no. 11) or footprint area (no. 12). Searching for the largest collision distance (no. 10) will produce solutions that are more likely to be easily assembled, will be less dangerous in human–robot collaboration [98], and will not suffer from self-collision if parameters are changed. Some of these objectives may also conflict, e.g., a large collision distance may produce a robot with a large installation space.

When performing a structural synthesis with the framework, the single objective is

14. the rank deficiency of the manipulator Jacobian.

Other objectives present more mathematical than physical values, such as

15. maximum condition number of the manipulator Jacobian,
16. manipulability (based on the Jacobian, [108]),
17. smallest singular value of the Jacobian.

The correlation of the Jacobian condition number (objective 15) with physical performance criteria (e.g., position-error objective 1) is questionable [108]. Manipulability (no. 16) and single singular values (no. 17) are criteria specific to every structure but are used for optimization within the synthesis or the redundancy resolution by some authors. To summarize, comparing different robots or even robots with different types of actuation by their Jacobian matrices is not feasible. However, objectives 15–17 can still be used to produce sufficiently dimensioned academic examples or as a first iteration to explore the solution space for a dimensional synthesis. Optimizing the average ("global") condition number within the synthesis is not promising, as stated in [19], since single near-singular poses have a strong influence. In contrast, singular poses have no effect when optimizing the *minimum* condition number, as in [49]. A better alternative presents the use of the maximum condition number (objective 15) within the given reference points or workspace.

Again, the specific implementation of the objectives is omitted at this point, and the reader is referred to the fundamental textbooks referenced above and the open-source implementation [96] within the ydirectory `dimsynth/fitness` (files `cds_obj_...`).

3.3. Design Optimization

In some cases, a combined synthesis only leads to feasible results if one or more of the subsequent design-optimization problems of Section 2.6 is solved for every particle within the dimensional optimization. The optimization scheme from Figure 6 is extended, as shown in Figure 10. A cascaded structure is chosen, motivated by the examination of Censi's co-design framework of Figure 4, which shows the dependency of the optimization variables.

The phases of checking points-based constraints (numbers 5–16 in Section 3.2.2) and trajectory-based constraints (17–26) are separated in the detailed box for the fitness function. The design optimization in the lower part of Figure 10 corresponds to constraint 27 in Section 3.2.2, and the "force limits" are exemplary for constraints 28 and 29. Furthermore, a loop for initial values for the trajectory is added, corresponding to the robot's assembly modes (see Section 3.5).

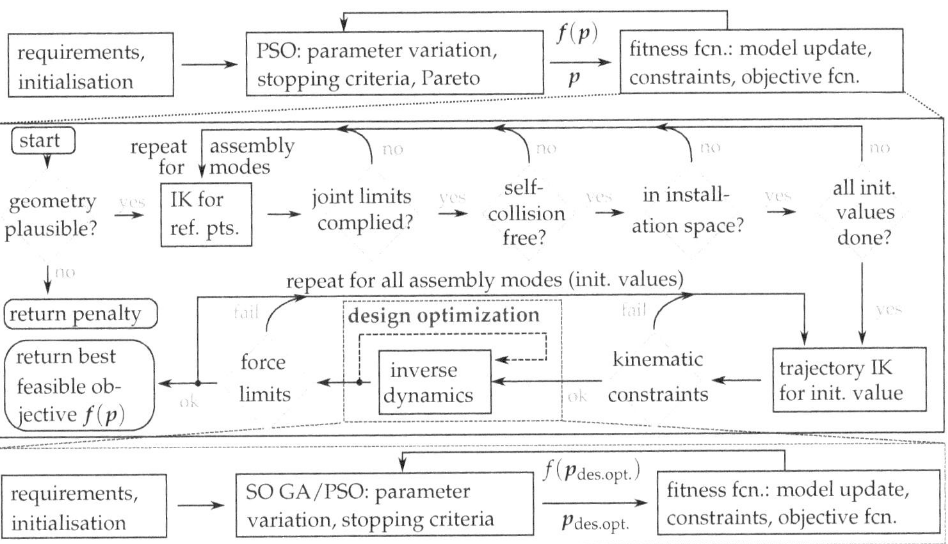

Figure 10. Extended dimensional-synthesis scheme with design optimization (Section 3.3) and loop over assembly modes (see Section 3.5), mod. from [97] (there published under CC-BY License).

3.3.1. Structure of the Design-Optimization Problem

The design optimization consists of several sub-problems that can be combined into one optimization, as introduced in Section 2.6.

The *link-design problem* gains importance if a lightweight (objective 7 in Section 3.2.3) or stiff structure (objective 6) is required. It is further relevant if dynamics are high (possibly combined with objective 2) or if material stress is considered (constraint 28), e.g., due to high payloads. If link-design optimization is involved, internal forces have to be regarded (via constraint 28 or objective 8). In most cases with small or medium payloads, a pre-defined moderate link dimensioning is sufficient and does not strongly influence the overall mass and dynamics when comparing parallel robots. The continuous optimization parameters are diameter d_{link} and strength e_{link} of hollow cylinders used for the links. The diameter is limited by the self-collision constraint 14, checked within the design optimization with a memory to avoid redundant checks if a larger diameter already succeeded. The strength is limited by the diameter when the hollow cylinder is filled. If a catalog of semi-finished parts (e.g., aluminum tubes) is regarded, not all continuous diameters and wall thicknesses are available, and a discrete design problem arises. When choosing the same dimensioning for all links, only two continuous parameters are subject to optimization.

The *drive-train selection problem* is mainly relevant if actuators are not mounted at the base, and their mass influences the overall dynamics. In the case of base-mounted drives, the design optimization specifically helps sort out dimensional parameters for which no suitable drives are available. The same effect is possible in principle by performing a Pareto optimization with objectives actuator force and speed (numbers 2 and 3) or by setting thresholds for force and speed as constraints (29 and 23). One discrete optimization parameter is sufficient for selecting existing gearbox–motor combinations for a symmetric parallel robot. Two discrete variables are optimized if the motor and gearbox are chosen separately.

The *passive-joint design problem* also has little impact on the overall mass and dynamics of the parallel robot. Its primary purpose is sorting out infeasible solutions for which no passive joints are available. This can also be achieved approximately by the corresponding joint-range constraints number 10 and 21 or by optimizing the joint-range objective 13 and only selecting solutions from the Pareto front below a specific value. The number of discrete variables for a symmetric robot (for catalog selection) corresponds to the number of universal and spherical joints, at most two, due to the number of joint DoFs. Revolute joints are assumed to provide infinite rotation, and a selection only has to be performed based on required bearing forces and tilting moments if internal forces are high enough to create relevant restrictions. Using fork joints would encounter this at the price of limiting the possible joint range.

If only continuous variables are optimized, the standard implementation of the *single-objective PSO* is chosen within this work as the *optimization algorithm for design optimization*. However, the GA could be used as well. Using MOO is impractical since selecting from the Pareto front of design solutions would be necessary to continue with the outer loop of dimensional synthesis optimization. If a mixed discrete–continuous problem has to be solved, e.g., due to consideration of the drive train, the PSO does not offer a dedicated approach other than rounding the design variables. Instead, a problem-specific modification of the genetic algorithm or a discrete algorithm like the Complex-RD could be used [63], which was, however, not investigated further in this work.

The design-optimization problem is structured with *hierarchical constraints* as the dimensional synthesis. First, the feasibility of the parameters $\boldsymbol{p}_{\text{des.opt.}}$ is checked to avoid, e.g., a wall thickness greater than the hollow cylinder's radius (for link design). Then, after an updated computation of the actuator and internal force dynamics with the new design parameters, the material stress is checked similar to constraint 28 from Section 3.2.2. Further constraints are task-based thresholds on the mass, actuator force, or stiffness, like constraint 29 from Section 3.2.2.

Similarly, the possible *objectives* are those from Section 3.2.3, which vary with the design parameters: mass (no. 7), energy (no. 5), power (no. 4), actuator force (no. 2), stiffness (no. 6), or material stress (no. 8). The objective functions are checked in the order above for congruence with the objective(s) of the dimensional synthesis, and the first match is selected as the SO criterion for the design optimization. Thereby, the order of objectives within the MOO in the dimensional synthesis presents a prioritization and changes not only the visualization of the Pareto diagram.

The selected approach corresponds to the nested strategy of bilevel optimization [5], where the upper level (dimensional synthesis) and the lower level (design optimization) are both solved with evolutionary algorithms.

3.3.2. Dynamics Regressor Form Within the Design Optimization

Figure 10 shows the design optimization's dependency on the robot's inverse dynamics. The dynamics should be implemented as efficiently as possible to reduce the computational costs of the overall optimization. One method to improve the computation time spent on inverse-dynamics calculations is to exploit the linearity of the dynamics equation with respect to the dynamics parameters, which is introduced in [100]. With the regressor form, it is possible to split the dynamics calculation into two steps: Calculating the regressor $\boldsymbol{\Phi}_{\mathrm{a}}$ and multiplying the regressor matrix with the dynamics parameter vector $\boldsymbol{p}_{\mathrm{dyn}}$.

Under certain assumptions, the dynamics calculation in the design-optimization process can then be performed with the second step. The first step can be done outside the optimization loop since the regressor matrix depends on the robot's kinematics, which stay the same in the inner loop within this work.

The approach is sketched in Figure 11 in a simplified block diagram with the computed terms and parameters as signal lines and computation steps as blocks. The internal forces can be calculated in the same way.

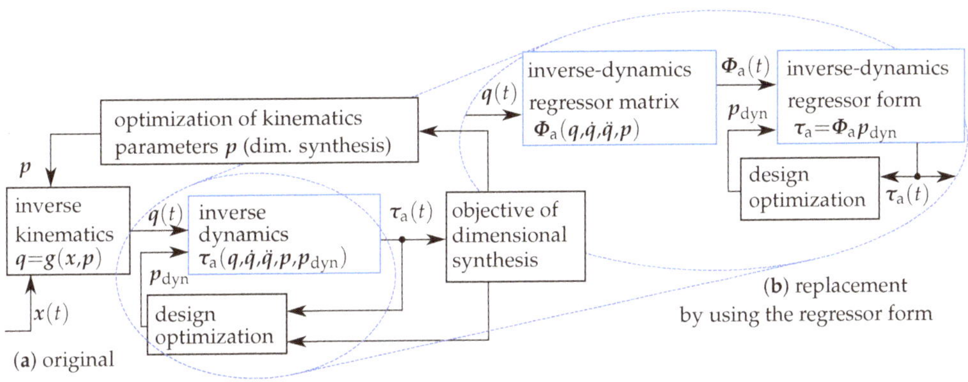

Figure 11. Block diagram of extending the existing optimization scheme (**a**) by the dynamics regressor form (**b**), mod. from [100]. The dynamics parameters $\boldsymbol{p}_{\mathrm{dyn}}$ may be in inertial- or base-parameter form.

This approach does not allow the use of point-to-point cycle time as a performance measure, as investigated by [52,61,63].

3.4. Discrete Redundancy: Assembly Modes

The inverse-kinematics problem contains *discrete redundancies* due to different possible configurations of the parallel robot's leg chains, also called *assembly modes* [1], or—in certain conditions—working modes [110], as the mode may also change after the assembly. The evaluation of a particle within the dimensional synthesis has to regard all possible assembly modes to avoid discarding a particle due to self-collisions or installation-space violations, which only result from looking at the "wrong" mode. A general selection of only the "right" mode is impossible in a combined synthesis since there is no prior knowledge of all the robot structures. The assembly modes are regarded by randomly generating initial values for the gradient-based IK and testing the point constraints 7–16 of Section 3.2.2 on them, as shown by the corresponding loops in Figure 10.

3.5. Continuous (Functional) Redundancy

A *continuous redundancy* common for parallel robots is the *functional redundancy* of full-mobility manipulators (3T3R) in tasks with rotationally symmetric tools (3T2R) [22] or of 3T1R parallel robots in 3T0R tasks with arbitrary planar rotation [101], as elaborated in Section 2.1. However, the following approach also holds for intrinsic redundancy. With the same argumentation as for the discrete assembly modes, the redundant degree of freedom must also be regarded within the dimensional synthesis.

3.5.1. Optimization Structure for Dimensional Synthesis and Redundancy Resolution

Due to the observed strong influence of motion-related constraints and the more straightforward implementation, a sampling-based approach is preferred over the gradient-based approach favored in [77] for the bilevel optimization problem of dimensional synthesis (there termed design) and motion planning. A discretization within the reference trajectory is not feasible due to the continuous nature and high computational requirements since redundancy resolution needs to be performed for every particle within the optimization. Therefore, the redundancy resolution by optimization is performed either by nullspace projection [111] or by global optimization using dynamic programming [29]. This can be regarded as another optimization loop within the dimensional synthesis but before the design optimization. If there is an interaction of design optimization and redundancy resolution, an iterative approach can be used. The necessity depends on the chosen objectives, and the topic remains out of this paper's scope.

The redundancy resolution is placed within the dimensional-synthesis algorithm as shown in Figure 12.

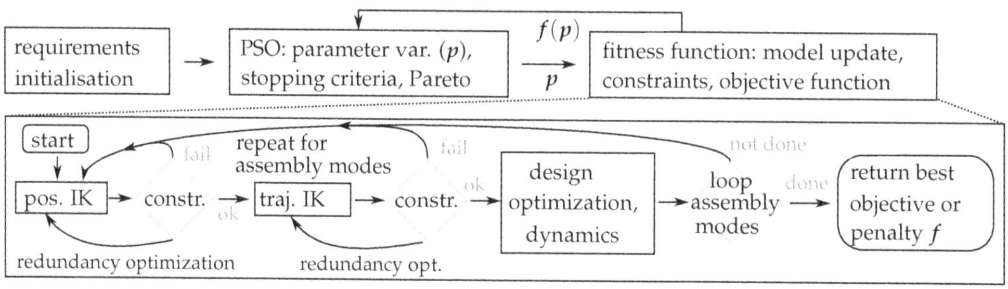

Figure 12. Dimensional synthesis scheme with functional redundancy. Mod. from [101].

First, it is used to generate more solutions to the position-level inverse kinematics for the reference points (based on the assembly-mode iterations) with the approach presented in [22]. These solutions then deliver optimal initial configurations for the trajectory inverse-kinematics algorithm. There, the methods from [29] are used to obtain optimal performance characteristics during the reference trajectory. In the concept of [5], this trajectory approach corresponds to the "optimistic version" of the bilevel optimization problem solved by a nested evolutionary algorithm (PSO) at the upper level and a classical algorithm at the lower level. Although ref. [5] states the nested strategy to be computationally expensive in general, the existence of an efficient problem-specific lower-level optimization favors the approach.

3.5.2. Selection of the Objective Function

Similar to the cascading of dimensional synthesis and design optimization, the *objective function of the dimensional synthesis and the redundancy optimization* should be the same to allow for each particle to reach its maximum performance. Thereby, the best convergence of the dimensional optimization can be expected. The objectives of the redundancy resolution that have been implemented, as detailed in [29], are

1. the soft (parabolic) joint-limit criterion from [22] (equation 44),
2. the Jacobian condition number as part of the singularity criterion [29] (equation 17),
3. the self-collision-distance criterion of [29] (equation 18) using a parabolic function,
4. a similar parabolic installation-space criterion,
5. the minimum actuator speed, and
6. the minimum actuator force or torque.

3.5.3. Redundancy Resolution and Limits of the Robot

Hard (hyperbolic) criteria for joint limits, singularities, collisions, and installation-space violations are active and overrule the soft criteria in case of a constraint violation in the IK. As a simple heuristic, the Jacobian condition number (IK objective 2) is optimized for the reference points unless a joint-limit, self-collision, or installation-space constraint is violated in the first iteration without optimization. In that case, the corresponding IK objective 1, 3, or 4 from the list above avoids the constraint violation in another iteration. The same holds if the synthesis objective already is one of these (numbers 10, 11, or 13 in Section 3.2.1). In trajectory optimization, the same principle is pursued. Since the correlation of the Jacobian condition number to the physical criteria is at most weak [1], a moderate activation threshold (of 250, empirically determined) is used for the condition-number criterion, which leads to a free motion (with minimal position change or acceleration) in most cases.

3.5.4. On the Necessity of Global Optimization over the Trajectory

When the dimensional synthesis energy objective (number 5) is optimized, the energy consumption during the trajectory can be used as a global criterion for the redundancy resolution. In this case, the design optimization would also need to be performed before the redundancy resolution to consider feasible dynamics parameters. A nullspace-projection approach for the trajectory IK is not suitable for this objective since established methods for reducing energy consumption show that a complete trajectory has to be regarded to optimally transfer between potential and kinetic energy, and local optimization thereof, e.g., via joint torques, is not constructive [112]. A similar argumentation can be used for the objective of minimizing the actuator torques (number 2) in case of significant dynamics effects through the relation of inertial and gravitational forces.

3.6. *Summary*

The presented synthesis framework can be used for structural *and* dimensional synthesis, which is another justification for the name "combined structural and dimensional synthesis". Application examples for the synthesis framework are presented in the following Sections 4 and 5. These case studies cover the first part of the design procedure of parallel robots as summarized by [1] (chapter 11) with the steps of structural synthesis, modeling, and geometrical (dimensional) synthesis. This provides a profound qualitative comparison of the different structures, which is quantitative regarding the level of abstraction. Subsequently, mechanical design must be considered apart from geometric parameters. Therefore, the combined structural and dimensional synthesis is only the first step toward creating a working prototype.

4. Case Study: High-Payload Parallel Robot for Naval Experiments

In this section, the synthesis framework is applied to a task requiring high payloads with high additional external and inertial forces. A naval testbed from the Ludwig Franzius Institute of Hydraulic, Estuarine, and Coastal Engineering (LuFI) at the Leibniz University Hannover (LUH) shall be equipped with a six-DoF parallel robot to measure external hydrodynamic forces during forced motion of mooring systems or ship models. The task is introduced in Section 4.1, detailed in Section 4.2, and transferred to the synthesis framework in Section 4.3. The results are shown in Sections 4.4 and 4.5, in Appendix A, and summarized in Section 4.6. A hexapod-robot prototype was constructed based on the results within the thesis [113].

4.1. Introduction and Related Work

Tasks that require high payloads and a limited range of motion are predestined for Gough–Steward-type parallel robots, such as flight simulators or devices for applying forces on mechanical parts in multiple directions (e.g., historically for tire testing) [1]. *Devices for moving attached objects* within water or wind belong to the latter category. They are required for identifying parameters and validating (numerical) fluid-dynamics models (e.g., from computational fluid dynamics or lumped-parameter models). This, combined with additional end-effector force measurements, helps improve simulations and the hydrodynamic or aerodynamic design of structures [114].

The 6-$\underline{P}$US structure with parallel coplanar rails (HexaGlide) from a milling machine was chosen for moving a down-scaled model for flying objects within a *wind tunnel* [115], termed "model-positioning mechanism". The dimensional synthesis of different (non-parallel) rail alignments of the 6-$\underline{P}$US (then termed HexaSlide) was, e.g., investigated in [12] in a multi-objective optimization using SQP. This was taken on by [114] to design a hybrid testbed for hardware-in-the-loop simulation of water-floating structures in a wind tunnel. The dimensional synthesis based on a single-objective GA showed better performance in that task for the HexaSlide than the hexapod [116] or HexaGlide [117].

In contrast to the wind tunnel, the installation space for a sole *hydrodynamic testbed* is less restricting, as free space exists above the water level, where the robot is mounted from above. As a result, hexapods are used for multi-axis naval experiments such as forced-oscillations tests, which were, e.g., performed in [118] for the down-scaled model of a subsea blowout preventer using a specialized hexapod from Symétrie [119]. A similar parallel-robot setup for a wave basin at LUH's LuFI is subject to the following combined structural and dimensional synthesis since the optimal structure depends on the specific requirements at the site, and a custom solution was preferred.

4.2. Task Description and Requirements

The *mechanical requirements* are defined by the dimensions of the cross-section of the 110 m long basin, as shown in Figure 13a for the empty channel with a width of 2.2 m, a height of 2 m, a regular water depth of 1.0 m, and waves with an amplitude of up to 0.5 m. The *installation space* of the parallel robot above the basin's upper edge should be kept relatively small to simplify construction and operation, as an indoor crane can only partially reach the basin. A safety margin of 0.3 m is defined relative to the basin's walls, which leaves an allowed installation space with a width of 1.6 m and a height of 1.5 m above normal water level. To avoid *design* complications, the robot's moving parts should not pass the robot's base frame, which should be compatible with the geometry of the basin's upper edge. Further, the robot shall also be used in an open wave basin at a second location of the LuFI and, therefore, has to be transportable, facing up and down, and be easy to assemble. The open wave basin produces 3D waves, which are also taken into account for the force requirements.

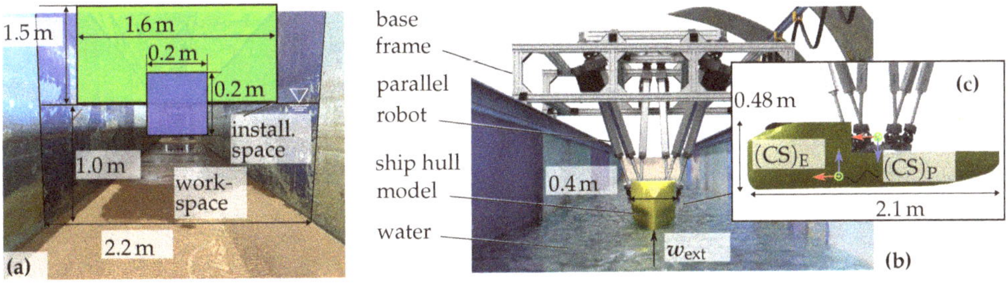

Figure 13. Photograph of the LuFI wave basin with sketched dimensions (not true to scale) of the allowed robot installation space (green) and workspace (blue) (**a**), rendering of the robot (**b**) with detail on the end effector (**c**). Geometrical relations and perspectives are depicted qualitatively. Modified from [113].

In operation, a *ship model* (dimensions 0.4 m, 0.48 m, 2.1 m, mass 54.2 kg) with a T-beam mechanical interface is mounted at the robot's end effector as shown in Figure 13b,c, which exerts external forces and moments w_{ext} by hydrostatic and hydrodynamic effects. The ship's center of mass lies in the origin of the end-effector frame $(CS)_E$. The interface forces are measured by a force-torque sensor mounted at the platform with frame $(CS)_P$ shown in Figure 13c, which requires a platform diameter of at least 80 mm. The *process requirements* are defined by the forced motion of the ship in waves with amplitudes up to 25 mm. The wave basin's full amplitude is not used due to the focus on forced oscillation and the down-scaled model. Separate harmonic oscillations with a frequency of 1.25 Hz and amplitude of 100 mm or 15° in all six DoFs are considered. Based on existing hydrodynamic model parameters, the maximal values of the external force in this motion are used for the synthesis. The respective maximal forces from different effects (i.e., wave force, forced oscillation, damping, hydrostatic forces) are superposed to obtain safe dimensioning. The demanded *accessible workspace* is defined as a cube with sides of 200 mm to allow vertical or horizontal motion of ±100 mm amplitude. Within the position workspace, tilting and rotation angles 15° against the neutral (untilted) pose must be achievable.

4.3. Transfer to the Combined Structural and Dimensional Synthesis

The task requirements from above are transferred to the proposed synthesis framework of Section 3. The constraints, parameters, and load details are summarized in Table 3 with reference to the numbered lists in Section 3.2.2 (for constraints) and Section 3.2.1 (for parameters). Installation space (row 1), base/platform diameters (rows 2 and 3), and base position (row 6) can be taken from the requirements with minor adaptions. Some parameters are not subject to optimization, such as the horizontal base position (row 7, thereby maintaining a symmetric alignment) or the chosen platform adapter for the ship model (row 9). Only a vertical rotation of the base relative to the task (row 8) and of the end effector relative to the platform (row 10) is considered for optimization, not the tilting angles, to simplify manufacturing and assembly. The payload and the worst-case external forces are given values (rows 11–14).

The *workspace requirements* are translated into 342 different poses ("reference points") at 27 positions that sample the cube in steps of 100 mm. Each position is tested with varying combinations of X-Y'-Z'' Euler angles for the end-effector frame, $\varphi_x, \varphi_y, \varphi_z \in \{-15°, 0°, 15°\}$, similar to [10] (p. 126), leading to tilting angles of up to 21°. To reduce the number of poses, symmetry of the workspace is assumed, and the full orientation variation is used only for outer points. The *reference trajectory* between the 342 poses has 7 k samples (with sampling time 50 ms, trapezoidal acceleration profile, $a_{\max} = 3\,\mathrm{m/s^2}$, and $v_{\max} = 1\,\mathrm{m/s}$, chosen based on previous one-axis forced-oscillation experiments).

A threshold for the robot Jacobian condition number (row 4) avoids structures with an unbalanced transmission ratio between the actuators and, therefore, asymmetric load. The material stress of the robot links is checked without additional safety factors (row 5) since the assumed external forces already include a safety margin. A link design optimization is performed to obtain a dimensioning that withstands the high loads.

The list of potential structures is not limited by specific design requirements. Therefore, all 234 of the 3T3R parallel robots from a structure database may be tested. It is based on a geometric permutation of the alignment of base- and platform-coupling joints from Section 3.2.1 and the leg chains of [30], which are based on the permutation of Denavit–Hartenberg parameters. Some structural filters are applied, nevertheless. By using only active revolute first or second joints, larger motion of the electric motors, which are expected to be bulky regarding the payload, is avoided.

Table 3. Constraints and parameter limits for the naval-testbed task.

#	Constraint	Section 3.2.2	Value
1	installation space (see Figure 13)	no. 15	max. width 1.6 m, height 1.5 m
2	base diameter	no. 2	800–1500 mm
3	platform diameter	no. 2	200–800 mm
4	Jacobian condition number	no. 26	max. 500 (mixed units)
5	material stress	no. 28	max. 100 % (of yield strength)
	Parameter	**Section 3.2.1**	**Value**
6	vertical base position in $_{(\mathrm{W})}\boldsymbol{r}_0$	no. 10	1.1–2.5 m
7	base position in x-y-plane	no. 10	0 (no optimization)
8	base rotation ${}^{\mathrm{W}}\boldsymbol{R}_0$	no. 11	$\varphi_{\mathrm{b},x} = \varphi_{\mathrm{b},y} = 0$ and $\varphi_{\mathrm{b},z} \in [-180°, 180°]$
9	end-effector translation $_{(\mathrm{P})}\boldsymbol{r}_{\mathrm{E}}$	no. 12	$[0, 0, 200]$ mm (schematic in Figure 13)
10	end-effector rotation ${}^{\mathrm{P}}\boldsymbol{R}_{\mathrm{E}}$	no. 13	$\varphi_{\mathrm{p},x} = \varphi_{\mathrm{p},y} = 0$ and $\varphi_{\mathrm{p},z} \in [-180°, 180°]$
	Load		**Value**
11	payload mass (ship hull)		54.2 kg
12	payload inertia $_{(\mathrm{P})}\boldsymbol{I}^{(\mathrm{C_P})}$		$[I_{xx}, I_{yy}, I_{zz}] = [0.4, 8.3, 8.3]\,\mathrm{kg\,m^2}$
13	maximal external forces		$_{(\mathrm{W})}\boldsymbol{f}_{\mathrm{ext}}^{\mathrm{T}} = [34, 451, 659]\,\mathrm{N}$
14	maximal external moments		$_{(\mathrm{W})}\boldsymbol{m}_{\mathrm{ext}}^{\mathrm{T}} = [67, 608, 162]\,\mathrm{N\,m}$

In summary, nearly six thousand different structure variants are considered due to the high number of base- and platform-coupling joint alignment combinations as discrete design parameters. For each variant, a PSO is performed on the LUH computing cluster [120], using 100 iterations and 200 individuals in multiple subsequent trials, where the initial values are obtained partly random, partly from previous results. Parameters from different coupling-joint alignments of a variant are also transferred between the consecutive trials to increase the population's diversity. The optimization was aborted after two to eight hours as this led to adequately feasible solutions for iteratively adjusting the settings.

The computation time per particle is about 60 s if no early abortion occurs. If multiple assembly modes are valid, the time further increases to evaluate all and find the best one. The number of optimization variables is up to 21, depending on the structure.

4.4. Synthesis Results for the Naval-Testbed Robot

The link design optimization is first omitted to investigate its necessity. Objectives for the mechanical design of the robot are evaluated for an initial estimate of possible solutions. The objectives are the collision distance (objective 10 from Section 3.2.3) and the material stress (obj. 8) for the relatively lightweight default link dimensioning (hollow cylinder with 13 mm diameter, 1 mm strength). The stress limit (row 5 in Table 3) is ignored at this stage. The Pareto diagram in Figure 14 summarizes the results for structures with three and four joints in the leg chains. Results for five and six joints are within appendix Figure A1. From the diagrams, it becomes clear that the high payload and external forces produce significant material stress, which can be held well by rod kinematics like the hexapod (6-UPS, Figure 15a), its kinematic inversion (6-SPU, Figure 15b), or variants of the two (6-RRPS in Figure 15c and 6-URPU in Figure 15d). Other feasible structures are another 6-URPU variant (Figure A8i) or the 6-RUPU (Figure 15e or Figure A8c) with short levers. Parallel robots with the same joint order can differ by the parallelism of their joints or additional skew angles, visible in the Denavit–Hartenberg parameters given in the tables in Appendix A. The kinematic inversions (e.g., 6-SPU instead of 6-UPS) were investigated since R, U, and S joints have different ranges of motion. The geometric dimensioning of the robot then has to be different, which is visible in the lower performance of the 6-SPU. Other structures exceed the material-stress limit and require stronger dimensioning of the links by the design-optimization loop, which is included in the following.

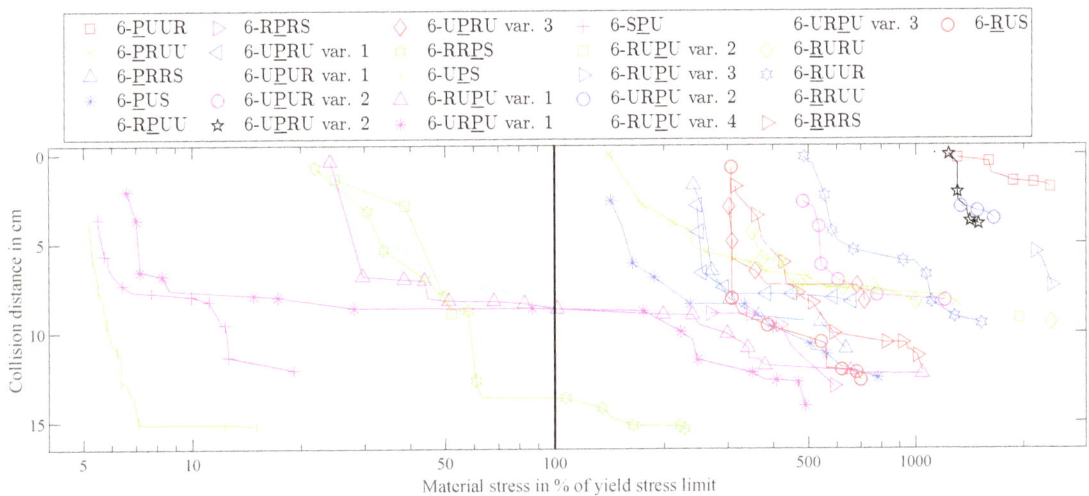

Figure 14. Pareto fronts for design-oriented objectives for chains with three and four joints with lightweight link dimensioning without the link-design optimization loop.

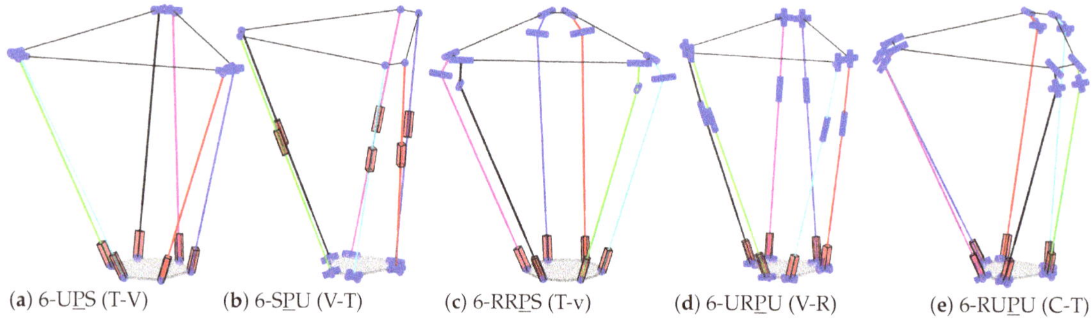

Figure 15. Visualization of valid solutions for the naval-testbed robot from Figure 14. The fixed base is at the top, and the moving platform is at the bottom. Red cuboids mark active prismatic joints, and blue cylinders or spheres mark passive joints.

In another run of the optimization, the objectives were selected as the moving parts' mass (excluding motors and joints, objective 7 from Section 3.2.3), the actuators' rated power (obj. 4), and again, the collision distance (obj. 10). Figure 16 shows the Pareto diagrams of the first two objectives for the combination of base- and platform-joint alignments for a hexapod (a), a Hexa (b), and a HexaSlide (c). The mass is influenced mainly by the length of the links and the force distribution within the structure, as high material stress (resulting from bending moments from levered forces) requires stronger dimensioning. The Hexa and the HexaSlide have higher stress in Figure 14, and also, after the design optimization, a higher moving structural mass than the hexapod. However, no dominant structure can be stated regarding the neglected mass of the hexapod's moving actuators and the mass difference of less than 10% (regarding 60 kg).

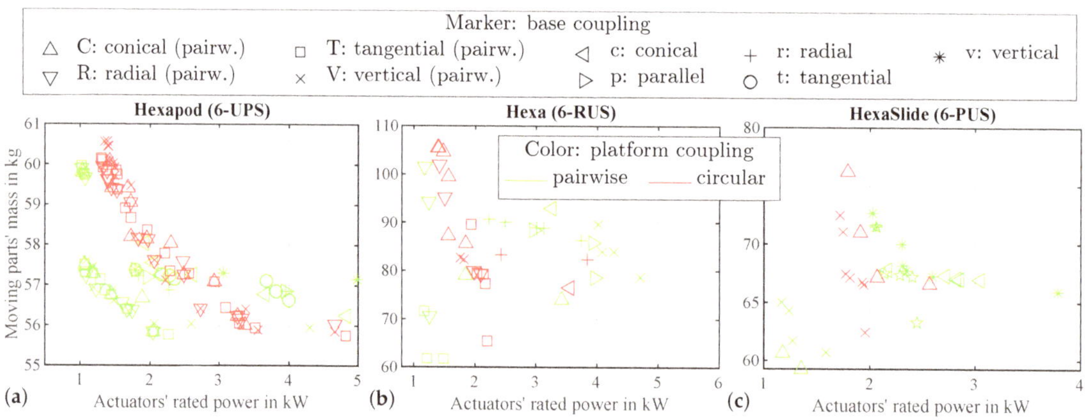

Figure 16. Pareto fronts for combinations of coupling-joint alignments of selected parallel robots: (**a**) Hexapod, (**b**) Hexa, and (**c**) HexaSlide.

Performing the dimensional synthesis for several coupling-joint alignments without preliminary fixing allows for comparing their performance. The HexaGlide discussed in Section 4.1 (6-PUS with parallel horizontal base coupling, "p") does not provide any solutions due to collisions and the limited available installation space. The actuator's rated power is approx. 20% lower for the hexapod (800 W compared to 1 kW). Since a spherical joint is used for the platform joints of the three structures, only pairwise and regular circular distributions are distinguished with two colors in Figure 16, in contrast to the base joints with nine possible markers. The pairwise alignment of both base and platform joints (shifted by one) outperforms other possible alignments for all structures. Thereby, the optimization algorithm obtains the established alignments from the literature without prior specific knowledge, showing the feasibility of the permutational approach for parallel-robot synthesis.

Next, all robot structures are compared regarding the three criteria relevant to the mechanical design and technical realization, summarized in the two Pareto diagrams in Figure 17. Only the Pareto-dominant coupling-joint alignments are included, as well as the custom hexapod engineering solution from [113] (see Section 4.5) based on the synthesis. By comparing moving mass and actuator power (upper part in Figure 17), the same robot assemblies as in Figure 14 are dominant since the necessary stronger link dimensioning of the previously underdimensioned structures increases their mass. Most of these structures lead to valid solutions, meaning that the link-design optimization found a stronger dimensioning that was also free of self-collisions despite the increased link diameters. The total of 40 structures are listed in Figure 17, as opposed to 26 in Figure 14 and 17 in Figure A1, with only a few with stress below 100%.

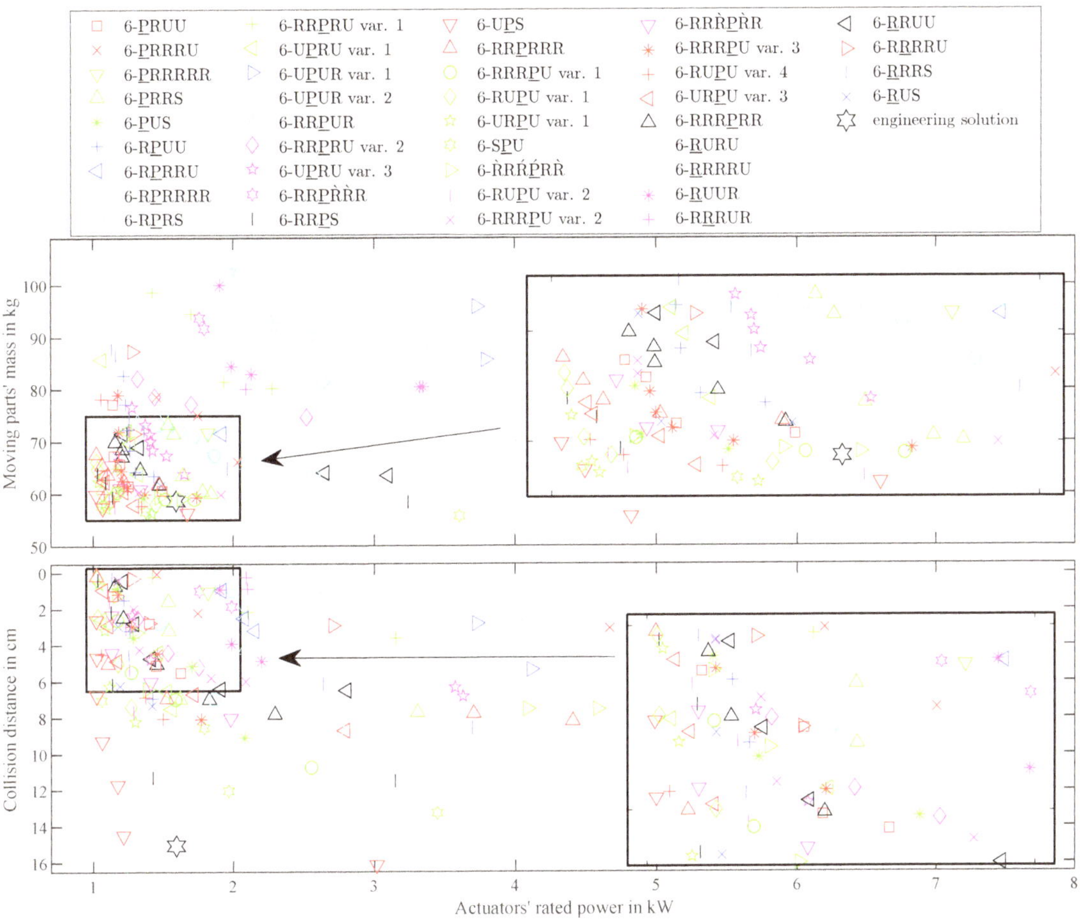

Figure 17. Grouped Pareto fronts of all results of the naval-testbed robot.

The collision criterion is investigated in the lower part of Figure 17. The dominance of the hexapod becomes visible for this task, as it provides the largest possible distances between the leg chains due to its compact construction. This also presents a safety margin for motion outside of the prescribed workspace and leaves room for detailed construction since collision bodies for the actuators are not included in the simplified model in the synthesis. For the design of the system, not only the actuators' rated power but also their torque and speed have to be compared to select from catalogs of existing motor-gear combinations. Together with the collision distance, these objectives (numbers 2 and 3 in Section 3.2.3) were optimized in another run, resulting in Pareto diagrams in Figures A2–A4 in the appendix. For prismatic actuation, these drive-train characteristics constitute one dense Pareto front, showing an equivalence of most structures regarding these criteria, which does not hold for revolute actuation. The stronger dimensioning of some of the structures as the result of the design optimization is visible as well (Figure A2 and Tables A1 and A3). Further robot sketches (Figures A7–A9) and tables with characteristic values and dimensional parameters (Tables A1–A4) are given in Appendix A for compactness.

4.5. Engineering Solution

A hexapod was selected in [113] as the most feasible solution for the task. The final dimensioning was determined based on several runs of the dimensional synthesis and as a compromise solution for multiple objectives (force, velocity, collision, precision, workspace, stiffness, workspace size) and computed for the 3D Pareto front of the synthesis with objectives of force, velocity, and collision distance. The workspace volume was evaluated by a discretization approach from [121,122], which is not feasible to be performed within the dimensional synthesis due to the computation time. The objectives were normalized to their best possible value and weighted using a paired comparison based on economic, practical, and task considerations. Therefore, the solution is not dominant in Figures 17, A3, and A4, i.e., it is above the Pareto front for the 6-U$\underline{P}$S. An overview of the resulting construction is given in Figure 18. The geometric parameters and performance characteristics are shown in the last row of appendix Table A1, together with values for other structures.

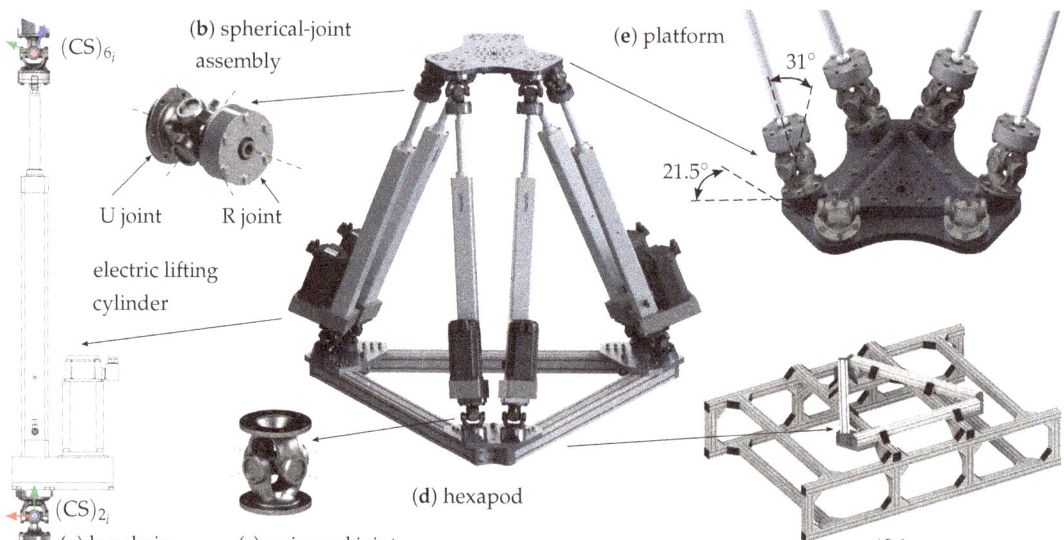

Figure 18. Details of the engineering solution; modified from [113]: (**a**) leg chain (modified from the catalog "Electromechanical cylinders EMC" from Bosch Rexroth AG, Lohr am Main, Germany), (**b**) spherical joint, (**c**) universal joint (derived from a CAD file of Elso Elbe GmbH & Co. KG., Hofheim, Germany), (**d**) hexapod assembly, (**e**) moving platform, and (**f**) fixed base.

The found structure (cf. Figure 18d) is a classical hexapod with pairwise alignments of base- and platform-coupling joints with correspondence shifted by one. Electric lifting cylinders based on a spindle and axis-parallel motor alignment (cf. Figure 18a) were used directly for the leg chains, dimensioned regarding the minimal and maximal stroke and the required actuator force and speed from the simulation. Further specific constraints had to be respected (i.e., inability to use the full stroke under load, cylinder length offset to account for joint and adapter dimensions, and ability to support the total system's mass when the robot stands on the mobile platform). The same standard universal joints (cf. Figure 18c) were used for the base and platform coupling due to their availability. An additional revolute joint was added for the kinematic spherical joint (cf. Figure 18b), and no ball-and-socket joints were used. A tilt adapter for the base- and platform-coupling joints (cf. 21.5° in Figure 18e) was chosen to reduce the joints' tilting angle to 31° (of maximal 35°) during the reference motion. The base-frame design (Figure 18f) fulfills the requirements regarding interface compatibility, transportability, and parts availability.

The mechanical design of the base showed that the resulting parameter for base-rotation offset is not feasible, and a parallel alignment of one base-joint pair to the frame's edge by $\varphi_{\mathrm{b},z}=180°$ was used instead (Figure 18f), which only led to negligible degradation of the system performance. The same holds for the platform rotation offset parameter $\varphi_{\mathrm{p},z}$, which now can be set in discrete values of 15° with the circular flange (Figure 18e) and makes the robot adaptable to different tasks and ship models.

4.6. Summary of the Case Study

For the naval-testbed task, the combined structural and dimensional synthesis found valid solutions for the majority of available kinematic structures. The best performance of the design-oriented objectives was achieved by widely-used structures with three joints per chain. For these standard solutions (see, e.g., [103]), the framework presents an efficient approach for dimensional synthesis. Other solutions were not able to outperform the standard designs. The reasons are the rather nonrestrictive ratio of workspace to installation space and the moderate required tilting angles already achievable by the hexapod. Furthermore, the payload produces high bending moments for structures with multiple joints and longer levers.

However, results for structures with more joints are still feasible regarding their performance despite their high design complexity. The acceptable computational effort shows the *feasibility of the approach for the link-design optimization*, which had to be employed for the multi-joint leg chains to obtain valid solutions. The case study further indicates that after some iterations of the requirements and the synthesis results, the kinematic structure and the drive characteristics were a helpful basis for the design and construction of a robot prototype within the time limitation of a master's thesis.

After exploring the diversity of parallel robots with 3T3R end-effector DoFs that the proposed synthesis framework can produce, the subsequent case study focuses on robots with reduced mobility and functional redundancy. An example of using functional redundancy with 3T3R robots with the framework is given in [68].

5. Translational Parallel Handling Robots with Arbitrary Planar Rotation

A general handling task is taken as a second case study to investigate structures with 3T0R and 3T1R motion exemplarily. Arbitrary planar rotation is allowed in the 3T0R task, which produces functional redundancy for 3T1R PRs. The inverse-kinematics model for this case used in the dimensional synthesis is elaborated on in [101]. The case study presents a simulative investigation without reference to an actual robot prototype, unlike in the previous section. A similar evaluation has been published in [101] in preliminary form.

In Section 5.1, the requirements, based on the paper [123], are summarized and formalized within the synthesis framework. The results are introduced in Section 5.2, followed by the evaluation of functional redundancy in Sections 5.4 and 5.5, a comparison to the reference work [123] in Section 5.3 and the summary in Section 5.6.

5.1. Task Requirements

By reproducing the requirements of the case study from [123], a validation of this thesis' synthesis framework against the one from [20] is performed (cf. Section 2.4). The requirements are summarized in Table 4. *The task* is specified as a *reachable cuboid workspace* of 200 mm × 200 mm × 100 mm. A visualization is given in Figure 19 with some of the robot structures. By sampling each dimension with three values, 27 *reference points* are defined. The orientation is not specified for 3T1R robots, which permits choosing it by optimization.

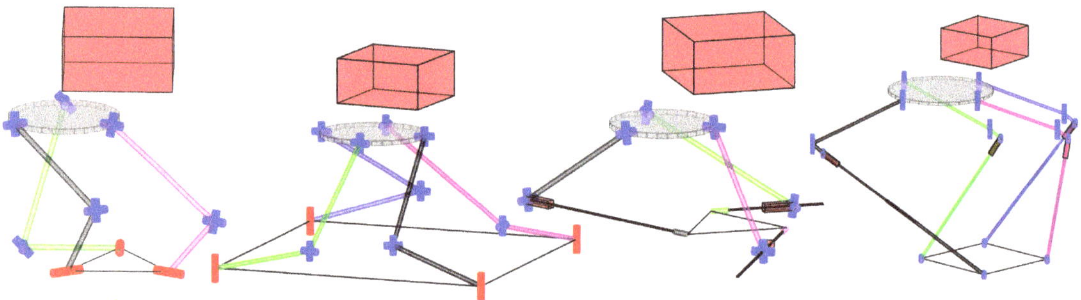

(a) 3-R̲UU △ **(b)** 4-R̲UU × **(c)** 3-P̲UU ☆ **(d)** 4-R̀P̲R̀R̀Ŕ

Figure 19. Visualization of selected results for the handling task from the Pareto diagrams in Figure 20.

Table 4. Constraints and parameter limits for the handling task with reference to numbers in Section 3.

#	Constraint	Section 3.2.2	Value
1	installation space (depending on joint index in the chain)	no. 15	joints 1–3 not above lower workspace limit, joints 4–5 not above upper workspace limit
2	prismatic joint stroke length	no. 10	max. 400 mm
3	length of a leg chain	no. 10	max. 1000 mm
4	base diameter	no. 2	200–1000 mm
5	platform diameter	no. 2	200–1000 mm, smaller than base
6	precision (position error)	no. 29	max. 0.5 mm (for objective 1 in Section 3.2.3)
7	Jacobian condition number	no. 8	max. 10^4 (IK and manipulator Jacobian)
	Parameter	**Section 3.2.1**	**Value**
8	vertical base position in $_{(\mathrm{W})}\boldsymbol{r}_0$	no. 10	1–0.1 m below the task
9	base position in x-y-plane	no. 10	0 (no optimization)
10	base rotation $^{\mathrm{W}}\boldsymbol{R}_0$	no. 11	$\varphi_{\mathrm{b},x} = \varphi_{\mathrm{b},y} = 0$ and $\varphi_{\mathrm{b},z} \in [-180°, 180°]$
11	base-joint inclination γ_b	no. 1	$5° < \gamma_\mathrm{b} < 30°$ (for conic alignment)
12	end-effector translation $_{(\mathrm{P})}\boldsymbol{r}_\mathrm{E}$	no. 12	$[0, 0, 0]$ (no optimization)
13	end-effector orientation $^{\mathrm{P}}\boldsymbol{R}_\mathrm{E}$	no. 13	$\varphi_{\mathrm{p},x} = \varphi_{\mathrm{p},y} = \varphi_{\mathrm{p},z} = 0$ (no opt.)
	Load		**Value**
14	maximal external forces		$_{(\mathrm{W})}\boldsymbol{f}_{\mathrm{ext}}^{\mathrm{T}} = [2, 2, -10]$ N

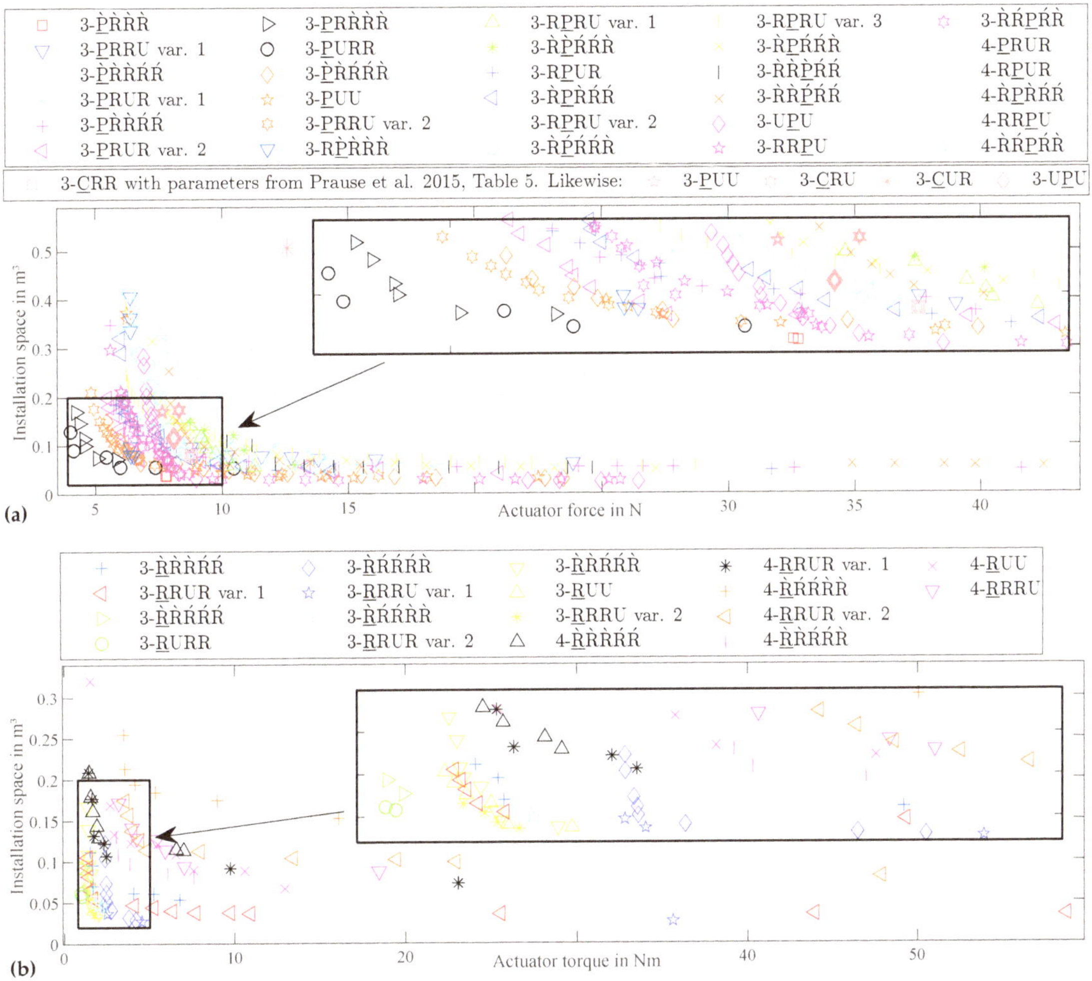

Figure 20. Pareto fronts for robots with (**a**) prismatic and (**b**) revolute actuation. The notation of [6] is used for distinguishing kinematic structures, where two R̀ denote two parallel revolute joints and Ŕ denotes a joint with a different axis. In addition to [6], prismatic joints with an axis parallel to a revolute R̀ joint are denoted by P̀.

The combined structural and dimensional synthesis in [123] is based on the 27 reference points, and results for 3T0R PRs with prismatic actuation are obtained using a single-objective optimization. There, a detailed investigation of the best parameterization for all structures is performed with a finer discretization of 10 mm in each dimension and, thereby, 4851 points. Since the proposed framework can handle this complexity, a *reference path* connecting the 4851 points in a zigzag motion (by the same number of samples) is used directly within the synthesis of this study after the check of the 27 reference points. Thereby, the risk of unreachable parts within the workspace decreases. For 3T1R robots, dynamic programming (cf. [29]) is used for optimizing the redundant rotational DoF with twelve stages and five states and a limitation of $\pm 60^{\circ}$ around the initial value, resulting from the IK optimization.

Only the force transmission, without statics or dynamics, is considered, as in [123]. The *robot and payload mass are neglected*, which allows the use of the simplified *trajectory* with discontinuous velocity and acceleration profiles by numeric differentiation. The actuator force is computed based on a given external force (corresponding to a payload of 1 kg, cf. row 14 in Table 4) and used as an objective (number 2 in Section 3.2.3), as in [123]. As a second additional objective (number 11 in Section 3.2.3), the installation space is used to be able to select robots with a good ratio of (fixed) workspace to (optimized) installation space. The precision (number 1 in Section 3.2.3) was selected as a third criterion for further exploration of the solution space as it conflicts with the actuator force by lever effects.

Several geometric *constraints* are set to increase the practicability of the solutions. A floor-mounted PR is assumed to simplify the visualizations contrary to the more practical existing ceiling-mounted robots. The robot structure should be below the workspace (row 1) to avoid impractical crane-like designs. Further constraints are taken from [123], such as a limited prismatic joint stroke (row 2), chain length (row 3), and base and platform diameters (rows 4 and 5). The required joint-coordinate range of prismatic joints in the chain (realized by lifting cylinders) results in 400–800 mm by constraint no. 13 from Section 3.2.2. To avoid singularities and to obtain feasible handling performance, e.g., in industrial pick-and-place tasks, the end-effector precision (row 6) was additionally limited to 500 μm, which is achieved by all results from [123] (Figure A17d in Appendix B) and does not deteriorate the comparability. The encoder error of 52 μm from [123] was used for prismatic actuation and 7″ from [124] for revolute actuation. For further exclusion of near-singular PRs (types I and II), Jacobian condition-number limits are set (row 7). The robot base is assumed below the center of the workspace (rows 8 and 9) with further optimization of the base rotation (row 10). No end-effector transformation is assumed (rows 12 and 13), and only an offset distance is used for visualization, e.g., in Figure 19.

In [123], only robots with tangential ("triangular") and conical ("star-shaped") *base-joint alignment* are considered as showcases. This restriction is lifted in the following, and all other structures (radial and vertical base joints) are allowed to increase the number of results and show the variety of produced solutions by the proposed framework. Only the restriction on the conical base joints' elevation of [123] is maintained (row 11) for compatibility.

The synthesis is performed for all 3T0R and 3T1R PRs from a previous structural synthesis with actuation of the first revolute or single prismatic joint. In total, 329 variants (combinations of leg chains and base- and platform-coupling-joint alignments) are investigated, 239 for 3T0R and 90 for 3T1R. Of these, 226 match the requirements (200 for 3T0R and 26 for 3T1R). The results are summarized into 36 groups of 3T0R (25 prismatic and 11 revolute actuation) and 12 groups of 3T1R (5 prismatic, 7 revolute actuation) by their respective leg chain. The results for all coupling-joint alignments for one leg chain are summarized into one Pareto front, similar to the case study in Section 4.

The particle swarm optimization for 7 to 13 parameters (depending on the structure) is configured similarly as before, with 400 individuals and a time limit of 10 h with a computation time of up to 20–40 s for 3T0R and 2 min for 3T1R robots, depending on the abortion due to constraint violation. The longer computation time for 3T1R robots results from the more complex model (one additional chain), more possibility for assembly modes, and especially, the redundancy resolution. Several optimization iterations are performed in which constraints are added and refined, and intermediate results are taken as initial values for the successive iterations.

5.2. Overview of the Synthesis Results

The optimization results are depicted in Figure 20 in Pareto fronts separately for prismatic and revolute actuation due to the different units and clarity. Four exemplary robot structures are shown in Figure 19. The remainder of the structures are visualized in Appendix B in Figures A10–A16 with parameters in Tables A5–A8.

For *prismatic actuation* in Figure 20a, many different structures show a comparably good performance; however, the 3-P̲URR (Figure A12h), 3-P̲UU (linear Delta, Figure 19c), and 3-P̲RRU (variant 2, Figure A12k) clearly outperform the other solutions regarding the selected criteria. The 3-P̲RR̀R̀R̀ (Figure A12g) and 3-P̲̀R̀ŔŔR̀ (Figure A12i) show similar performance, as they correspond to the same robots but with universal joints expanded to single revolute joints. The 3T1R structures require a significantly higher actuator force of about 10 N (e.g., 4-R̀P̲R̀ŔŔ, Figure 19d) compared to 4 N (Figure 20a) for a similar installation space.

For the *revolute actuation* in Figure 20b, the 3-R̲URR (Figure A10g) performs best, followed by the Delta robot structure (3-R̲UU, Figure 19a) and variants thereof. The Delta's base-joint alignment is chosen unconventionally by a radial drive axis. The ability of this alignment to support the established Delta robot construction using a spatial parallelogram needs further investigation before eventual realization. The relative disadvantage of 3T1R structures is similar to prismatic actuation of about 100% regarding the actuator force/torque of the best 3T0R structure. The 3T1R 4-R̲UU (Figure 19b), for instance, and the 4-R̲RUR (Figure A11e), e.g., require 3 N m and 2 N m actuator torque compared to 1 N m for the 3T0R 3-R̲URR.

5.3. Comparison to Results from the Literature and Evaluation of the Framework

The structures from [123] are evaluated in Figure 20a as an additional benchmark if they exist in the database. Structures with both prismatic (P) and cylindrical (C) joints are omitted since this corresponds to two prismatic DoFs (one passive) within one leg chain. A rather exact reproduction of the actuator forces from [123] was possible for the 3-C̲RR, 3-P̲UU, and 3-UP̲U using the parameters from the paper, as is visible in a comparison of the box plots from [123], cf. Figure A17. For the 3-C̲UR and 3-C̲RU, the values were in the same range but with a different distribution over the 4851 trajectory samples. The difference is already contained within the Jacobian, which differs between the robot assembly modes, visible in Figure A17e,f. In conclusion, the *kinetostatic model* is likely consistent. Still, the reference's collision model (if any) and the selection strategy of the assembly mode may explain the deviation using the same kinematic parameters.

The markers from [123] are dominated by the corresponding Pareto fronts (using the same marker symbol but a different color). Due to the complexity of the *optimization problem*, a possible reason is hard to determine. The additional use of the reference trajectory does not explain the difference from [123], where results are only based on the 27 reference points, as the optimization only based on these 27 reference points leads to qualitatively similar results as Figure 20a, shown in appendix Figure A15. Next to the differences mentioned above, one other reason could be an undocumented further constraint in the paper or the underlying framework of the PhD thesis [20]. An influence that is not less likely is the insufficient exploration ability of the employed single-objective genetic algorithm from [82] instead of the PSO, which was also observed (for the multi-objective variant) in the case study [97].

5.4. Evaluation of Functional Redundancy of 3T1R Parallel Robots

Although symmetric 3T1R PRs do not reach as low actuator forces as their 3T0R counterparts, their principal operational capability presents an interesting result compared to the state of the art, where mostly theoretical structures are shown. They benefit from the redundancy resolution using dynamic programming (see [29]).

In Figure 21b, an illustrative performance map is given for the 4-RP̲UR (Figure A14b). This heat-map representation of the redundant degree of freedom over the trajectory was introduced by Wenger (see [110]) and is discussed in [29] in more detail. The actuator-force representation in the map color corresponds to the objective of the IK optimization. It provides a direct interpretation in this case study since only the transmission of the given external force is considered, and rigid-body dynamics are neglected. Within the optimization of the planar rotation coordinate, the maximal force of any of the actuators is minimized to 5.3–12.2 N by avoiding orientations with higher forces. The upper bound is taken as objective (number 2 in Section 3.2.3) in Figure 20. For reference, the position trajectory is given in Figure 21a, showing the zigzag motion of the x and y coordinates while ascending in the z coordinate and, thereby, sampling the cuboid workspace. The actuator-force/-torque performance maps of the other 3T1R PRs are given in appendix Figures A18 and A19. There, avoiding constraint violations (singularities, collisions, accuracy) dominates the force objective within the motion due to the relatively homogeneous pose-independent transmission ratio of the given external force (regarding the most-stressed actuator). The evaluation already shows that the redundancy-resolution scheme from [29] also provides plausible results for 3T1R PRs regarding the optimality of the trajectory of the redundant coordinate.

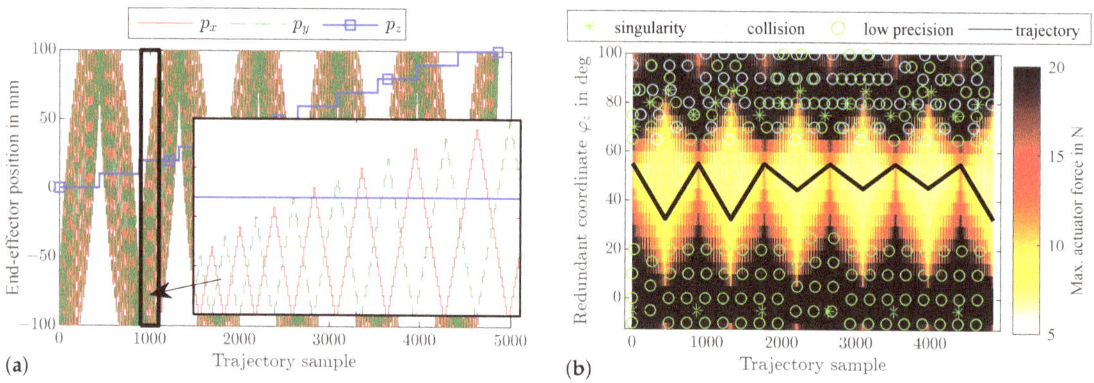

Figure 21. Trajectory of end-effector position (**a**) and of the planar rotation (redundant coordinate) within the force performance map with markers for constraint violation (**b**) for the 4-RPUR.

The inspection of performance maps for other criteria in Figures A20–A23 shows possible disadvantages of the symmetric 3T1R PRs compared to 3T0R PRs, even though all constraints are met within the results of the Pareto front. The position error is much higher for prismatic actuation, except for the 4-PRUR (Figure A14a), cf. Figure A20a. Many 3T1R PRs with revolute actuation exhibit IK singularities within the workspace, which is not possible to detect or avoid numerically with the 10 mm trajectory sampling and the chosen approach of discretely evaluating the condition number, as, e.g., visible in Figure A23b,f,g). Therefore, special attention must be paid to this aspect to investigate the structures further.

The objective of the inverse kinematics influences the motion of the platform rotation, provided that the constraints leave sufficient space for optimization. An example of the trajectories resulting from different IK objectives is given in Figure 22a for a parameterization of the 4-RPUR with radially aligned base-coupling joints (unlike in Figure A14b). The precision is chosen as map color as it also highlights singularities. Repeating the variation of the IK objective for each point of the robot's Pareto front from the optimization (termed "heuristic") and evaluating the objective functions leads to a new Pareto front for each of the IK objectives, shown in Figure 22b,c. Therefore, the selection of the IK objective indirectly influences the performance of the dimensional synthesis. Using the same objectives for the IK redundancy resolution and the dimensional synthesis provides the most promising results and corresponds best to the principle of bilevel optimization. In the specific example of Figure 22b, the precision can be reduced by approximately 5% against the constant-orientation case and 10% against the heuristic approach of a weighted sum of several objectives, depending on the objectives and constraints of the dimensional synthesis. The approach does not provide significant improvement for the installation-space objective, as visible in Figure 22c. Other examples (case studies or robots) show qualitatively similar results. The extent of possible improvement depends on the objectives of the dimensional synthesis and the IK settings that were used to obtain the results, as well as on the robot and task requirements.

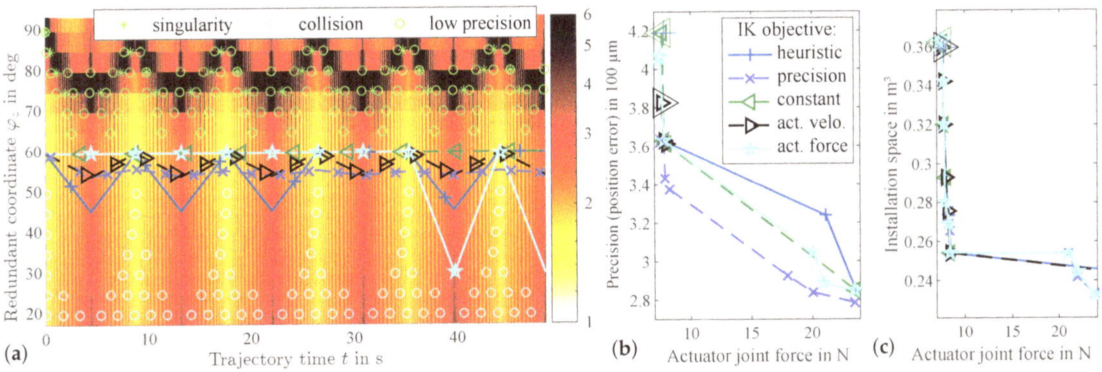

Figure 22. Redundant-coordinate trajectory (**a**) and Pareto fronts (**b**,**c**) for another 4-RPUR structure based on the same parameters with different inverse-kinematics optimization objectives. The method legend in (**b**) holds for all (**a**–**c**). Large markers in (**b**,**c**) represent parameters for the performance map in (**a**).

5.5. Discussion of Redundancy Resolution Within the Dimensional Synthesis

The downside of possible improvements by redundancy resolution is that the (multi-objective) dimensional synthesis optimization results are based on the (single-objective) redundancy-resolution method. The performance of the synthesis with one setting can then not be obtained by other IK settings. This overfitting can be avoided by using generic or no IK objectives (e.g., constant orientation or minimal platform motion) with little degree of freedom (e.g., few states in the dynamic programming) at the cost of lower performance in the synthesis objectives. By these results and considerations, the two underlying assumptions of the paper regarding dimensional synthesis and functional redundancy can be addressed:

Assumption 1. *Optimizing the objective of the dimensional synthesis in an inner-loop redundancy resolution improves the synthesis results in the case of functional redundancy.*

An improvement of the synthesis objective by up to 10% was shown for some examples. Therefore, the proposed inner loop "*is likely to improve the synthesis results*". As the effect varies, the assumption may not hold in all cases or can not be tested generally (for the complete Pareto front, all tasks, and all robot structures).

A thorough statistical evaluation for greater generality would require a metric for comparison of Pareto fronts. This is aggravated by the fact that the density of particles on the front is imbalanced, and gaps in the front are sometimes due to impossibility by constraint violations, sometimes by insufficient exploration. The true form of the front is thereby unknown and only sampled incompletely by the heuristic optimization algorithm. Quantifying the effect in a single-objective optimization would be easier from a methodological point of view. However, it would also be less valuable, as a trade-off between multiple objectives has to be found for an actual robot design. This leads to

Assumption 2. *For a meaningful comparison including functionally redundant robots, the redundancy resolution has to be performed within the dimensional synthesis. Only then can the dominance (outperformance) of one structure compared to another be determined.*

The redundancy resolution should be used to avoid constraint violations and receive feasible results in the first place. Without using the redundancy at all, too many particles (i.e., sets of kinematic parameters) would be unnecessarily discarded for constraint violation, which would deteriorate the results. The assumption can be *argued to be true if redundancy resolution is only defined as avoiding constraint violation* and having no influence otherwise (i.e., leaving the redundant coordinate constant or using some shortest-distance metric for optimization beyond the constraints).

Furthermore, optimizing for a general objective for *the initial poses of the reference trajectory* proved to be a promising approach. This may be influenced by the fact that most of the Pareto-optimal performance maps could be traversed well with constant orientation, and the performance distribution over the redundant coordinate did not change much. It can be assumed that this was not induced by the synthesis (i.e., self-fulfilling by overfitting) since the redundancy resolution was also proven to cope with other, more complicated cases. For the synthesis of other trajectories, this could already be different, and the initial pose may have less influence as extensive nullspace motion would be necessary regardless. Thereby, from the results obtained so far, the assumption *is not likely valid for understanding redundancy resolution as initial-pose optimization.*

The assumption also does *not hold when understanding redundancy resolution as optimization within the reference trajectory* since several further assumptions have to hold, depending on the use case. In multi-objective optimization, the non-redundancy-optimized solution may present a compromise solution, and the optimized solution may be better in one objective but worse in the other objectives and, therefore, not Pareto-dominant. The extent of this effect depends on the selection of objectives and their respective opposition. Optimizing the inverse kinematics for one objective in the design phase may lead to overfitting, and using the structure later with a different IK objective may present worse results. Therefore, the priority of the objectives and the aversion to this overfitting would have to be defined beforehand for the assumption to be true without restrictions.

5.6. Summary of the Case Study

The numerous theoretical results show the efficiency of the synthesis approach for this general 3T0R task. The results for functional redundancy show, in principle, that reasonable solutions exist and a comparison is feasible. However, the performance of redundant structures is weaker than that of non-redundant. The 3T1R structures may still perform better for other objectives, which should be evaluated if a task-specific synthesis is performed. However, due to the simplicity of the 3T0R structures, the advantage of the 3T1R structures to avoid collisions and singularities may not be relevant since 3T0R robots, such as the Delta robot, do not suffer widely from singularities or self-collisions within the workspace. By direct comparison to the results of [123], parts of the methodology could be validated for the 3T0R case.

6. Discussion

The results obtained can answer the paper's *research question*: "How should the optimization problem of the dimensional synthesis be structured if functional redundancy and design optimization are considered?" Solving the problem by nested optimization loops is shown to be feasible due to the reduced number of optimization variables in the single steps. By this, the most efficient method can be used at each stage of the optimization. Design optimization and redundancy resolution should be executed consecutively as far as the problem interconnections allow, which holds for the selected evaluation examples.

Within this paper, the optimization framework for the dimensional synthesis of parallel robots is explained in detail, focusing on several aspects, such as a complete description of the optimization problem for later reference in terms of variables, objectives and constraints, design optimization, and redundancy resolution. A comprehensive overview of the literature on dimensional synthesis provides applicable approaches that are further developed for the requirements of this paper. The concept of hierarchical constraints with continuous penalties presents a general contribution to the state of the art on nonlinear optimization. The field of dimensional synthesis is furthered by promoting multi-objective particle swarm optimization with a bilevel strategy and using the dynamics regressor form for the design optimization. The extensions lead to improved computational efficiency in contrast to existing methods, making the combined synthesis practically possible.

Further gains in efficiency are possible in the non-redundant case by using analytic solutions instead of the general numeric iterative approach for solving the inverse-kinematics problem. However, this has not yet been implemented for the parallel-robot database due to the large number and variety of structures. On the other hand, the iterative solution to the inverse-kinematics problem is necessary to perform gradient-descent optimization in the case of functional redundancy, which is used to generate more valid and generally better solutions for dimensional synthesis.

Several limiting assumptions from the literature regarding included robot structures in the synthesis have been lifted; some more prevent a completely comprehensive robot synthesis. The approach of permutational synthesis is also promising for *non-symmetric alignments of coupling joints* for parallel robots with identical leg chains. The complexity of the optimization for *non-symmetric parallel robots* (i.e., with different leg chains) should be subject to further investigation, as the robots' performance (especially in non-symmetric tasks) may be further improved, as already shown in the literature for some manually designed structures. In principle, the proposed framework would be efficient enough to increase the number of structures by one order of magnitude. Still, the complexity of non-symmetric robots in a geometric permutation could be more extensive than that. For such numbers of structures, the optimization algorithm has to be improved by, e.g., exchanging parameter values between different structures, corresponding to sub-populations in the particle-swarm optimization, and more efficient reuse of previous results of other use cases. Realizing this would require multiple iterative runs of the optimization or synchronization of the currently independent parallel optimizations.

Another direct application of the developed optimization scheme can be found by lifting the restriction of fully parallel robots by allowing *more than one active joint per leg chain* and reducing the number of legs. This step could, in principle, *combine some advantages of serial and parallel robots*, provided that the dimensioning is optimal. For instance, the workspace could be larger than that of a parallel robot of comparable size while still being more energy efficient than serial robots by low moving masses. Concepts for manipulators of this type already exist, but their design was optimized manually or only for one structure. Exploring this relation to both sides presents an interesting research question that requires a *drive-train optimization*, which has not yet been implemented in the proposed framework and would require corresponding component databases or scaling methods already available in the literature. Serial robots are principally supported by the framework—in the theoretical framework as well as in the implementation.

7. Conclusions

The ambition of fast computation regarding the large number of structural permutations from a previous structural synthesis characterizes the proposed algorithm extensions for the *dimensional synthesis*. Compliance with this goal can already be seen by the application within structural synthesis, where several hundred to thousand iterations must be performed for each of the thousands of structural permutations. The proposed concept of *hierarchical constraints* with *continuous penalties* and *early abortion* has not yet been explicitly demonstrated within dimensional synthesis. In combination with particle swarm optimization, also uncommon in the literature on robot synthesis, fast convergence and good exploration were achieved simultaneously despite the high number of optimization variables. Additions to the framework regarding reusing results for initial values, also across similar structures, further improved the computational efficiency, which is crucial to solving a task-specific synthesis problem. Incorporating *design optimization* into the concept was possible by separating independent optimization problems efficiently. *Functional redundancy* was integrated similarly, again by following the principle of bilevel optimization, which can be termed nested or *cascaded optimization*.

The ability of the combined structural and dimensional synthesis to generalize on practical problems is shown in two case studies. The examples show how to transfer technical requirements to the parameters of the synthesis framework and, used as a template, simplify solving new problems.

Funding: This research is based on works that received funding from the Deutsche Forschungsgemeinschaft (DFG) under grant number 341489206. Computation for the case studies was performed on the computing cluster of LUH funded by DFG (project number INST 187/592-1 FUGG).

Data Availability Statement: The original data presented in the study are openly available in the research data repository of the Leibniz Universität Hannover at https://doi.org/10.25835/00p7zgvk for Section 4 and https://doi.org/10.25835/76g8d6a8 for Appendix B. The framework described in Section 3 is published open-source at https://github.com/SchapplM/robsynth-structdimsynth (accesed on 21 February 2025) and the case studies of this paper (including figures and tables) can be reproduced by the MATLAB code provided in https://github.com/SchapplM/robotics-paper_mdpi2025_dimsynth (accesed on 21 February 2025).

Acknowledgments: The author thanks the following persons for proof-reading earlier versions of the manuscript: Tobias Ortmaier (Sections 2–7), Tim-David Job (Sections 2 and 3), Tim-Lukas Habich (Sections 2 and 5–7, Appendix B), Tim Sterneck (Section 4, Appendix A), and Jannik Fettin (Section 4, Appendix A). Contributions to the case study of Section 4: The collection of requirements and computation of external forces in Section 4.2 was carried out majorly by Jannik Fettin in collaboration with Jannik Meyer (LuFI). The prototype realization of Section 4.5 was performed by Jannik Fettin in cooperation with Tim Sterneck (Institute for Mechatronic Systems, IMES.)

Conflicts of Interest: The author declares no conflicts of interest. The funders had no role in the design of the study, in the collection, analyses, or interpretation of data, in the writing of the manuscript, or in the decision to publish the results.

Abbreviations

The following abbreviations are used in this manuscript:

3T0R	three translations and no rotation (also for 3T1R, 3T2R, and 3T3R)
C	cylindrical joint
CAD	computer-aided design
DH	Denavit–Hartenberg
DoFs	degrees of freedom

GA	genetic algorithm
IK	inverse kinematics
MOO	multi-objective optimization
P	prismatic joint
PR	parallel robot
PSO	particle swarm optimization
R	revolute joint
S	spherical joint
SOO	single-objective optimization
U	universal joint

Appendix A. Further Results for the Case Study of Section 4

In the following, further results for the high-payload robot within the naval testbed are shown, as initially discussed in Section 4. The Pareto diagram for collision distance versus material stress from Figure 14 showing parallel robots with three- and four-joint leg chains is extended in Figure A1 for the cases with leg chains with five and six joints.

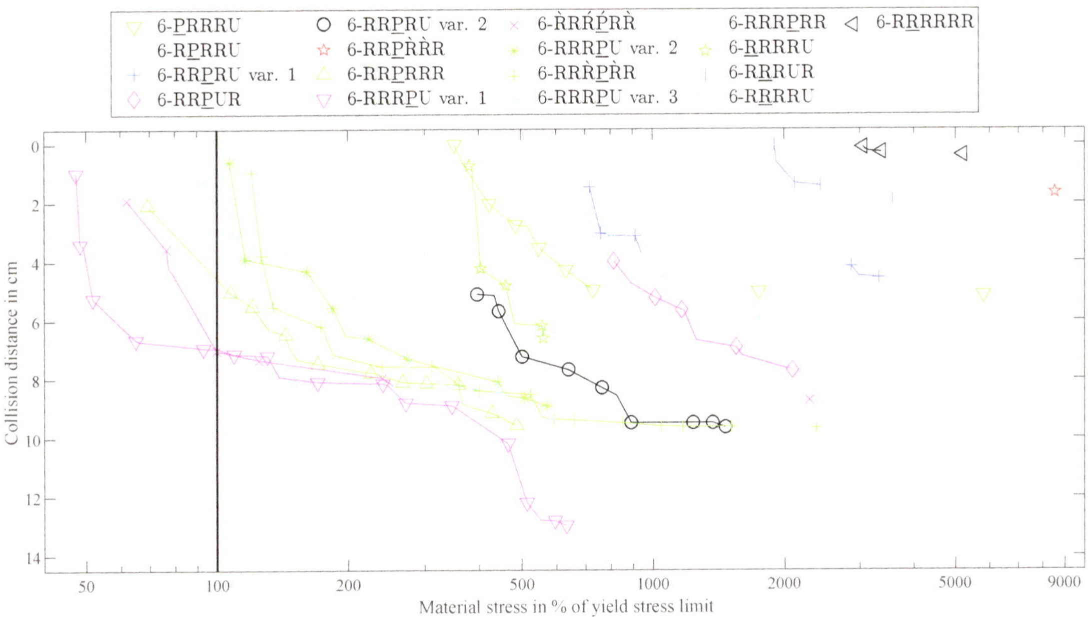

Figure A1. Pareto fronts for the design-oriented objectives for chains with five and six joints and fixed-dimension lightweight links without link design optimization.

The Pareto diagrams for the drive-train objectives velocity and force are separated into revolute joints in Figure A3 and prismatic joints in Figure A4 since the units are different. However, the required rated power of 1–2 kW for many structures is similar for both actuation types. The stronger link dimensioning, visible in Figure A2, can partly explain the markers above the ideal hyperbola. Assumptions regarding transmission ratios (by gear pitch of a spindle or radius of a belt disk), which would allow a unit conversion between rotational and translational quantities, are not included in the synthesis framework.

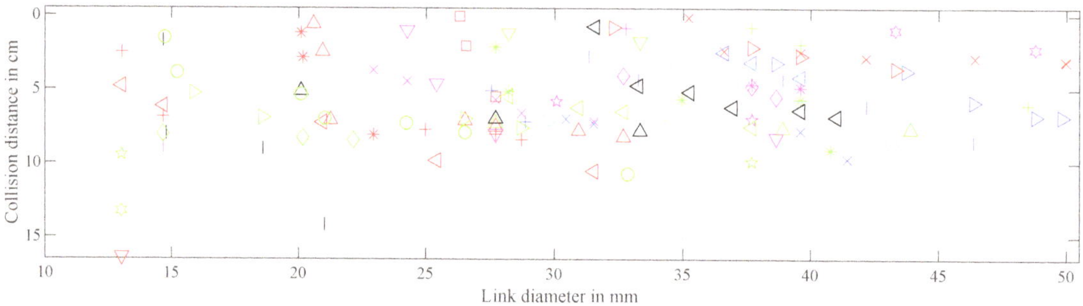

Figure A2. Pareto fronts for the resulting link dimensioning of the results above together with the collision distance as third optimization objective. Only Pareto-dominant particles in these two criteria are shown. The legend is identical to that of Figures A3 and A4.

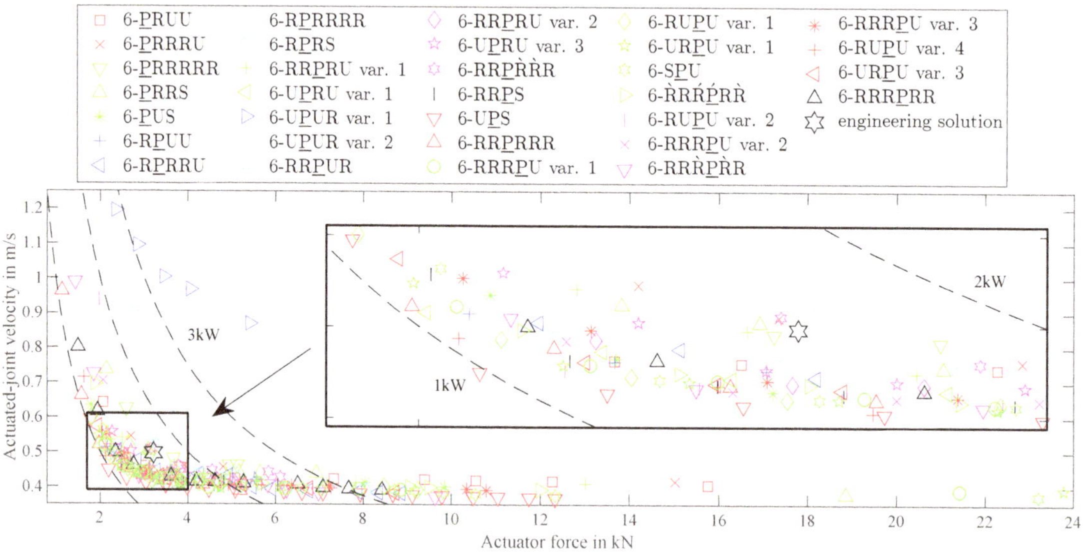

Figure A3. Pareto fronts for the actuator-oriented objectives for prismatic actuation.

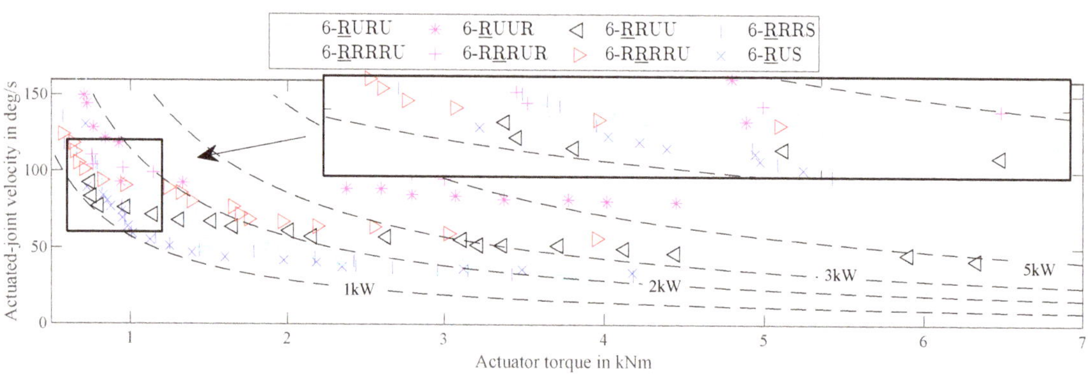

Figure A4. Pareto fronts for the actuator-oriented objectives for revolute actuation.

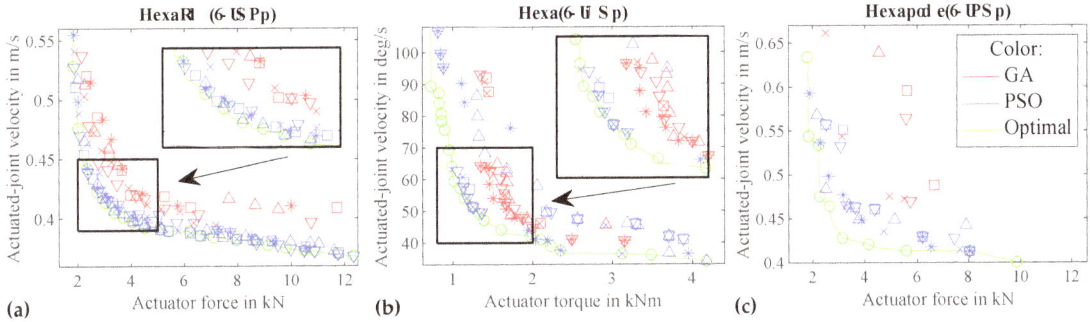

Figure A5. Comparison of particle swarm optimization (blue) and genetic algorithm (red) for the same settings as Figures A3 and A4 for three different parallel robots (**a**–**c**). Each independent repetition has its own marker. The optimal solution after several iterations (from Figures A3 and A4) is marked in green for reference. Identical markers were thinned out to improve visibility.

The same drive-train objectives are used for an evaluation of the optimization algorithm. The synthesis was performed with the multi-objective genetic algorithm and particle swarm optimization with a random initial population without prior results, set up equally for both. Default values of the algorithms [82,102] were used with 200 individuals and a time limit of eight hours of computation time on the same partition of an Intel Xeon computing cluster [120] with MATLAB version 2024b. The experiment was repeated five times with random seeds. The results are depicted in Figure A5 for the 6-UPS, 6-RUS, and 6-PUS. The genetic algorithm shows a worse exploration and performance than the PSO, which clearly dominates the GA's results. For this reason, only the PSO is used within the paper.

The feasibility of the proposed concept of hierarchical constraints from Section 3.1 is evaluated based on the computation times of the single particles of the synthesis above. All successful parallel robots (187 structures with different chains and coupling-joint alignments) were included with a total of 1.2 million particles. A box plot of the computation times over the resulting fitness value (increasing from left to right) is depicted in Figure A6. The fitness values correspond to the constraints from Section 3.2.2 that were violated and led to the abortion. The targeted increasing relation of computation time and validity is evident in the first constraints 1–9 that account for the plausibility by only checking parameters (10 ms in median) or reference points (up to 5 s in median). An abortion at the trajectory-related constraints 17–26 (which includes computing all previous constraints) exhibits a required time of 5–20 s. Reaching the design optimization (constraint 27) takes about 100 s. The outliers mainly result from structures with many assembly modes (due to multi-joint leg chains) that are all checked. Better handling of this aspect may present a possibility for increasing efficiency in the future. The evaluation has to be regarded rather qualitatively, as different robots are combined in the box plots, which presents a bias and explains the low median time of the "success" case of only 40 s.

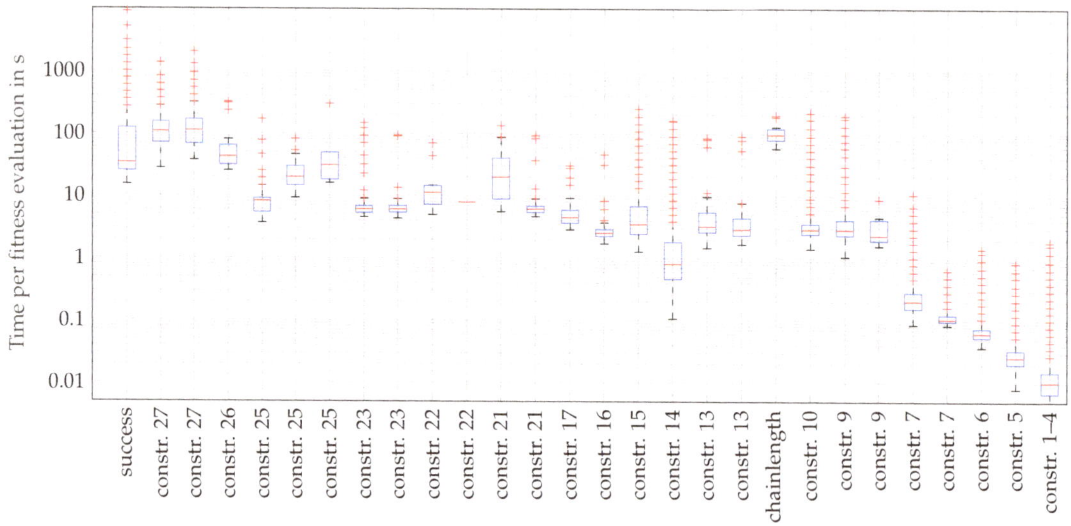

Figure A6. Computation time of the fitness function (logarithmic scale) in relation to the number of the violated constraint (from Section 3.2.2) that led to abortion. The class "chainlength" comes from checking if the length of a leg chain that includes a prismatic joint exceeds the maximum allowed length. Outlier markers were thinned out (193 of 97 k left over) to increase visibility of the data. Duplicate constraints result from multiple similar checks, e.g., for different Jacobian matrices in constraint 9.

One parameterization is chosen for each of the valid parallel robots from the Pareto diagrams. The selection is based on the Pareto diagram of collision distance vs. actuator power in Figure 17. Both criteria are normalized to their best-achieved values (for all structures), and the best particle for equal weighting is chosen to illustrate the results' appearance. The performance values and kinematic parameters of these reference solutions are summarized in Table A1 for prismatic and Table A3 for revolute actuation. Kinematic sketches are given in Figures A7–A9.

Table A1. Summary of one typical particle for each robot with *prismatic actuation*, visualized in Figures A7 and A8. Abbreviations: "coll." (collision distance), "velo." (actuator velocity), "cond." (largest Jacobian condition number), n (number of optimization variables), r_b (base radius), d_b (base-joint pair distance), γ_b (base-joint inclination), r_p (platform radius), d_p (platform-joint pair distance), e_{link} (material strength of the links), d_{link} (diameter of the links), "Eng. Sol." engineering solution. Continued horizontally in Table A2.

Robot	Performance					Kinematic and Design Parameters							
	coll. mm	Power kW	Force kN	velo. m/s	cond.	n	r_b mm	d_b mm	γ_b deg	r_p mm	d_p mm	e_{link} mm	d_{link} mm
6-PRUU, C-R	18	1.15	2.17	0.53	25	16	613	219	37	165	346	11	27
6-PRRRU, V-R	0	1.45	2.73	0.53	24	17	495	618	0	212	219	10	40
6-PRRRRR, C-T	10	1.82	3.47	0.52	50	21	410	255	173	152	275	14	31
6-PRRS, V-V	69	1.62	2.89	0.56	23	16	551	254	0	194	273	10	36
6-PUS, V-V	35	1.28	2.46	0.52	23	13	534	257	0	174	320	4	33

Table A1. *Cont.*

Robot	Performance					Kinematic and Design Parameters							
	coll. mm	Power kW	Force kN	velo. m/s	cond.	n	r_b mm	d_b mm	γ_b deg	r_p mm	d_p mm	e_{link} mm	d_{link} mm
6-RPUU, R-R	69	1.42	3.03	0.47	25	13	722	159	0	183	287	12	32
6-RPRRU, C-R	14	1.94	4.06	0.48	30	16	536	269	33	210	233	10	37
6-RPRRRR, C-c	11	2.57	5.92	0.43	116	19	419	146	10	148	—	18	43
6-RPRS, C-V	27	1.47	2.84	0.52	20	15	407	126	83	166	297	10	36
6-RRPRU var. 1, T-T	2	1.43	2.66	0.54	31	14	502	203	0	157	305	19	38
6-UPRU var. 1, T-T	25	1.06	2.06	0.52	17	11	691	196	0	171	330	15	33
6-UPUR var. 1, R-T	35	3.73	3.08	1.21	21	11	403	1112	0	216	186	8	46
6-UPUR var. 2, V-T	3	1.87	2.61	0.71	34	10	401	80	0	176	326	18	43
6-RRPUR, C-R	11	1.99	3.19	0.62	35	13	398	203	14	191	275	20	49
6-RRPRU var. 2, V-T	25	1.32	2.69	0.49	19	13	411	230	0	177	324	10	40
6-UPRU var. 3, C-T	48	1.41	2.93	0.48	37	11	470	237	14	170	289	16	32
6-RRPR̀R̀R, V-T	10	1.76	3.31	0.53	35	15	408	217	0	164	243	16	43
6-RRPS, T-v	113	1.43	3.05	0.47	28	13	398	1097	0	231	—	1	39
6-UPS, T-V	155	1.24	2.64	0.47	21	11	399	1219	0	212	226	1	13
6-RRPRRR, R-R	50	1.10	1.30	0.85	15	18	408	1069	0	184	201	1	36
6-RRRPU var. 1, C-R	55	1.27	2.69	0.47	32	16	406	154	72	185	274	3	35
6-RUPU var. 1, C-T	74	1.27	2.47	0.51	28	13	615	186	111	181	308	1	23
6-URPU var. 1, V-R	74	1.15	2.40	0.48	26	12	557	111	0	171	306	1	14
6-SPU, V-T	84	1.15	2.31	0.50	23	8	651	130	0	175	301	1	13
6-R̀RŔPRR̀, R-V	64	1.39	3.01	0.46	25	17	700	219	0	176	273	7	31
6-RUPU var. 2, T-T	76	1.29	2.77	0.46	23	11	725	225	0	188	292	1	32
6-RRRPU var. 2, C-t	42	1.33	3.03	0.44	17	13	399	660	45	185	—	15	30
6-RRR̀PR̀R, C-R	49	1.15	1.64	0.70	15	16	428	807	81	161	198	3	35
6-RRRPU var. 3, V-R	29	1.27	2.35	0.54	27	14	454	240	0	167	344	5	24
6-RUPU var. 4, C-T	45	1.06	2.38	0.45	19	12	658	177	114	176	310	7	31
6-URPU var. 3, R-T	48	1.17	2.54	0.46	22	11	442	207	0	172	308	14	28
6-RRRPRR, C-c	31	1.23	2.20	0.56	21	17	465	224	50	141	—	3	35
Eng. Sol.	151	1.59	3.21	0.50	24	11	670	156	0	185	275	1	13

The kinematic complexity of the solutions remains manageable for the structures with three or four joints per chain. For some structures with five and six joints, like 6-RRRPRR (Figure A8j), feasible solutions exist with enough space between components for design and assembly. The same holds for 6-RRRPU (Figure A8k); however, the small angle between a joint axis and a link complicates the mechanical design (visible by the relation of the d and a parameters in Table A2).

The necessary stronger dimensioning of structures with longer levers (e.g., 6-RPRRRR, 6-RRPRRR, 6-RRPUR), mentioned in Section 4.4, can be observed by the higher values in Table A1 for material strength e_{link} and link diameter d_{link} resulting from the design optimization.

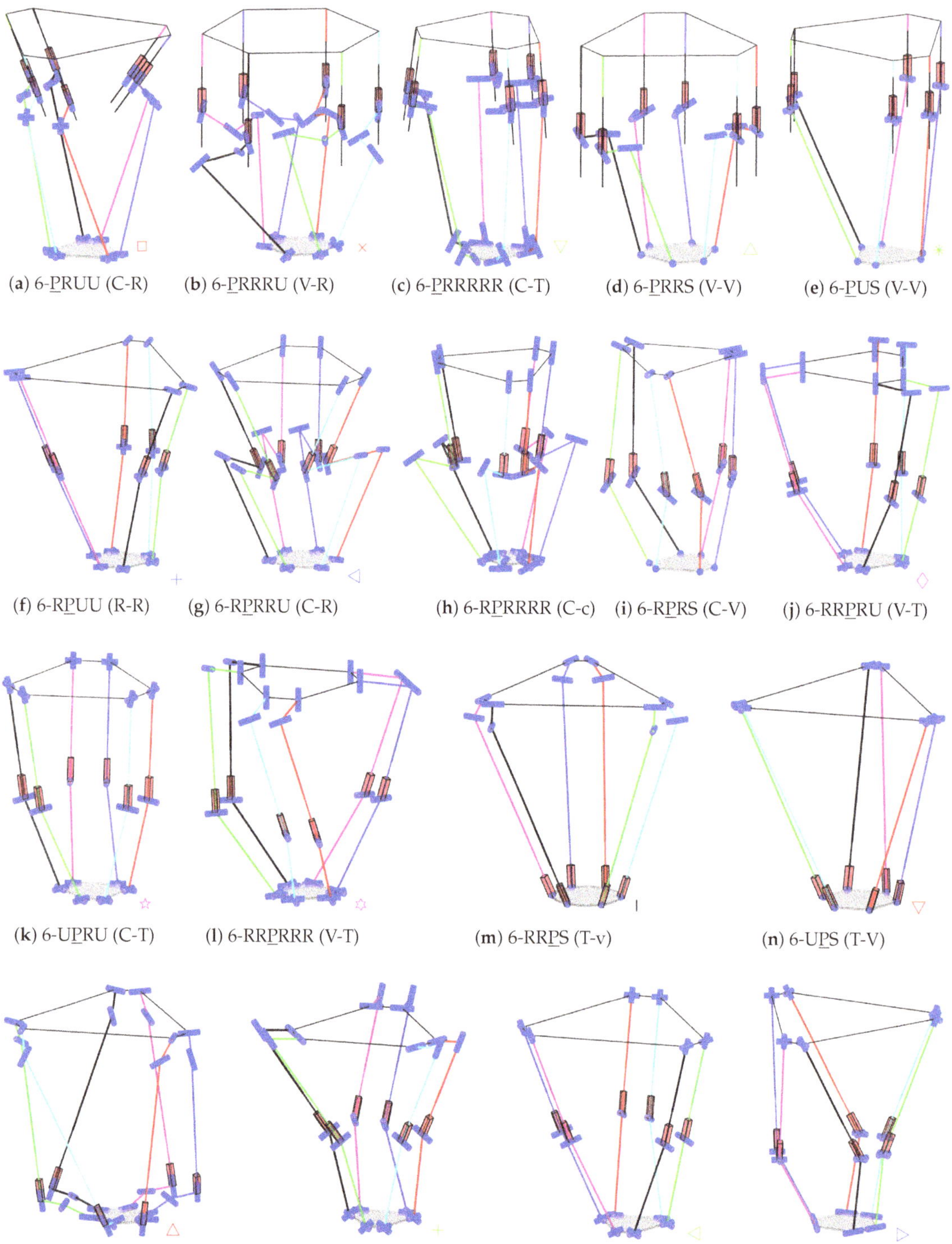

(**a**) 6-PRUU (C-R) (**b**) 6-PRRRU (V-R) (**c**) 6-PRRRRR (C-T) (**d**) 6-PRRS (V-V) (**e**) 6-PUS (V-V)

(**f**) 6-RPUU (R-R) (**g**) 6-RPRRU (C-R) (**h**) 6-RPRRRR (C-c) (**i**) 6-RPRS (C-V) (**j**) 6-RRPRU (V-T)

(**k**) 6-UPRU (C-T) (**l**) 6-RRPRRR (V-T) (**m**) 6-RRPS (T-v) (**n**) 6-UPS (T-V)

(**o**) 6-RRPRRR (R-R) (**p**) 6-RRPRU (T-T) (**q**) 6-UPRU (T-T) (**r**) 6-UPUR (R-T)

Figure A7. Visualization of the results with *prismatic actuation* from Table A1 with corresponding marker from Figures 17 and A3—part 1.

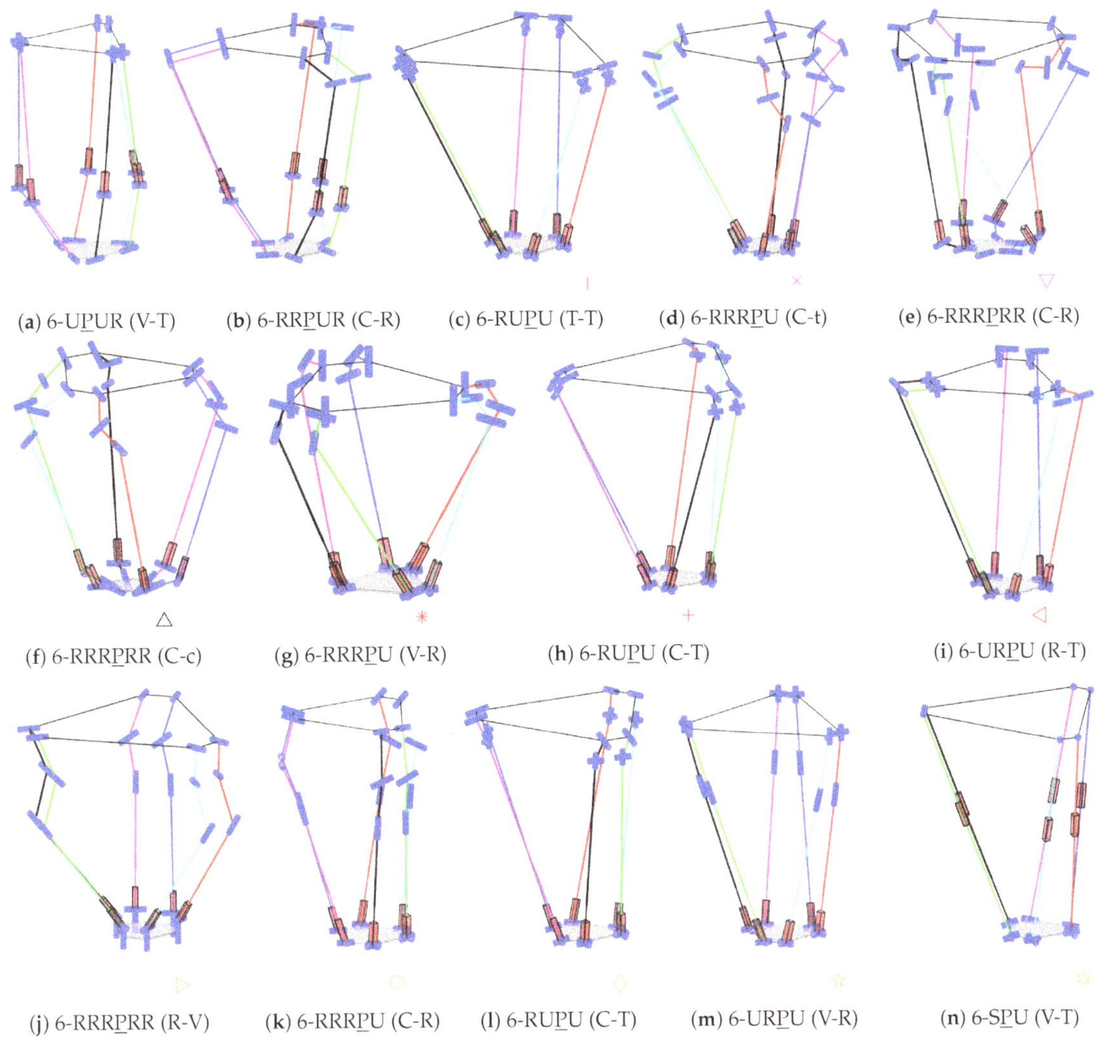

Figure A8. Visualization of the results with *prismatic actuation* from Table A1 with corresponding marker from Figures 17 and A3—part 2.

Fewer results with *revolute actuation* are available due to the lower number of possible combinations and their lower success rate. In total, nine robots with revolute actuation fulfill the kinematic requirements (self-collision free, in installation space) when using lightweight limbs: five in Figure 14 and four in Figure A1. Due to the high material stress, the strong required dimensioning leads partly to self-collisions and partly still does not lead to valid material stress (for 70 out of 926 investigated variants). For prismatic actuation, this is only the case for 85 out of 4944 variants.

Table A2. Leg chain kinematic optimization parameters for results of Table A1, using modified DH notation. Fixed structural parameters are highlighted in light gray color.

PR	Leg-Chain Kinematic Parameters																
	d_1 mm	α_2 deg	a_2 mm	θ_2 deg	d_2 mm	α_3 deg	a_3 mm	θ_3 deg	d_3 mm	α_4 deg	a_4 mm	θ_4 deg	d_4 mm	a_5 mm	d_5 mm	a_6 mm	d_6 mm
6-PRUU, C-R	act.	2		—		79	55	—	167			—		982	195		
6-PRRRU, V-R	act.	18		—		68	123	—	217		123	—	246	837	23		
6-PRRRRR, C-T	act.	8		—		84	38	—	92	87	12	—	125	602	723	93	5
6-PRRS, V-V	act.	58		—		76	36	—	168	51	649	—	455				
6-PUS, V-V	act.	86		—				—		90	1023	—	205				
6-RPUU, R-R		68		90	act.	4		—				—		748	10		
6-RPRRU, C-R		7		3	act.	49		—			422	—	79	866	83		
6-RPRRRR, C-c		0		59	act.	78		—		73	204	—	188	806	122	12	39
6-RPRS, C-V		81		90	act.	41		—		3	512	—	275				
6-RRPRU var. 1, T-T		86	168	—	135			25	act.			—		535	192		
6-UPRU var. 1, T-T				—				80	act.			—		682	40		
6-UPUR var. 1, R-T				—				45	act.			—				535	1
6-UPUR var. 2, V-T				—					act.			—				475	33
6-RRPUR, C-R			373	—	4				act.			—				360	238
6-RRPRU var. 2, V-T		85	290	—	123				act.			—		597	182		
6-UPRU var. 3, C-T				—					act.			—		617	131		
6-RRPR̀R̀R, V-T		72	305	—	174				act.			—		561	379	7	14
6-RRPS, T-v		86	136	—	79	77		90	act.	75		—					
6-UPS, T-V				—		89		90	act.	82		—					
6-RRPRRR, R-R		82	97	—	161	55		82	act.	0		—		187	81	74	187
6-RRRPU var. 1, C-R		79	283	—	122	73	30	—	318			46	act.				
6-RUPU var. 1, C-T		74	156	—	40			—				90	act.				
6-URPU var. 1, V-R				—		89	16	—	429			31	act.				
6-SPU, V-T				—				—					act.				
6-R̀RR̀ṔRR̀, R-V		65	182	—	237	58	275	—	201			27	act.			6	80
6-RUPU var. 2, T-T		87	39	—	70			—					act.				
6-RRRPU var. 2, C-t		86	280	—	27		270	—	75				act.				
6-RRR̀PRR, C-R		73	180	—	169		173	—	54				act.			74	163
6-RRRPU var. 3, V-R		73	101	—	29	87	85	—	113				act.				
6-RUPU var. 4, C-T		33	23	—	126			—					act.				
6-URPU var. 3, R-T				—		86	232	—	5				act.				
6-RRRPRR, C-c		90	181	—	139	90	105	—	151				act.			28	170
Eng.				—		90		18	act.	88		—					

The eight remaining structures are shown in Figure A9 and Table A3. All of them require a significantly stronger link design than the robots with prismatic actuation from Table A1. The structures with an actuated second joint (6-RRRS, 6-RRRRU, and 6-RRRUR) are not suitable for realization since the moving mass and volume of a motor of up to 2.1 kW rated power will likely lead to self-collisions and degrade dynamic performance criteria in the selected base-joint alignment (pairwise-conical, tangential, or radial), cf. Figure A9a,d,f. The motor selection was excluded from the synthesis and design optimization to focus on the link design. The Hexa robot in Figure A9b presents the classical design already reported in the literature [8], which has been found without prior detailed knowledge within the synthesis algorithm and could be constructed and realized with moderate effort, provided joints for the given high material stresses are available. The other structures could at least be built in theory; however, their increased complexity is not countered by any advantages, giving them the character of an academic example.

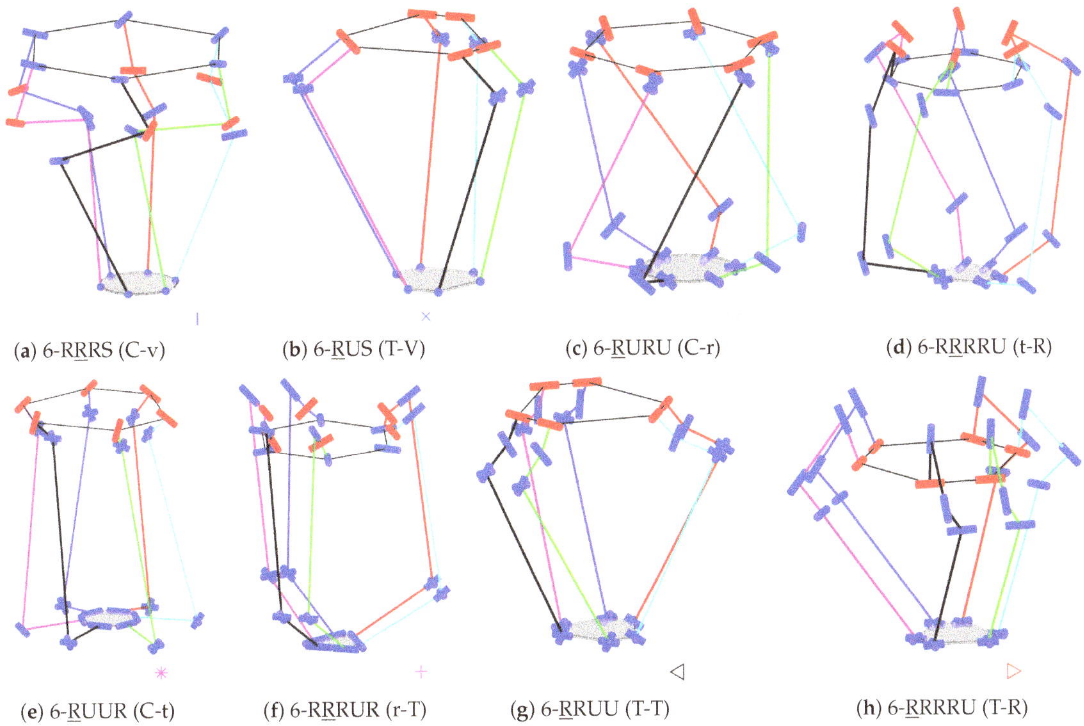

Figure A9. Visualization of the results with *revolute actuation* from Table A3 with corresponding marker from Figures 17 and A4.

Table A3. Summary of the naval-testbed results (part 2 of Table A1) for structures with *revolute actuation*, continued horizontally in Table A4).

Robot	Performance					Kinematic and Design Parameters							
	coll. mm	Power kW	Torque Nm	velo. deg/s	cond.	n	r_b mm	d_b mm	γ_b deg	r_p mm	d_p mm	e_{link} mm	d_{link} mm
6-RURU, C-r	59	1.60	927	99	17	15	401	457	66	236	—	10	40
6-RRRRU, t-R	29	1.41	781	104	12	16	419	—	0	177	147	10	33
6-RUUR, C-t	41	1.99	1170	97	26	15	399	560	131	173	—	10	40
6-RRRUR, r-T	8	2.10	1022	118	12	16	408	—	0	114	164	13	33
6-RRUU, T-T	47	1.42	998	81	28	16	403	234	0	177	282	7	39
6-RRRRU, T-R	27	1.40	774	104	20	18	398	326	0	168	288	10	40
6-RRRS, C-v	45	1.25	1331	54	36	17	537	661	77	229	—	10	40
6-RUS, T-V	67	1.21	855	81	19	14	398	208	0	184	297	9	32

Table A4. Leg chain kinematic optimization parameters for results of Table A3, using modified DH notation.

PR	Leg-Chain Kinematic Parameters												
	d_1 mm	α_2 deg	a_2 mm	d_2 mm	α_3 deg	a_3 mm	d_3 mm	α_4 deg	a_4 mm	d_4 mm	a_5 mm	d_5 mm	a_6 mm	d_6 mm
6-RURU, C-r		88	71	9					1080	256	375	13		
6-RRRRU, t-R		71	163	112		276	317		962	283	426	104		
6-RUUR, C-t		90	162	81					963	834			383	70
6-RRRUR, r-T		86	105	12		0	158		913	553			545	89
6-RRUU, T-T		77	149	68	85	234	2				1074	182		
6-RRRRU, T-R		90	18	243	60	9	434		204	99	803	115		
6-RRRS, C-v		60	346	62	34	585	146	3	801	554				
6-RUS, T-V		37	343	31				0	1129	431				

Appendix B. Further Results for the Case Study of Section 5

Further results for the case study of Section 5 are given below. Sketches of each possible resulting structure are shown in Figures A10–A14 for redundant and prismatic actuation and for 3T0R and 3T1R motion. The kinematic parameters are selected from the Pareto diagrams in Figure 20 based on an equal weight for both criteria, normalized to the best-found value of each criterion. The corresponding parameters are given in Tables A5 and A6 for prismatic and Tables A7 and A8 for revolute actuation.

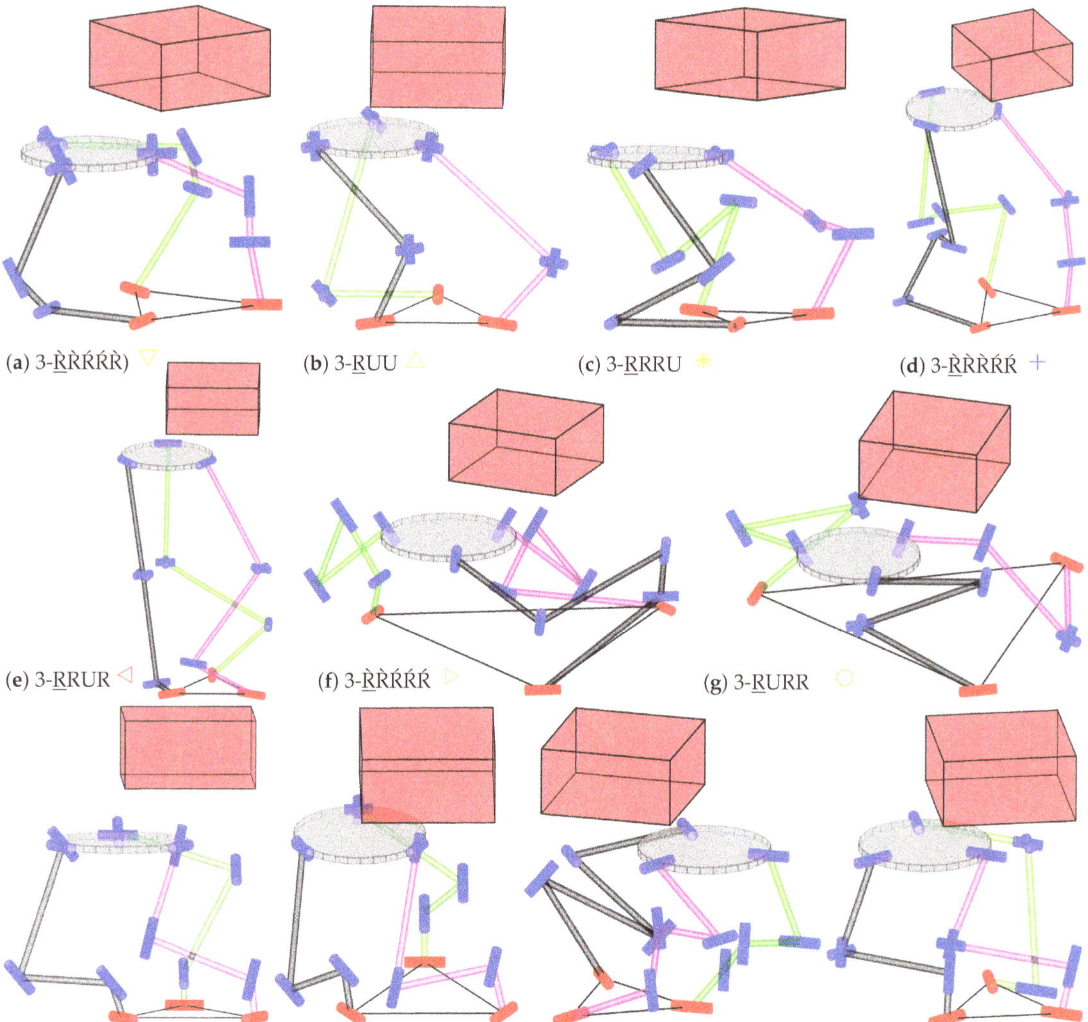

(**a**) 3-R̲̀R̀ŔŔR̀) ▽ (**b**) 3-R̲UU △ (**c**) 3-R̲RRU ✳ (**d**) 3-R̲̀R̀R̀ŔŔ +

(**e**) 3-R̲RUR ◁ (**f**) 3-R̲̀R̀ŔŔŔ ▷ (**g**) 3-R̲URR ○

(**h**) 3-R̲̀ŔŔŔR̀ ◇ (**i**) 3-R̲RRU ☆ (**j**) 3-R̲̀ŔŔR̀R̀ (**k**) 3-R̲RUR

Figure A10. Visualization of *3T0R* parallel robots for the handling task with *revolute* actuation. Markers are consistent with Figure 20.

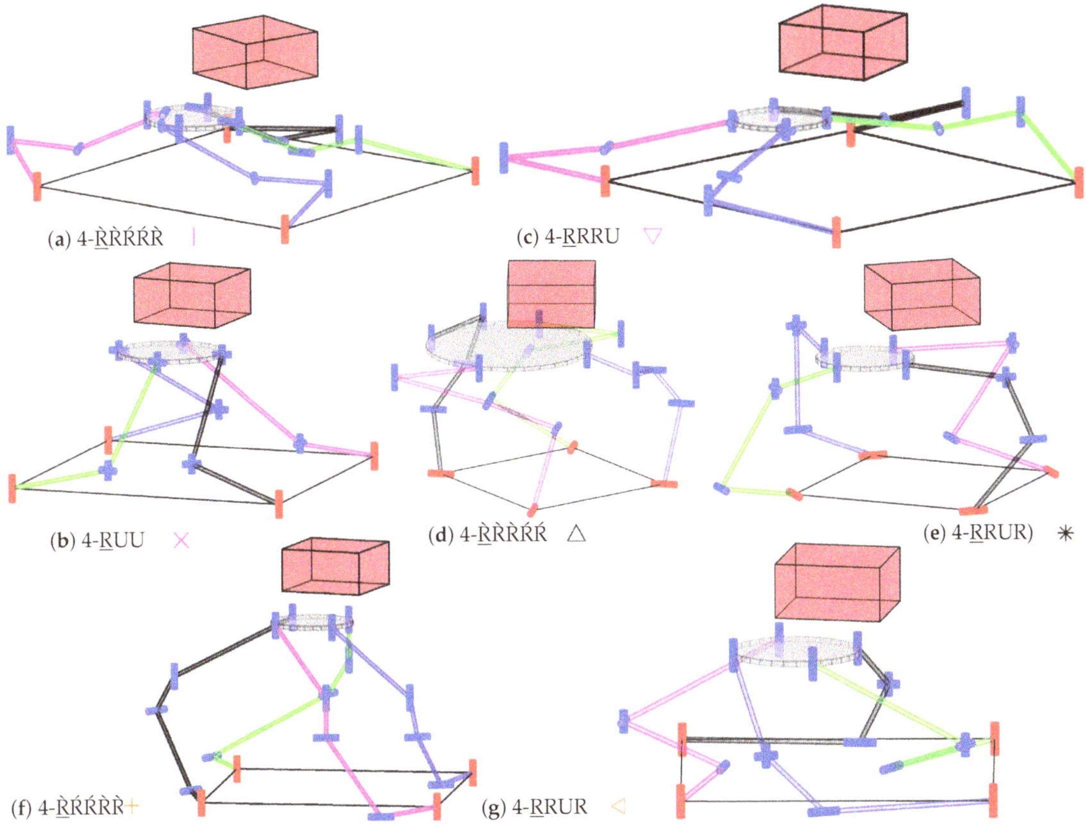

Figure A11. Visualization of *3T1R* parallel robots for the handling task with *revolute* actuation.

The Pareto diagram in Figure A15 corresponds to the one in Figure 20a. It is based on the optimization only using the 27 reference points, not the reference trajectory with 4851 samples. Thereby, the results are based on the same assumptions as the study in [123].

For a direct comparison with [123], sketches of the optimal results from the paper are given in Figure A16. Especially for 3-PUU (Figure A16b), 3-CRU (Figure A16c), and 3-CUR (Figure A16d), large structures have been found by [123] that are outperformed by the results of this replication of the study. The box plots from the paper (Figure A17b,d,e) are compared to the reproduction by the proposed synthesis toolchain (in Figure A17a,c,e) for different criteria, as explained in Section 5.2.

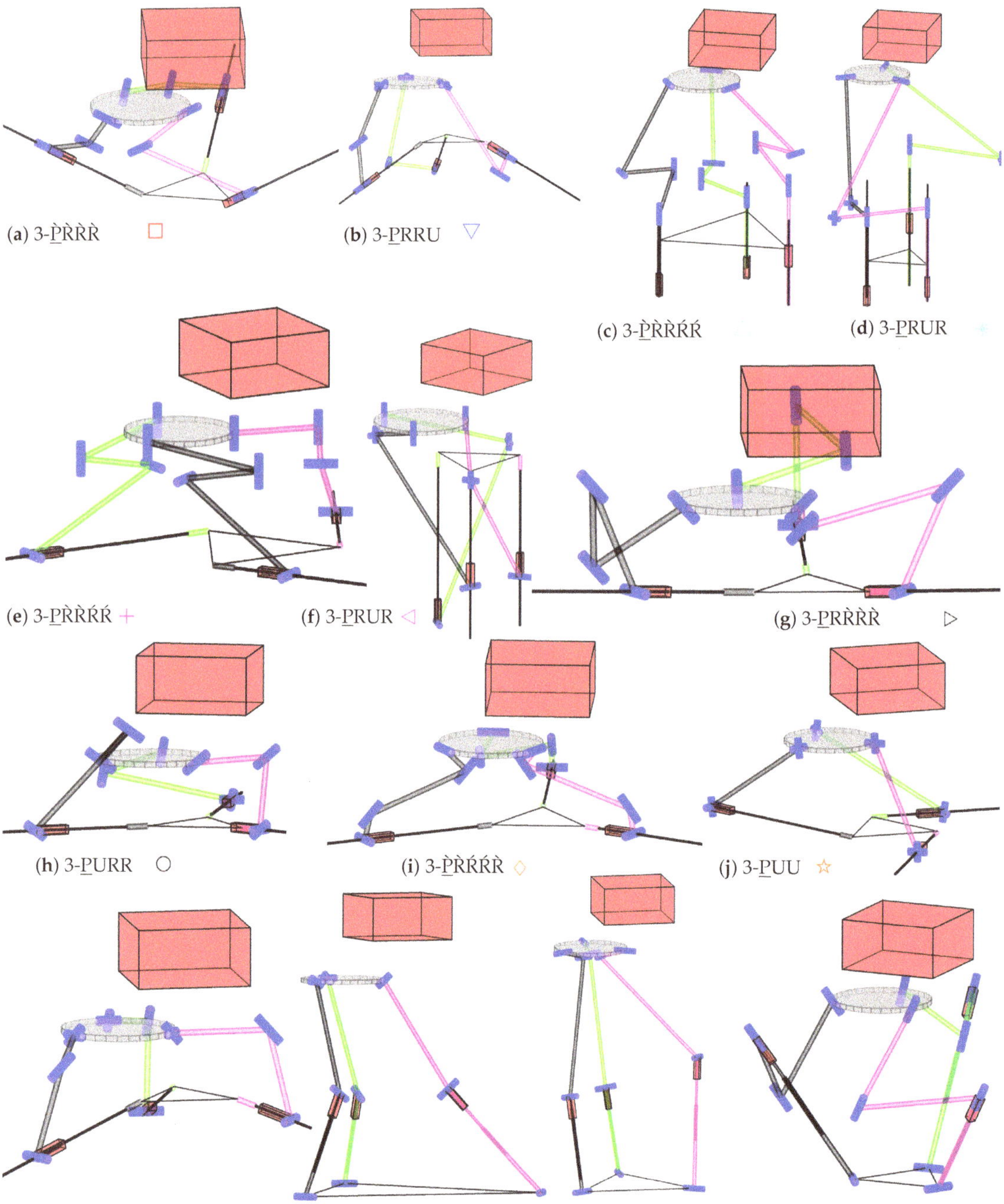

(**a**) 3-P̲̀R̀R̀R̀ □ (**b**) 3-P̲RRU ▽ (**c**) 3-P̲̀R̀R̀ŔŔ (**d**) 3-P̲RUR

(**e**) 3-P̲R̀R̀ŔŔ + (**f**) 3-P̲RUR ◁ (**g**) 3-P̲RR̀R̀R̀ ▷

(**h**) 3-P̲URR ○ (**i**) 3-P̲̀R̀ŔŔR̀ ◇ (**j**) 3-P̲UU ☆

(**k**) 3-P̲RRU ✡ (**l**) 3-RP̲RU | (**m**) 3-R̀P̲ŔŔR̀ × (**n**) 3-RP̲̀R̀R̀R̀ ▽

Figure A12. Visualization of *3T0R* parallel robots for the handling task with *prismatic* actuation. Markers are consistent with Figure 20. Continued in Figure A13.

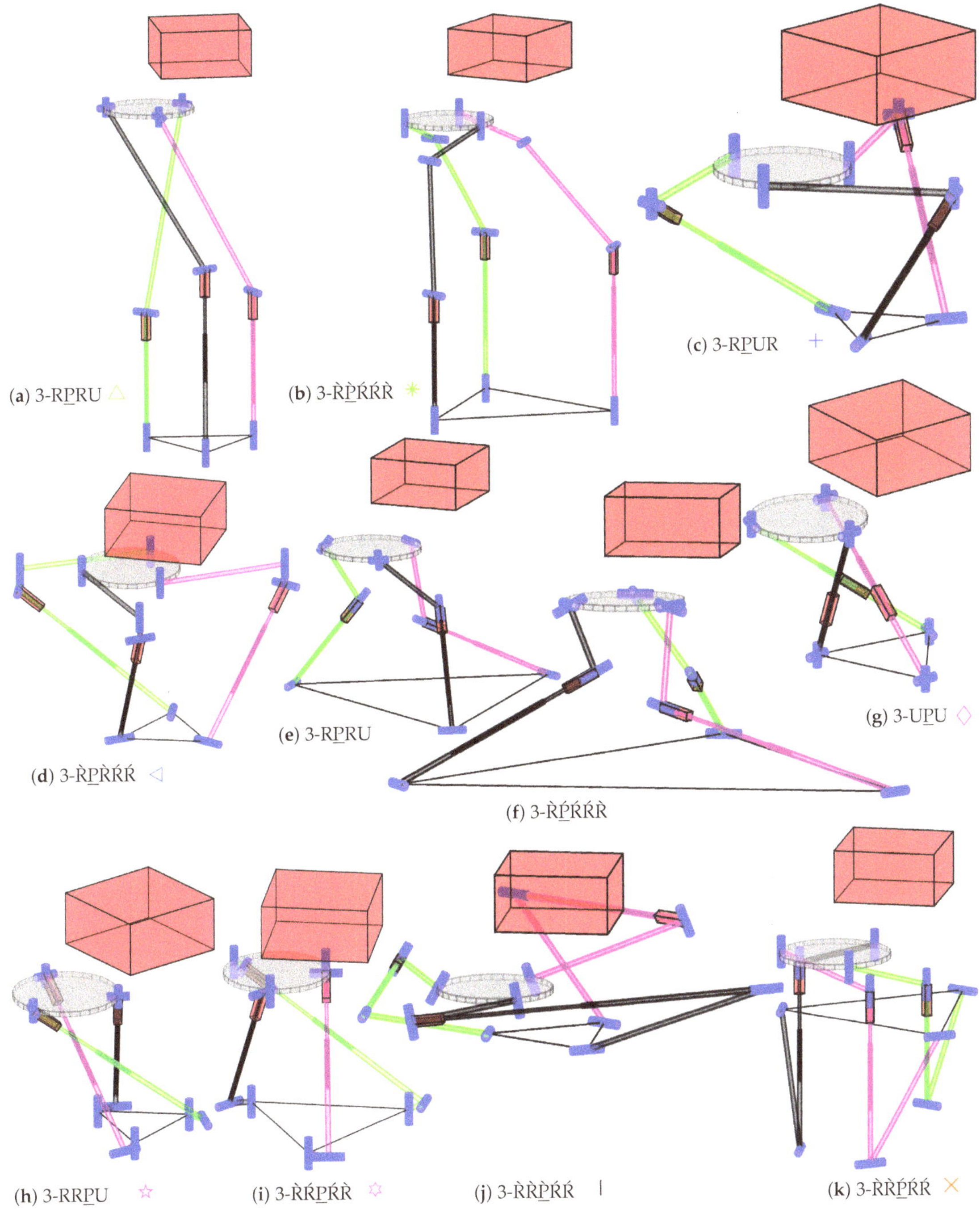

(**a**) 3-RP̲RU △ (**b**) 3-R̀P̲̀ŔŔR̀ ✳ (**c**) 3-RP̲UR + (**d**) 3-R̀P̲R̀ŔŔ ◁ (**e**) 3-RP̲RU (**f**) 3-R̀Ṕ̲ŔŔR̀ (**g**) 3-UP̲U ◇ (**h**) 3-RRP̲U ☆ (**i**) 3-R̀ŔP̲ŔR̀ ✡ (**j**) 3-R̀R̀P̲̀ŔŔ | (**k**) 3-R̀R̀Ṕ̲ŔŔ ×

Figure A13. Visualization of *3T0R* parallel robots (part 2), continuation of Figure A12.

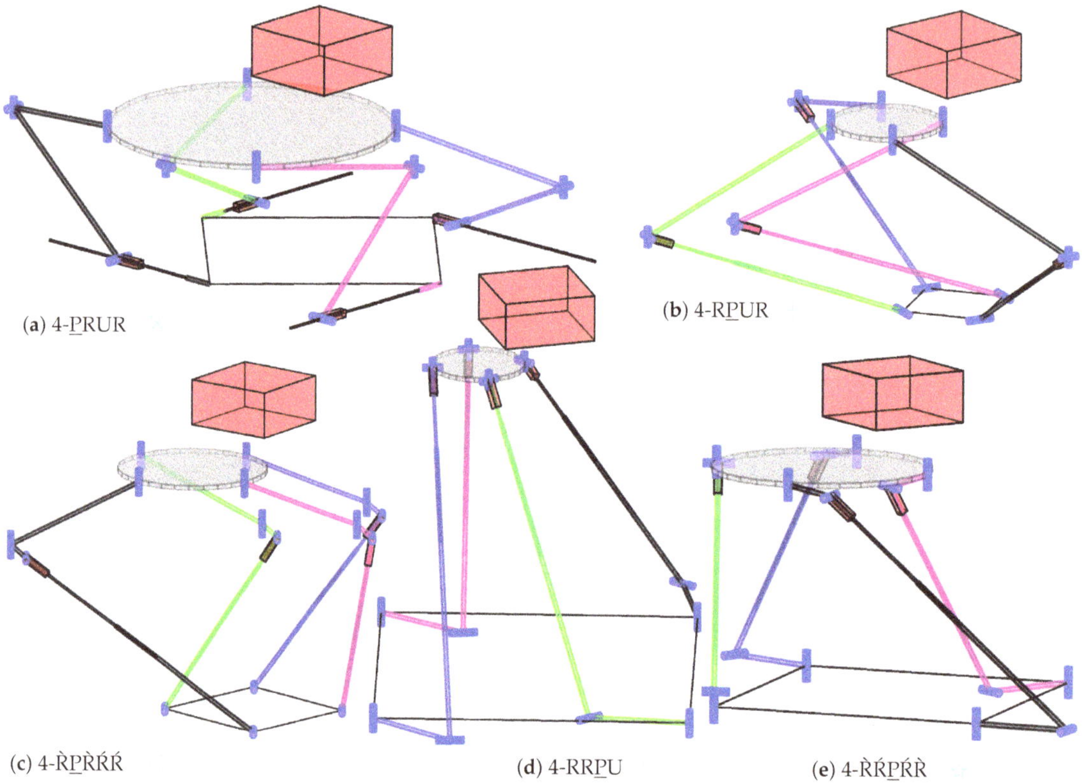

Figure A14. Visualization of *3T1R* parallel robots for the handling task with *prismatic* actuation.

Table A5. Summary of the handling robot results (part 1, prismatic actuation).

Robot	Performance							Kinematic Parameters				
3T0R:	**coll.** mm	**inst.spc.** m^3	**footprt.** m^2	**Power** W	**Force** N	**velo.** m/s	**cond.**	n	r_b mm	γ_b deg	r_p mm	γ_p deg
3-P̲̀R̀R̀R̀, c-m	1	0.04	0.23	8.3	7.8	61	1.2	11	106	120	101	120
3-P̲RRU var. 1, c-t	0	0.08	0.29	11.5	9.4	70	1.4	11	100	63	100	0
3-P̲̀R̀R̀ŔŔ, v-t	0	0.08	0.21	34.1	13.4	146	6.0	12	207	0	100	0
3-P̲RUR var. 1, v-r	0	0.11	0.21	26.6	8.7	176	7.2	10	110	0	100	0
3-P̲R̀R̀ŔŔ, t-v	4	0.07	0.27	17.8	10.2	100	4.6	11	156	0	100	0
3-P̲RUR var. 2, v-v	3	0.07	0.19	14.8	6.7	126	6.3	10	122	0	121	0
3-P̲RR̀R̀R̀, c-c	26	0.07	0.44	6.4	5.1	72	1.4	13	100	95	101	45
3-P̲URR, r-c	5	0.06	0.39	8.2	6.0	78	1.0	10	100	0	101	36
3-P̲̀R̀ŔŔR̀, r-t	1	0.05	0.30	14.9	10.0	86	2.6	11	103	0	100	0
3-P̲UU, t-r	0	0.05	0.24	19.7	8.1	140	7.3	7	113	0	100	0
3-P̲RRU var. 2, c-t	4	0.06	0.23	12.7	7.7	94	3.6	10	100	73	100	0
3-RP̲̀R̀R̀R̀, t-c	42	0.08	0.31	5.3	6.4	47	2.0	11	108	0	100	55
3-RP̲RU var. 1, v-v	42	0.06	0.12	16.7	11.3	85	3.4	8	136	0	100	0
3-R̀P̲̀ŔŔR̀, v-v	28	0.09	0.15	23.6	13.6	99	5.6	10	236	0	100	0
3-RP̲UR, r-v	49	0.05	0.24	12.8	10.5	70	3.9	7	100	0	100	0
3-R̀P̲R̀ŔŔ, r-v	1	0.06	0.21	12.8	10.9	68	5.6	9	104	0	100	0

Table A5. *Cont.*

Robot	Performance							Kinematic Parameters				
3T0R:	coll. mm	inst.spc. m^3	footprt. m^2	Power W	Force N	velo. m/s	cond.	n	r_b mm	γ_b deg	r_p mm	γ_p deg
3-RP̱RU var. 2, t-t	8	0.06	0.16	11.5	10.7	62	3.3	8	330	0	111	0
3-R̀P̱ŔŔŔ, t-t	6	0.08	0.32	11.2	8.7	74	1.6	10	500	0	102	0
3-RP̱RU var. 3, t-t	70	0.08	0.17	12.2	12.3	56	6.6	7	347	0	100	0
3-R̀P̱ŔR̀Ŕ, r-r	6	0.07	0.12	20.6	14.0	84	4.8	9	184	0	103	0
3-R̀R̀P̱ŔŔ, t-c	6	0.06	0.46	22.4	12.1	106	3.4	11	127	0	102	−60
3-R̀R̀P̱ŔŔ, t-v	0	0.07	0.14	25.6	12.6	117	4.5	10	224	0	110	0
3-UP̱U, v-v	11	0.03	0.12	10.8	9.4	66	2.8	5	127	0	102	0
3-RRP̱U, v-v	4	0.03	0.12	11.1	9.7	65	3.0	7	100	0	100	0
3-R̀ŔP̱ŔR̀, v-v	1	0.04	0.12	10.3	8.9	66	2.5	9	172	0	100	0
3T1R:												
4-P̱RUR, t-v	6	0.19	0.73	16.1	12.1	77	4.4	9	327	0	282	0
4-RP̱UR, t-v	32	0.20	0.44	17.3	12.3	80	31.2	7	223	0	171	0
4-R̀P̱R̀ŔŔ, p-v	1	0.18	0.42	15.3	10.2	86	28.5	9	202	0	169	0
4-RRP̱U, v-v	10	0.27	0.61	15.4	14.3	62	73.1	7	483	0	101	0
4-R̀R̀P̱ŔR̀, v-v	7	0.23	0.51	12.7	13.3	55	36.5	9	384	0	216	0
[123]:									From reference			
3-C̱RR	54	0.08	0.33	10.4	8.8	68	1.2	11	300	120	110	120
3-P̱UU	315	0.17	0.71	13.5	7.6	101	2.8	8	250	−85	230	0
3-C̱RU	4	0.17	0.54	0.7	8.3	5	8.0	10	300	115	420	0
3-C̱UR	158	0.51	0.78	33.4	12.6	152	4.6	11	200	120	150	0
3-UP̱U	150	0.12	0.32	8.8	8.1	63	2.1	5	500	0	120	0

Table A6. Summary of the handling robot results (continuation of Table A5, leg-chain parameters).

Robot	Leg-Chain Kinematic Parameters										
3T0R:	α_2 deg	a_2 mm	d_2 mm	α_3 deg	a_3 mm	d_3 mm	α_4 deg	a_4 mm	d_4 mm	a_5 mm	d_5 mm
3-P̀R̀R̀R̀, c-m					183	154		134	41	—	—
3-P̱RRU var. 1, c-t					131	25		165	236		
3-P̀R̀R̀ŔŔ, v-t					159	80		149	3	274	1
3-P̱RUR var. 1, v-r					284	33				454	61
3-P̀R̀ŔŔŔ, t-v					271	33		151	30	227	27
3-P̱RUR var. 2, v-v					493	0				320	80
3-P̱ŔR̀R̀R̀, c-c					61	170		148	174	165	57
3-P̱URR, r-c								240	47	146	38
3-P̀R̀ŔR̀R̀, r-t					25	42		192	216	53	23
3-P̱UU, t-r								299	135		
3-P̱RRU var. 2, c-t					84	134		153	128		
3-RP̀R̀R̀R̀, t-c								213	177	210	114
3-RP̱RU var. 1, v-v								453	18		
3-R̀P̀ŔŔŔ, v-v								301	25	149	58
3-RP̱UR, r-v										281	2
3-R̀P̱R̀R̀R̀, r-v								3	44	274	4

Table A6. *Cont.*

Robot	Leg-Chain Kinematic Parameters										
3T0R:	α_2 deg	a_2 mm	d_2 mm	α_3 deg	a_3 mm	d_3 mm	α_4 deg	a_4 mm	d_4 mm	a_5 mm	d_5 mm
3-RP̱RU var. 2, t-t								157	83		
3-R̀P̱ŔŔR̀, t-t								161	45	3	0
3-RP̱RU var. 3, t-t								292	53		
3-R̀P̱ŔŔR̀, r-r								391	6	21	73
3-R̀R̀P̱ŔŔ, t-c		201	275							299	63
3-R̀R̀P̱ŔŔ, t-v		294	95							190	61
3-UP̱U, v-v											
3-RRP̱U, v-v		22	35								
3-R̀ŔP̱ŔR̀, v-v		21	37							20	8
3T1R:											
4-P̱RUR, t-v					455	135				419	74
4-RP̱UR, t-v										371	142
4-R̀P̱R̀ŔŔ, p-v								35	30	292	126
4-RRP̱U, v-v		153	155								
4-R̀ŔP̱ŔR̀, v-v		191	8							90	16
From [123]:	Kinematic parameters from reference										
3-C̱RR					290	0		250	0	—	—
3-P̱UU								450	0		
3-C̱RU					100	0		420	0		
3-C̱UR					490	0				490	0
3-UP̱U											

Table A7. Summary of the handling robot results (part 2, revolute actuation).

Robot	Performance							Kinematic Parameters				
3T0R:	coll. mm	inst.spc. m^3	footprt. m^2	Power W	Torque Nm	velo. deg/s	cond.	n	r_b mm	γ_b deg	r_p mm	γ_p deg
3-R̲̀R̀R̀ŔŔ, c-t	3	0.07	0.16	41.1	1.7	455	3.3	14	122	100	103	0
3-R̲RUR var. 1, r-t	8	0.05	0.13	41.4	1.8	447	2.7	11	100	0	100	0
3-R̲̀R̀ŔŔŔ, t-c	3	0.08	0.34	24.6	1.1	436	3.1	14	344	0	100	60
3-R̲URR, t-c	16	0.06	0.30	23.2	1.0	430	2.6	12	299	0	100	55
3-R̲̀ŔŔŔR̀, t-t	17	0.05	0.16	57.7	2.5	439	4.1	13	119	0	100	0
3-R̲RRU var. 1, t-t	2	0.04	0.14	56.9	2.5	433	4.2	11	117	0	100	0
3-R̲̀ŔŔR̀R̀, r-r	0	0.05	0.18	43.9	1.8	454	3.8	13	104	0	100	0
3-R̲RUR var. 2, r-r	7	0.04	0.17	45.6	2.0	439	4.2	11	100	0	100	0
3-R̲̀R̀ŔŔR̀, r-r	18	0.04	0.15	40.7	1.7	456	3.7	13	126	0	103	0
3-R̲UU, r-r	57	0.04	0.13	45.1	1.9	452	4.2	9	107	0	100	0
3-R̲RRU var. 2, r-r	12	0.04	0.13	40.0	1.7	450	3.1	11	152	0	104	0
3T1R:												
4-R̲̀R̀R̀ŔŔ, r-v	18	0.13	0.42	15.8	2.0	451	38.1	13	270	0	203	0
4-R̲RUR var. 1, t-v	0	0.11	0.46	74.5	2.4	453	90.3	11	349	0	109	0
4-R̲̀ŔŔR̀R̀, v-v	0	0.19	0.59	125.5	4.2	428	19.8	13	372	0	100	0
4-R̲RUR var. 2, v-v	0	0.16	0.83	375.8	14.0	384	30.7	11	403	0	122	0
4-R̲̀R̀ŔŔR̀, v-v	11	0.10	0.74	322.9	10.7	433	49.6	13	491	0	102	0
4-R̲UU, v-v	15	0.14	0.69	110.6	3.6	436	27.9	9	453	0	126	0
4-R̲RRU, v-v	7	0.12	0.62	400.4	12.7	451	46.7	11	468	0	147	0

Table A8. Summary of the handling robot results (part 2 continuation of Table A7, leg chain parameters).

Robot	Leg-Chain Kinematic Parameters										
3T0R:	α_2 deg	a_2 mm	d_2 mm	α_3 deg	a_3 mm	d_3 mm	α_4 deg	a_4 mm	d_4 mm	a_5 mm	d_5 mm
3-R̲̀R̀R̀ŔŔ, c-t		184	44		174	13		17	48	276	66
3-R̲RUR var. 1, r-t		197	44		225	45				204	47
3-R̲̀R̀ŔŔŔ, t-c		166	165		20	117		206	89	196	16
3-R̲URR, t-c		161	161					196	73	177	37
3-R̲̀ŔŔŔR̀, t-t		3	74		175	30		169	157	14	19
3-R̲RRU var. 1, t-t		36	26		161	21		146	161		
3-R̲̀ŔŔR̀R̀, r-r		164	48		196	52		175	41	180	31
3-R̲RUR var. 2, r-r		1	72		212	39				176	4
3-R̲̀R̀ŔŔŔ, r-r		175	25		22	63		181	108	5	7
3-R̲UU, r-r		173	0					235	126		
3-R̲RRU var. 2, r-r		175	53		50	158		170	35		
3T1R:											
4-R̲̀R̀R̀ŔŔ, r-v		213	53		167	100		217	51	225	2
4-R̲RUR var. 1, t-v		199	73		178	52				219	2
4-R̲̀ŔŔR̀R̀, v-v		91	113		290	164		59	80	274	125
4-R̲RUR var. 2, v-v		20	335		204	18				336	80
4-R̲̀R̀ŔŔŔ, v-v		257	41		172	64		250	30	21	5
4-R̲UU, v-v		274	45					384	131		
4-R̲RRU, v-v		265	74		69	167		237	47		

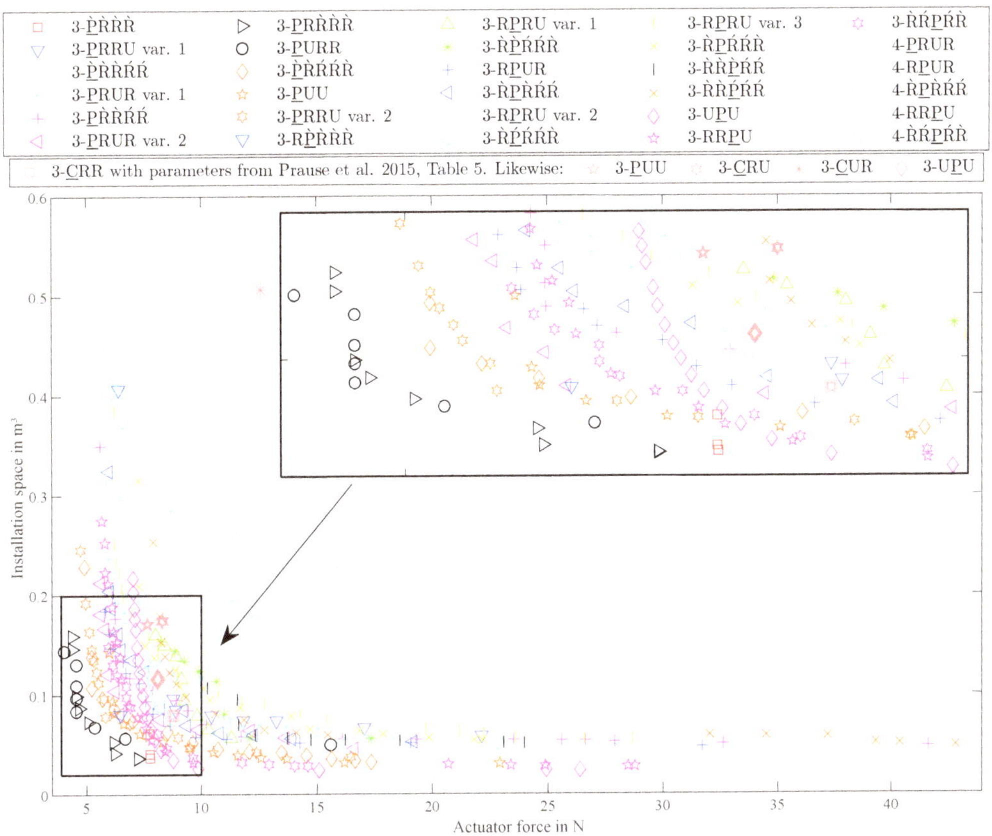

Figure A15. Pareto fronts for robots with prismatic actuation using only 27 reference points.

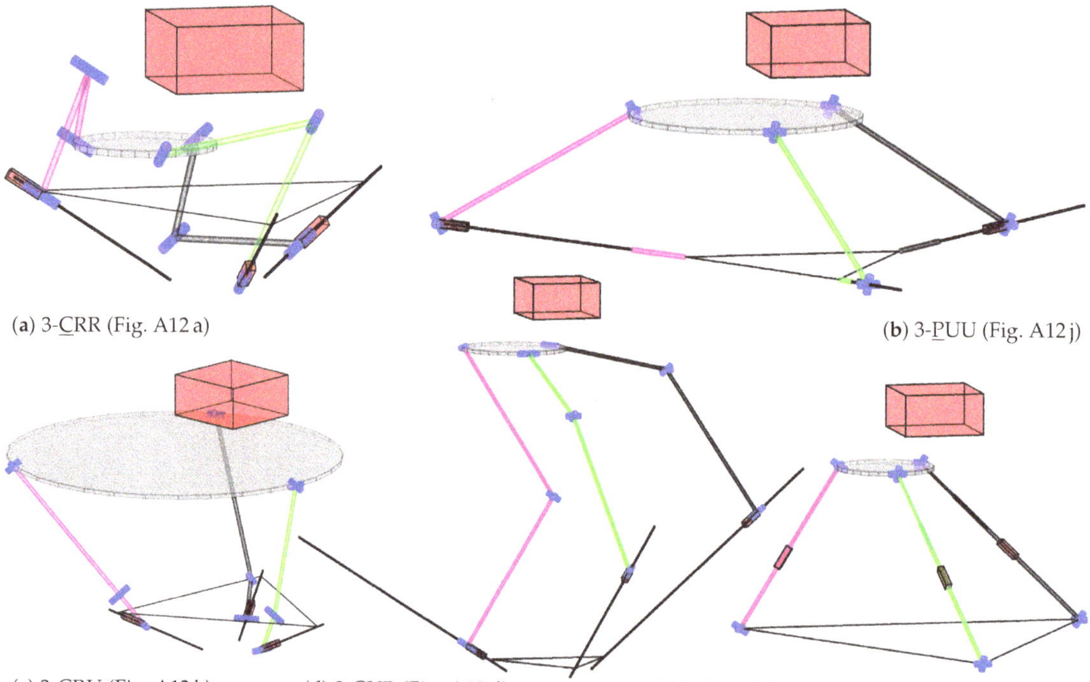

(**a**) 3-$\underline{\mathrm{C}}$RR (Fig. A12 a)
(**b**) 3-$\underline{\mathrm{P}}$UU (Fig. A12 j)
(**c**) 3-$\underline{\mathrm{C}}$RU (Fig. A12 k)
(**d**) 3-$\underline{\mathrm{C}}$UR (Fig. A12 d)
(**e**) 3-U$\underline{\mathrm{P}}$U (Fig. A13 g)

Figure A16. Visualization of the 3T0R parallel robots with parameters from [123], listed in the bottom of Tables A5 and A6.

The following pages give different performance maps for the resolution of the functional redundancy of the 3T1R parallel robots within the handling task. Each page contains the maps of all robots for one heat-map criterion. The always identical trajectory comes from the dimensional synthesis, where the minimization of actuator force or torque was selected as the IK objective. The robots with prismatic actuation are given at the top (Figures A18, A20, and A22), and with revolute actuation at the bottom of the pages (Figures A19, A21, and A23). The actuator-force/torque maps (Figures A18 and A19) show the ability of the redundancy resolution to minimize the actuator forces by avoiding areas of high required force (dark colors). For most structures, however, the force necessary in the reachable orientations is relatively uniform. The limitation of the end-effector precision to 500 µm is only influencing the 3T1R PRs with prismatic actuation, as is visible in Figure A20. The error is much lower for revolute actuation in Figure A21. Since the position error is proportional to the assumed encoder accuracy, principally better precision of the revolute actuation cannot be concluded, as a strictly equivalent selection of encoders was not subject to the case study. The heat maps in Figures A22 and A23 show the condition number of the full inverse-kinematics matrix "IK Jacobian", see [22], connected to the IK singularities within the workspace of 3T1R PRs, especially with revolute actuation.

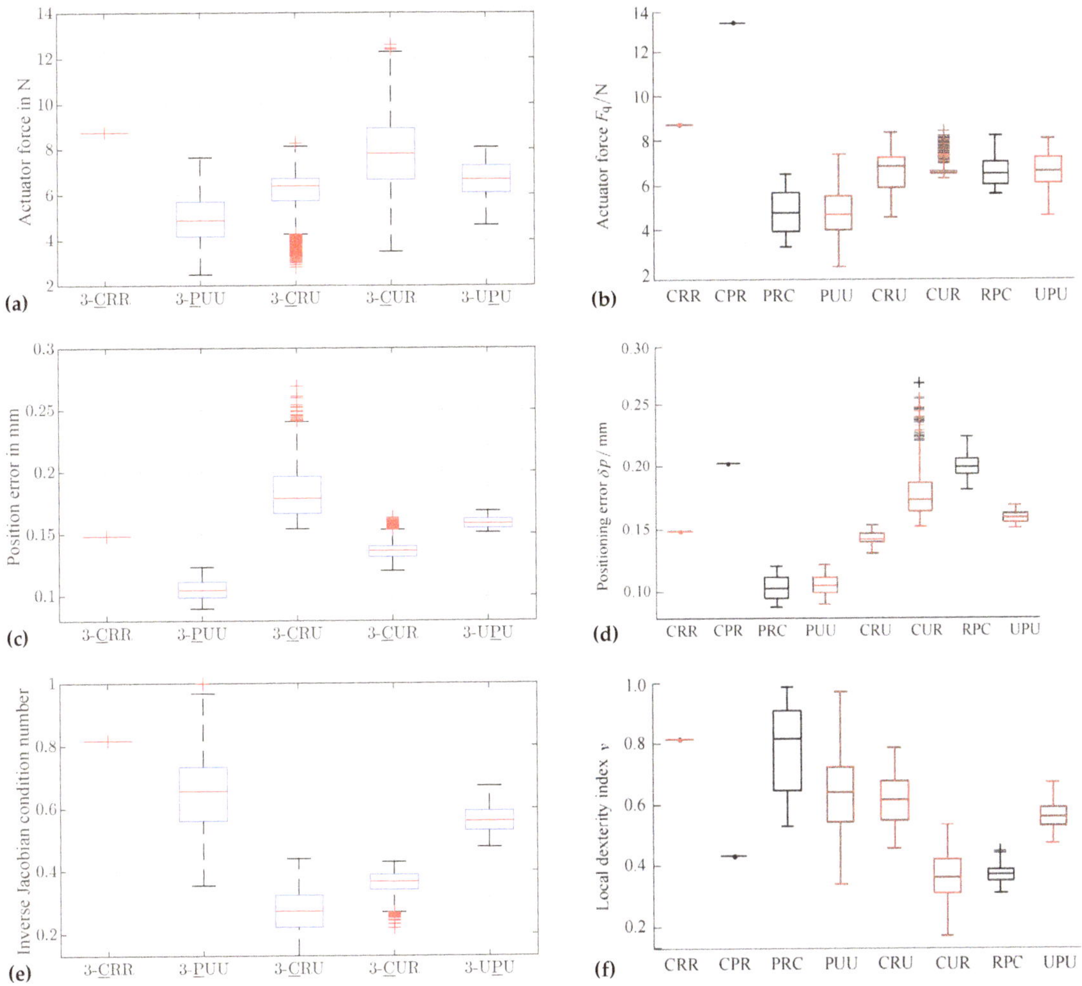

Figure A17. Distribution of actuator force (**a**,**b**), position error/precision (**c**,**d**), and dexterity (**e**,**f**) as performance criterion computed with parameters from Table 4 of [123] with this paper's implementation (left side) in comparison to the original results obtained from [123] (right side, with structures for comparison in red).

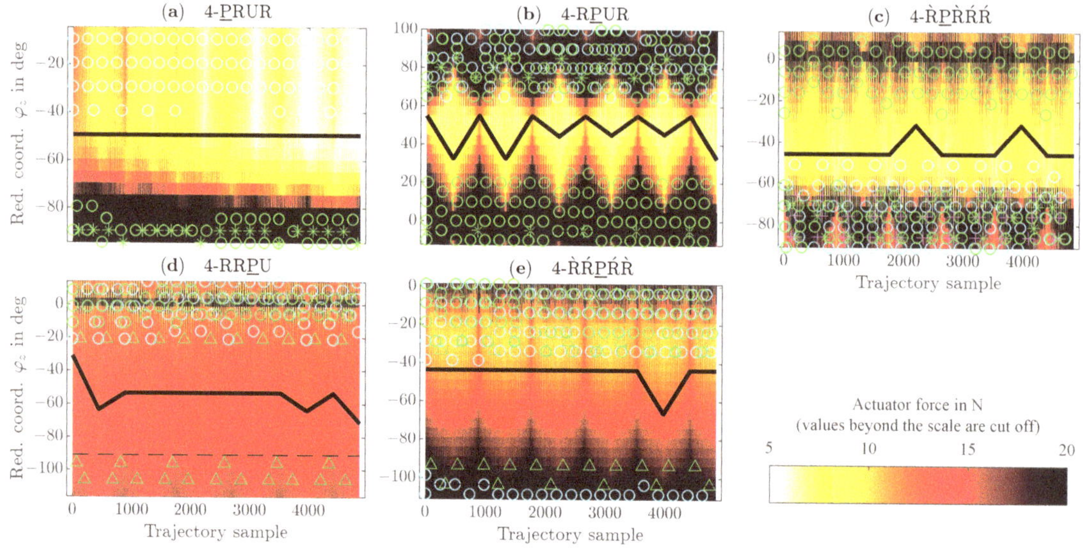

Figure A18. Performance maps for the prismatic-actuation parallel robots of Figure A14 with maximum actuator force as IK objective and heat-map criterion. For marker legend, see below in Figure A19.

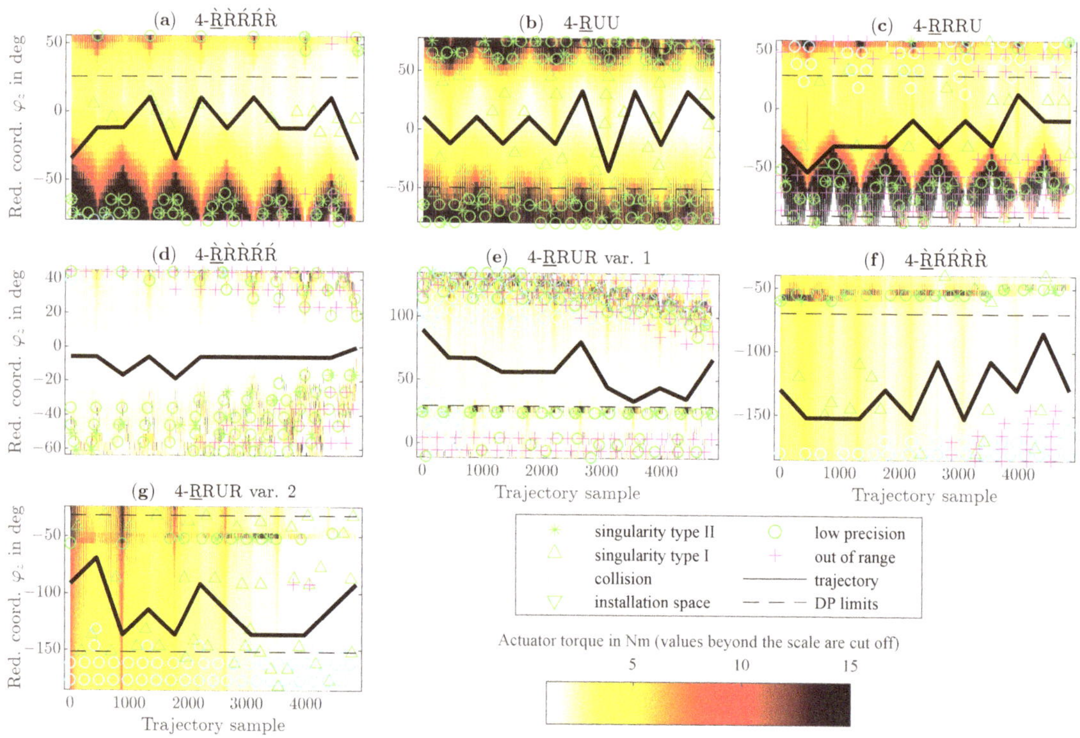

Figure A19. Performance maps for the revolute-actuation parallel robots of Figure A11 with maximum actuator torque as IK objective and heat-map criterion.

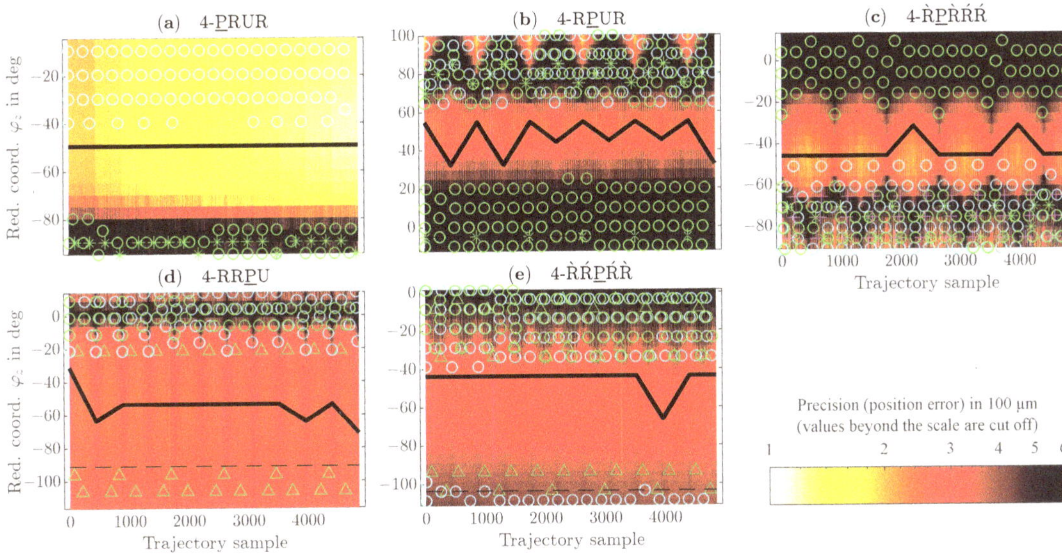

Figure A20. Performance maps for the prismatic-actuation parallel robots of Figure A14 with position error (precision) as heat-map criterion and trajectory from Figure A18. For marker legend, see below in Figure A21.

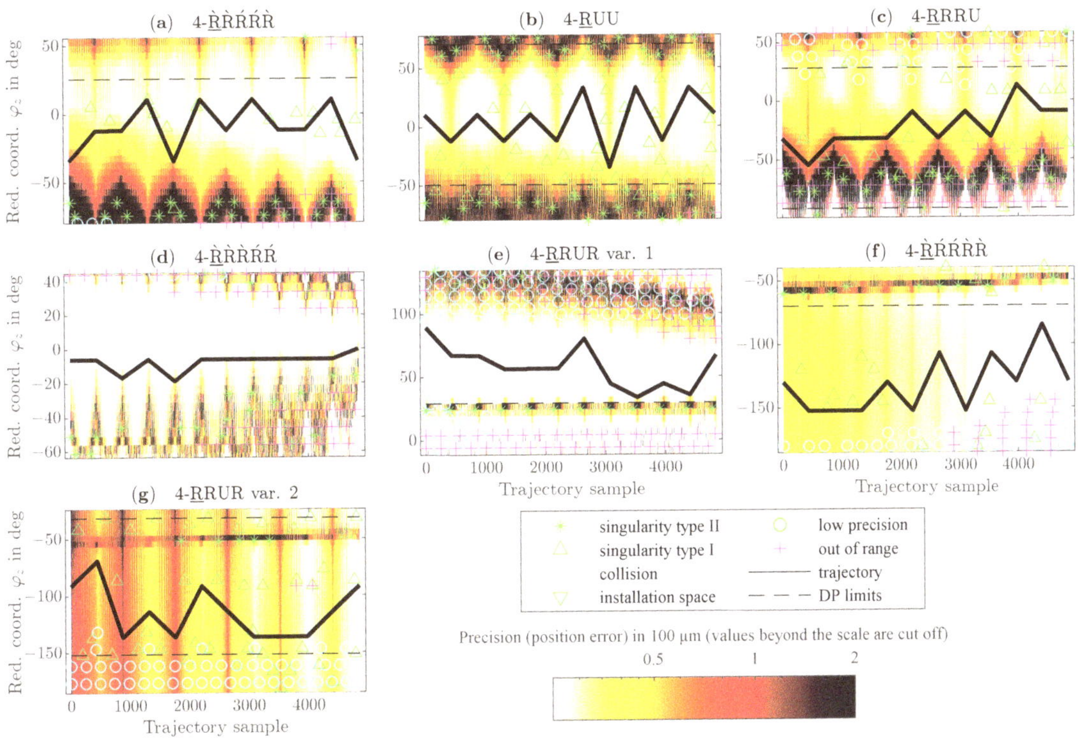

Figure A21. Performance maps for the revolute-actuation parallel robots of Figure A11 with position error as heat-map criterion and trajectory from Figure A19. Color scale is different than in Figure A20 above.

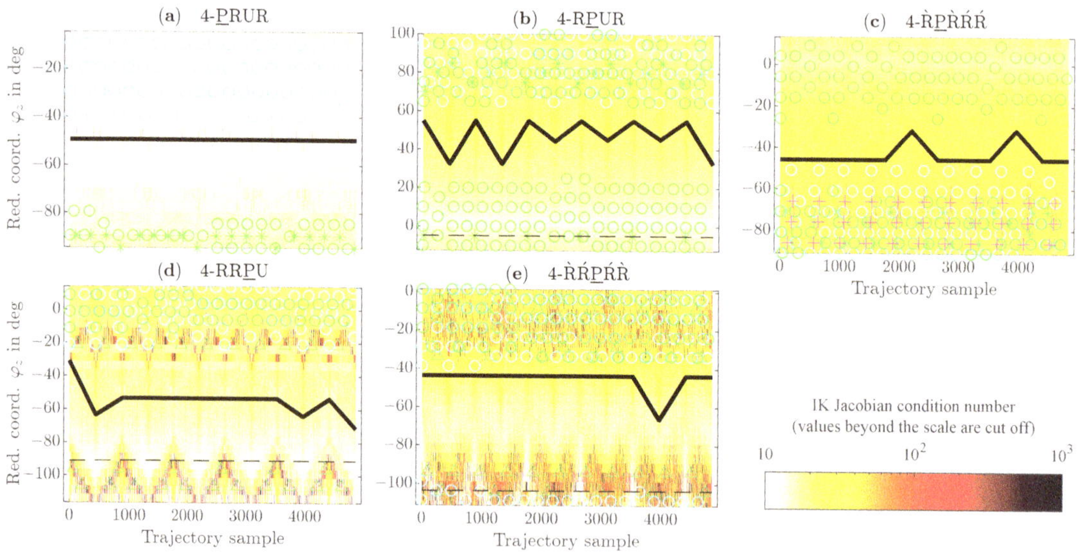

Figure A22. Performance maps for the prismatic-actuation parallel robots of Figure A14 with IK-Jacobian condition number as heat-map criterion and trajectory from Figure A18. For marker legend, see below in Figure A23.

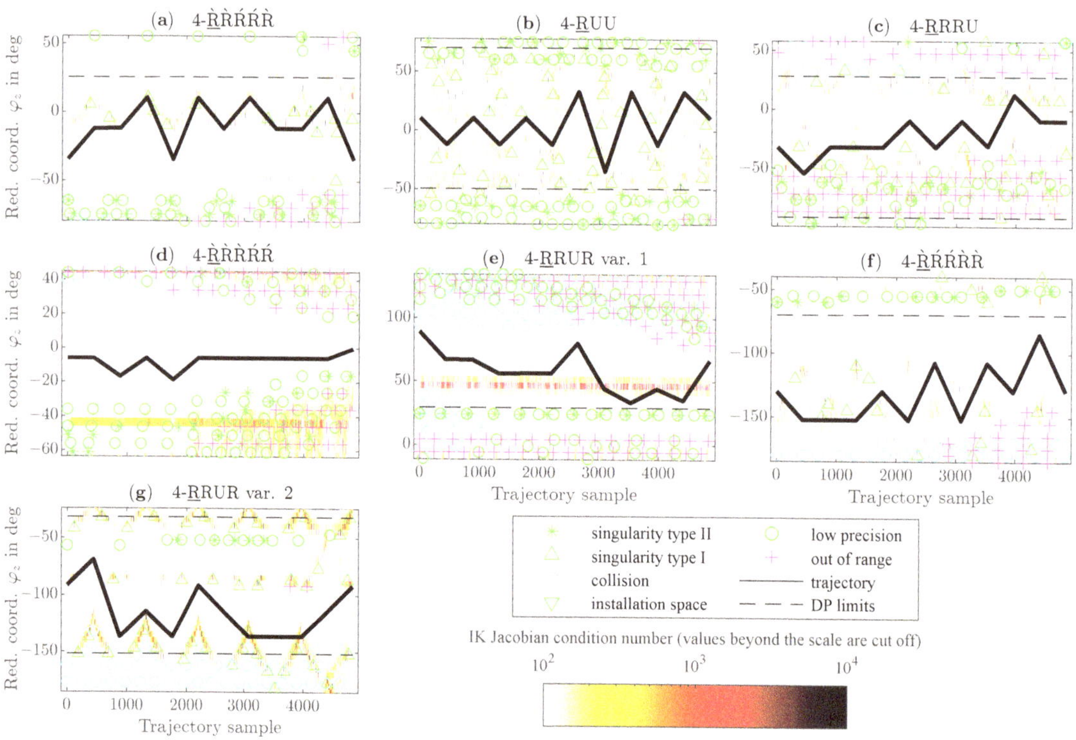

Figure A23. Performance maps for the revolute-actuation parallel robots of Figure A11 with IK-Jacobian condition number as heat-map criterion and trajectory from Figure A19. Color scale is different than in Figure A22 above.

References

1. Merlet, J.P. *Parallel Robots*, 2nd ed.; Solid Mechanics and Its Applications; Springer Science & Business Media: Berlin, Germany, 2006; Volume 128. [CrossRef]
2. Russo, M.; Zhang, D.; Liu, X.J.; Xie, Z. A review of parallel kinematic machine tools: Design, modeling, and applications. *Int. J. Mach. Tools Manuf.* **2024**, *196*, 104118. [CrossRef]
3. Briot, S.; Burgner-Kahrs, J. New Frontiers in Parallel Robotics. *Front. Robot. AI* **2023**, *10*, 1282798. [CrossRef]
4. Shao, Z.; Zhang, D.; Caro, S. New Frontiers in Parallel Robots. *Machines* **2023**, *11*, 386. [CrossRef]
5. Sinha, A.; Malo, P.; Deb, K. A review on bilevel optimization: From classical to evolutionary approaches and applications. *IEEE Trans. Evol. Comput.* **2017**, *22*, 276–295. [CrossRef]
6. Kong, X.; Gosselin, C.M. *Type Synthesis of Parallel Mechanisms;* Springer Publishing Company, Incorporated: New York, NY, USA, 2007. [CrossRef]
7. Gogu, G. *Structural Synthesis of Parallel Robots, Part 1: Methodology*; Springer: Berlin/Heidelberg, Germany, 2008; Volume 866. [CrossRef]
8. Frindt, M. Modulbasierte Synthese von Parallelstrukturen für Maschinen in der Produktionstechnik (Module-Based Synthesis of Parallel Structures for Machines in Production Technology). Ph.D. Thesis, Technische Universität Braunschweig, Braunschweig, Germany, 2001.
9. Krefft, M. Aufgabenangepasste Optimierung von Parallelstrukturen für Maschinen in der Produktionstechnik (Task-Specific Optimization of Parallel Structures for Machines in Manufacturing Technology). Ph.D. Thesis, Technische Universität Braunschweig, Braunschweig, Germany, 2006.
10. Kirchner, J. Mehrkriterielle Optimierung von Parallelkinematiken (Multi-Criterial Optimization of Parallel Kinematics). Ph.D. Thesis, Technische Universität Chemnitz, Zwickau, Germany, 2000.
11. Stock, M.; Miller, K. Optimal kinematic design of spatial parallel manipulators: Application to linear Delta robot. *J. Mech. Des.* **2003**, *125*, 292–301. [CrossRef]
12. Rao, A.B.K.; Rao, P.M.; Saha, S.K. Dimensional design of hexaslides for optimal workspace and dexterity. *IEEE Trans. Robot.* **2005**, *21*, 444–449. [CrossRef]
13. Carbone, G.; Ottaviano, E.; Ceccarelli, M. An optimum design procedure for both serial and parallel manipulators. *Proc. Inst. Mech. Eng. Part C J. Mech. Eng. Sci.* **2007**, *221*, 529–843. [CrossRef]
14. Kelaiaia, R.; Company, O.; Zaatri, A. Multiobjective optimization of a linear Delta parallel robot. *Mech. Mach. Theory* **2012**, *50*, 159–178. [CrossRef]
15. Miller, K. Optimal Design and Modeling of Spatial Parallel Manipulators. *Int. J. Robot. Res.* **2004**, *23*, 127–140. [CrossRef]
16. Laribi, M.; Romdhane, L.; Zeghloul, S. Analysis and dimensional synthesis of the DELTA robot for a prescribed workspace. *Mech. Mach. Theory* **2007**, *42*, 859–870. [CrossRef]
17. Liu, S.; Huang, T.; Mei, J.; Zhao, X.; Wang, P.; Chetwynd, D.G. Optimal Design of a 4-DOF SCARA Type Parallel Robot Using Dynamic Performance Indices and Angular Constraints. *J. Mech. Robot.* **2012**, *4*, 031005. [CrossRef]
18. von Daake, A. Beitrag zur Kalibrierung und Anforderungsbasierten Arbeitsraumoptimierung Parallelkinematischer Manipulatoren in Hexapodbauweise. Ph.D. Thesis, Technische Universität Clausthal, Clausthal, Germany, 2012.
19. Jamwal, P.K.; Hussain, S.; Xie, S.Q. Three-Stage Design Analysis and Multicriteria Optimization of a Parallel Ankle Rehabilitation Robot Using Genetic Algorithm. *IEEE Trans. Autom. Sci. Eng.* **2015**, *12*, 1433–1446. [CrossRef]
20. Prause, I. Task-Based Comparison of Parallel Translational Kinematic Structures. Ph.D Thesis, RWTH Aachen, Aachen, Germany, 2016. [CrossRef]
21. Sciavicco, L.; Siciliano, B.; Villani, L.; Oriolo, G. *Robotics: Modelling, Planning and Control*; Advanced Textbooks in Control and Signal Processing; Springer: Berlin, Germany, 2009. [CrossRef]
22. Schappler, M.; Tappe, S.; Ortmaier, T. Modeling Parallel Robot Kinematics for 3T2R and 3T3R Tasks Using Reciprocal Sets of Euler Angles. *Robotics* **2019**, *8*, 68. [CrossRef]
23. Agarwal, A.; Nasa, C.; Bandyopadhyay, S. Dynamic singularity avoidance for parallel manipulators using a task-priority based control scheme. *Mech. Mach. Theory* **2016**, *96*, 107–126. [CrossRef]
24. Merlet, J.P.; Perng, M.W.; Daney, D. Optimal trajectory planning of a 5-axis machine-tool based on a 6-axis parallel manipulator. In *Advances in Robot Kinematics*; Springer: Berlin/Heidelberg, Germany, 2000; pp. 315–322. [CrossRef]
25. Oen, K.T.; Wang, L.C.T. Optimal dynamic trajectory planning for linearly actuated platform type parallel manipulators having task space redundant degree of freedom. *Mech. Mach. Theory* **2007**, *42*, 727–750. [CrossRef]
26. Corinaldi, D.; Angeles, J.; Callegari, M. Posture optimization of a functionally redundant parallel robot. In *Advances in Robot Kinematics 2016*; Springer: Berlin/Heidelberg, Germany, 2016; pp. 101–108. [CrossRef]
27. Santos, J.C.; da Silva, M.M. Redundancy Resolution of Kinematically Redundant Parallel Manipulators via Differential Dynamic Programing. *J. Mech. Robot.* **2017**, *9*, 041016. [CrossRef]

28. Gao, Y.; Chen, K.; Gao, H.; Xiao, P.; Wang, L. Small-angle perturbation method for moving platform orientation to avoid singularity of asymmetrical 3-RRR planner [sic!] parallel manipulator. *J. Braz. Soc. Mech. Sci. Eng.* **2019**, *41*, 538. . [CrossRef]
29. Schappler, M. Pose Optimization of Task-Redundant Robots in Second-Order Rest-to-Rest Motion with Cascaded Dynamic Programming and Nullspace Projection. In *Proceedings of the Informatics in Control, Automation and Robotics*; Gusikhin, O., Madani, K., Nijmeijer, H., Eds.; Springer International Publishing: Cham, Switzerland, 2023; pp. 106–131. . [CrossRef]
30. Ramirez, D.A. Automatic Generation of Task-Specific Serial Mechanisms Using Combined Structural and Dimensional Synthesis. Ph.D. Thesis, Leibniz University Hannover, Hanover, Germany, 2018. [CrossRef]
31. Conkur, E.S.; Buckingham, R. Clarifying the definition of redundancy as used in robotics. *Robotica* **1997**, *15*, 583–586. [CrossRef]
32. Huo, L.; Baron, L. The joint-limits and singularity avoidance in robotic welding. *Ind. Robot. Int. J.* **2008**, *35*, 456–464. [CrossRef]
33. Léger, J.; Angeles, J. Off-line programming of six-axis robots for optimum five-dimensional tasks. *Mech. Mach. Theory* **2016**, *100*, 155–169. [CrossRef]
34. Luces, M.; Mills, J.K.; Benhabib, B. A Review of Redundant Parallel Kinematic Mechanisms. *J. Intell. Robot. Syst.* **2017**, *86*, 175–198. [CrossRef]
35. Gosselin, C.; Schreiber, L.T. Redundancy in Parallel Mechanisms: A Review. *Appl. Mech. Rev.* **2018**, *70*, 010802. [CrossRef]
36. Alba-Gomez, O.; Wenger, P.; Pamanes, A. Consistent kinetostatic indices for planar 3-DOF parallel manipulators, application to the optimal kinematic inversion. In Proceedings of the International Design Engineering Technical Conferences and Computers and Information in Engineering Conference, Brooklyn, NY, USA, 3–6 August 2008; Volume 47446, pp. 765–774. [CrossRef]
37. Kotlarski, J.; Do Thanh, T.; Heimann, B.; Ortmaier, T. Optimization strategies for additional actuators of kinematically redundant parallel kinematic machines. In Proceedings of the 2010 IEEE International Conference on Robotics and Automation (ICRA), Anchorage, AK, USA, 3–7 May 2010; pp. 656–661.
38. Briot, S.; Khalil, W. *Dynamics of Parallel Robots*; Mechanisms and Machine Science; Springer: Cham, Switzerland, 2015; Volume 35. [CrossRef]
39. *VDI 2206:2021*; Development of Mechatronic and Cyber-Physical Systems. Beuth Verlag: Berlin, Germany, 2021.
40. Maier, J.R.; Ezhilan, T.; Fadel, G.M.; Summers, J.D.; Mocko, G.M. A hierarchical requirements modeling scheme to support engineering innovation. In Proceedings of the DS 42: Proceedings of ICED 2007, the 16th International Conference on Engineering Design, Paris, France, 28–31 July 2007; pp. 839–840.
41. Stechert, C.; Franke, H.J. Managing requirements as the core of multi-disciplinary product development. *CIRP J. Manuf. Sci. Technol.* **2009**, *1*, 153–158. [CrossRef]
42. Stechert, C.; Franke, H.J.; Vietor, T. Knowledge-Based Design Principles and Tools for Parallel Robots. In *Robotic Systems for Handling and Assembly*; Springer: Berlin/Heidelberg, Germany, 2010; pp. 59–75. [CrossRef]
43. Neugebauer, R. (Ed.) *Parallelkinematische Maschinen: Entwurf, Konstruktion, Anwendung*; VDI-Buch, Springer: Berlin/Heidelberg, Germany, 2006. [CrossRef]
44. Censi, A. A mathematical theory of co-design. *arXiv* **2015**, arXiv:1512.08055. [CrossRef]
45. Baumgärtner, J.; Kanagalingam, G.; Fleischer, A.P.J. One Problem, One Solution: Unifying Robot and Environment Design Optimization. *arXiv* **2023**, arXiv:2310.05520. [CrossRef]
46. Ding, H.; Cao, W.; Cai, C.; Kecskeméthy, A. Computer-aided structural synthesis of 5-DOF parallel mechanisms and the establishment of kinematic structure databases. *Robotica* **2015**, *83*, 14–30. [CrossRef]
47. Mayer, M.; Külz, J.; Althoff, M. CoBRA: A Composable Benchmark for Robotics Applications. *arXiv* **2022**, arXiv:2203.09337. [CrossRef]
48. Salunkhe, D.H.; Michel, G.; Kumar, S.; Sanguineti, M.; Chablat, D. An efficient combined local and global search strategy for optimization of parallel kinematic mechanisms with joint limits and collision constraints. *Mech. Mach. Theory* **2022**, *173*, 104796. [CrossRef]
49. Su, Y.; Duan, B.; Zheng, C. Genetic design of kinematically optimal fine tuning Stewart platform for large spherical radio telescope. *Mechatronics* **2001**, *11*, 821–835. [CrossRef]
50. Zhang, D. *Parallel Robotic Machine Tools*; Springer Science & Business Media: New York, NY, USA, 2009. [CrossRef]
51. Ben Hamida, I.; Laribi, M.A.; Mlika, A.; Romdhane, L.; Zeghloul, S. Dimensional Synthesis and Performance Evaluation of Four Translational Parallel Manipulators. *Robotica* **2021**, *39*, 233–249. [CrossRef]
52. Tarkian, M.; Persson, J.; Ölvander, J.; Feng, X. Multidisciplinary design optimization of modular industrial robots. In Proceedings of the ASME 2011 International Design Engineering Technical Conferences & Computers and Information in Engineering Conference IDETC/CIE, Washington, DC, USA, 28–31 August 2011. [CrossRef]
53. Zhou, L.; Bai, S. A new approach to design of a lightweight anthropomorphic arm for service applications. *J. Mech. Robot.* **2015**, *7*, 031001. [CrossRef]
54. Tan, S.; Liang, F.; Fan, J.; Zhang, Y.; Lin, Z. Optimization Design for Structure Parameters of Six Degree-of-Freedom (DOF) Positioner of Secondary Mirror in a Space Optical Remote Sensor Based on ADAMS. In Proceedings of the Chinese Control Conference (CCC), Guangzhou, China, 27–30 July 2019; pp. 7201–7205. [CrossRef]

55. *VDI 2221:2019*; Design of Technical Products and Systems—Model of Product Design. Beuth Verlag: Berlin, Germany, 2019.
56. Singla, E.; Tripathi, S.; Rakesh, V.; Dasgupta, B. Dimensional synthesis of kinematically redundant serial manipulators for cluttered environments. *Robot. Auton. Syst.* **2010**, *58*, 585–595. [CrossRef]
57. Kivelä, T.; Mattila, J.; Puura, J. A generic method to optimize a redundant serial robotic manipulator's structure. *Autom. Constr.* **2017**, *81*, 172–179. [CrossRef]
58. Romiti, E.; Iacobelli, F.; Ruzzon, M.; Kashiri, N.; Malzahn, J.; Tsagarakis, N. An Optimization Study on Modular Reconfigurable Robots: Finding the Task-Optimal Design. In Proceedings of the 2023 IEEE 19th International Conference on Automation Science and Engineering (CASE), Auckland, New Zealand, 26–30 August 2023; pp. 1–8. [CrossRef]
59. Shiller, Z.; Sundar, S. Design of robotic manipulators for optimal dynamic performance. In Proceedings of the 1991 IEEE International Conference on Robotics and Automation, Sacramento, CA, USA, 9–11 April 1991; pp. 334–339. [CrossRef]
60. Chedmail, P.; Gautier, M. Optimum choice of robot actuators. *J. Eng. Ind.* **1990**, *112*, 361–367. [CrossRef]
61. Pettersson, M.; Ölvander, J. Drive train optimization for industrial robots. *IEEE Trans. Robot.* **2009**, *25*, 1419–1424. [CrossRef]
62. Zhou, L.; Bai, S.; Hansen, M.R. Design optimization on the drive train of a light-weight robotic arm. *Mechatronics* **2011**, *21*, 560–569. [CrossRef]
63. Pettersson, M.; Andersson, J.; Krus, P. Methods for discrete design optimization. In Proceedings of the ASME 2005 International Design Engineering Technical Conferences and Computers and Information in Engineering Conference, Long Beach, CA, USA, 24–28 September 2005; American Society of Mechanical Engineers: New York, NY, USA, 2005; pp. 295–303.
64. Padilla-García, E.A.; Cruz-Villar, C.A.; Rodriguez-Angeles, A. Multi-Objective Design/Control Optimization on the Power Train of Robot Manipulators using a Genetic Algorithm. In Proceedings of the 14th IFToMM World Congress, Taipei, Tajwan, 25–30 October 2015. [CrossRef]
65. Wang, H.; Zhang, L.; Chen, G.; Huang, S. Parameter optimization of heavy-load parallel manipulator by introducing stiffness distribution evaluation index. *Mech. Mach. Theory* **2017**, *108*, 244–259. [CrossRef]
66. Pastor, R.; Bobovský, Z.; Huczala, D.; Grushko, S. Genetic Optimization of a Manipulator: Comparison between Straight, Rounded, and Curved Mechanism Links. *Appl. Sci.* **2021**, *11*, 2471. [CrossRef]
67. Otremba, R. Systematische Entwicklung von Gelenken für Parallelroboter. Ph.D. Thesis, Technische Universität Braunschweig, Braunschweig, Germany, 2005.
68. Sterneck, T.; Fettin, J.; Schappler, M. Task-Specific Synthesis and Design of a Mobile Six-DoF Hexa Parallel Robot for Weed Control. In *Proceedings of the 2nd IFToMM for Sustainable Development Goals Workshop (I4SDG Workshop 2023), Bilbao, Spain, 22–23 June 2023*; Petuya, V., Quaglia, G., Parikyan, T., Carbone, G., Eds.; Springer: Cham, Switzerland, 2023; pp. 105–114. [CrossRef]
69. Vulliez, M.; Zeghloul, S.; Khatib, O. Design strategy and issues of the Delthaptic, a new 6-DOF parallel haptic device. *Mech. Mach. Theory* **2018**, *128*, 395–411. [CrossRef]
70. Brandstötter, M. Adaptable Serial Manipulators in Modular Design. Ph.D. Thesis, Tyrolean Private University UMIT, Institute of Automation and Control Engineering, Hall in Tirol, Austria, 2016. [CrossRef]
71. Whitman, J.; Choset, H. Task-specific manipulator design and trajectory synthesis. *IEEE Robot. Autom. Lett.* **2018**, *4*, 301–308. [CrossRef]
72. Ginnante, A.; Simetti, E.; Caro, S.; Leborne, F. Task priority based design optimization of a kinematic redundant robot. *Mech. Mach. Theory* **2023**, *187*, 105374. [CrossRef]
73. Maaroof, O.W.; Dede, M.I.C.; Aydin, L. A Robot Arm Design Optimization Method by Using a Kinematic Redundancy Resolution Technique. *Robotics* **2022**, *11*, 1.
74. Russo, M.; Raimondi, L.; Dong, X.; Axinte, D.; Kell, J. Task-oriented optimal dimensional synthesis of robotic manipulators with limited mobility. *Robot. Comput.-Integr. Manuf.* **2021**, *69*, 102096. [CrossRef]
75. Patel, S.; Sobh, T. Task based synthesis of serial manipulators. *J. Adv. Res.* **2015**, *6*, 479–492. [CrossRef]
76. Wan, J.; Ding, L.; Yao, J.; Wu, H. A hybrid CHAOS-PSO algorithm for dimensional synthesis of a redundant manipulator based on tracking trajectories without or with singularities. *Prod. Eng.* **2018**, *12*, 579–587. [CrossRef]
77. Dinev, T.C. Concurrent Design and Motion Planning in Robotics Using Differentiable Optimal Control. Ph.D. Thesis, University of Edinburgh, School of Informatics, Edinburgh, UK, 2023. [CrossRef]
78. Spielberg, A.; Araki, B.; Sung, C.; Tedrake, R.; Rus, D. Functional co-optimization of articulated robots. In Proceedings of the 2017 IEEE International Conference on Robotics and Automation (ICRA), Singapore, 29 May–3 June 2017; IEEE: Piscataway, NJ, USA, 2017; pp. 5035–5042. [CrossRef]
79. Xu, Q.; Li, Y. Error analysis and optimal design of a class of translational parallel kinematic machine using particle swarm optimization. *Robotica* **2009**, *27*, 67–78. [CrossRef]
80. Yun, Y.; Li, Y. Optimal design of a 3-PUPU parallel robot with compliant hinges for micromanipulation in a cubic workspace. *Robot. Comput.-Integr. Manuf.* **2011**, *27*, 977–985. [CrossRef]
81. Lara-Molina, F.; Rosario, J.; Dumur, D. Multi-objective design of parallel manipulator using global indices. *Open Mech. Eng. J.* **2010**, *4*, 37–47. [CrossRef]

82. MathWorks (The MathWorks, Inc.). Global Optimization Toolbox. Available online: https://mathworks.com/help/gads (accessed on 31 October 2024).
83. Wang, R.; Zhang, X. Optimal design of a planar parallel 3-DOF nanopositioner with multi-objective. *Mech. Mach. Theory* **2017**, *112*, 61–83. [CrossRef]
84. Sun, T.; Lian, B. Stiffness and mass optimization of parallel kinematic machine. *Mech. Mach. Theory* **2018**, *120*, 73–88. [CrossRef]
85. Qi, Y.; Sun, T.; Song, Y. Multi-Objective Optimization of Parallel Tracking Mechanism Considering Parameter Uncertainty. *J. Mech. Robot.* **2018**, *10*, 041006. [CrossRef]
86. Lian, B.; Wang, X.V.; Wang, L. Static and dynamic optimization of a pose adjusting mechanism considering parameter changes during construction. *Robot. Comput.-Integr. Manuf.* **2019**, *59*, 267–277. [CrossRef]
87. Lou, Y.; Zhang, Y.; Huang, R.; Chen, X.; Li, Z. Optimization algorithms for kinematically optimal design of parallel manipulators. *IEEE Trans. Autom. Sci. Eng.* **2013**, *11*, 574–584. [CrossRef]
88. Ceccarelli, M.; Lanni, C. A multi-objective optimum design of general 3R manipulators for prescribed workspace limits. *Mech. Mach. Theory* **2004**, *39*, 119–132. [CrossRef]
89. Yang, C.; Ye, W.; Li, Q. Review of the performance optimization of parallel manipulators. *Mech. Mach. Theory* **2022**, *170*, 104725. [CrossRef]
90. Hassan, R.; Cohanim, B.; De Weck, O.; Venter, G. A comparison of particle swarm optimization and the genetic algorithm. In Proceedings of the 46th AIAA/ ASME/ ASCE/ AHS/ ASC Structures, Structural Dynamics and Materials Conference, Austin, TX, USA, 18–21 April 2005; p. 1897. [CrossRef]
91. Coello, C.A.C.; Pulido, G.T.; Lechuga, M.S. Handling multiple objectives with particle swarm optimization. *IEEE Trans. Evol. Comput.* **2004**, *8*, 256–279. [CrossRef]
92. Mezura-Montes, E.; Coello, C.A.C. Constraint-handling in nature-inspired numerical optimization: Past, present and future. *Swarm Evol. Comput.* **2011**, *1*, 173–194. [CrossRef]
93. Jordehi, A.R. A review on constraint handling strategies in particle swarm optimisation. *Neural Comput. Appl.* **2015**, *26*, 1265–1275. [CrossRef]
94. Petuya, V.; Macho, E.; Altuzarra, O.; Pinto, C.; Hernandez, A. Educational software tools for the kinematic analysis of mechanisms. *Comput. Appl. Eng. Educ.* **2014**, *22*, 72–86. [CrossRef]
95. Andersson, J.A.E.; Gillis, J.; Horn, G.; Rawlings, J.B.; Diehl, M. CasADi—A software framework for nonlinear optimization and optimal control. *Math. Program. Comput.* **2019**, *11*, 1–36. [CrossRef]
96. Moritz Schappler. Matlab Toolbox for Structural and Dimensional Robot Synthesis. Available online: https://github.com/SchapplM/robsynth-structdimsynth (accessed on 21 February 2025).
97. Schappler, M.; Jahn, P.; Raatz, A.; Ortmaier, T. Combined Structural and Dimensional Synthesis of a Parallel Robot for Cryogenic Handling Tasks. In *Proceedings of the Annals of Scientific Society for Assembly, Handling and Industrial Robotics*; Schüppstuhl, T., Tracht, K., Henrich, D., Eds.; Springer: Berlin/Heidelberg, Germany, 2022. [CrossRef]
98. Mohammad, A.; Seel, T.; Schappler, M. Towards Optimized Parallel Robots for Human-Robot Collaboration by Combined Structural and Dimensional Synthesis. In Proceedings of the 2024 VDI-Mechatronik-Konferenz, Dresden, Germany, 14–15 March 2024. [CrossRef]
99. Schappler, M.; Ortmaier, T. Dimensional Synthesis of Parallel Robots: Unified Kinematics and Dynamics using Full Kinematic Constraints. In Proceedings of the 6. IFToMM D-A-CH Konferenz 2020, Lienz, Austria, 27–28 February 2020. [CrossRef]
100. Schappler, M.; Tappe, S.; Ortmaier, T. Exploiting Dynamics Parameter Linearity for Design Optimization in Combined Structural and Dimensional Robot Synthesis. In Proceedings of the 15th IFToMM World Congress, Krakow, Poland, 30 June–4 July 2019. [CrossRef]
101. Schappler, M. Inverse Kinematics for Task Redundancy of Symmetric 3T1R Parallel Manipulators using Tait-Bryan-Angle Kinematic Constraints. In *Proceedings of the International Symposium on Advances in Robot Kinematics*; Springer: Berlin/Heidelberg, Germany, 2022. [CrossRef]
102. Martínez, V. Matlab Implementation of the Multi-Objective Particle-SWARM Optimization from Mezura-Montes and Coello (2011). Available online: https://mathworks.com/matlabcentral/fileexchange/62074-multi-objective-particle-swarm-optimization-mopso (accessed on 21 February 2025).
103. Frindt, M.; Krefft, M.; Hesselbach, J. Structure and type synthesis of parallel manipulators. In *Robotic Systems for Handling and Assembly*; Springer: Berlin/Heidelberg, Germany, 2010; pp. 17–37. [CrossRef]
104. Huang, Z.; Li, Q.; Ding, H. *Theory of Parallel Mechanisms*; Springer Science & Business Media: Berlin/Heidelberg, Germany, 2012; Volume 6. [CrossRef]
105. Stechert, C.; Franke, H.J. Requirement-oriented Configuration of Parallel Robotic Systems. In *Proceedings of the Future of Product Development*; Krause, F.L., Ed.; Springer: Berlin/Heidelberg, Germany, 2007; pp. 259–268. [CrossRef]
106. Schütz, D.; Budde, C.; Raatz, A.; Hesselbach, J. Parallel Kinematic Structures of the SFB 562. In *Proceedings of the Robotic Systems for Handling and Assembly*; Schütz, D., Wahl, F.M., Eds.; Springer: Berlin/Heidelberg, Germany, 2011; pp. 109–124. [CrossRef]

107. Khalil, W.; Dombre, E. *Modeling, Identification and Control of Robots*; Hermes Penton Science: Stanmore, UK, 2002. [CrossRef]
108. Merlet, J.P. Jacobian, manipulability, condition number, and accuracy of parallel robots. *J. Mech. Des.* **2006**, *128*, 199–206. [CrossRef]
109. MathWorks (The MathWorks, Inc.). Alpha Shape Algorithm. Available online: https://mathworks.com/help/matlab/ref/alphashape.html (accessed on 31 October 2024).
110. Reveles R., D.; Pamanes G., J.A.; Wenger, P. Trajectory planning of kinematically redundant parallel manipulators by using multiple working modes. *Mech. Mach. Theory* **2016**, *98*, 216–230. [CrossRef]
111. Schappler, M.; Ortmaier, T. Singularity-Avoidance of Task-Redundant Robots in Pointing Tasks: On Nullspace Projection and Cardan Angles as Orientation Coordinates. In Proceedings of the 18th International Conference on Informatics in Control, Automation and Robotics (ICINCO), Paris, France, 6–8 July 2021. [CrossRef]
112. Hansen, C.; Öltjen, J.; Meike, D.; Ortmaier, T. Enhanced Approach for Energy-Efficient Trajectory Generation of Industrial Robots. In Proceedings of the 2012 IEEE International Conference on Automation Science and Engineering (CASE 2012), Seoul, Republic of Korea, 20–24 August 2012. [CrossRef]
113. Fettin, J. Entwurf und Konstruktion Einer Parallelkinematischen Maschine für Meerestechnische Anwendungen (Design and Construction of a Parallel Kinematic Machine for Applications in Ocean Engineering). Master's Thesis, Leibniz University Hannover, Hanover, Germany, 2023. [CrossRef]
114. Giberti, H.; Ferrari, D. A novel hardware-in-the-loop device for floating offshore wind turbines and sailing boats. *Mech. Mach. Theory* **2015**, *85*, 82–105. [CrossRef]
115. Loeser, T.; Bergmann, A. Capabilities of Deployment Tests at DNW-NWB. In *Fluid Dynamics of Personnel and Equipment*; Precision Delivery from Military Platforms; Meeting Proceedings RTO-MP-AVT-133, Paper 10; RTO: Neuilly-sur-Seine, France, 2006; pp. 10-1–10-12.
116. Fiore, E.; Giberti, H. Optimization and comparison between two 6-DoF parallel kinematic machines for HIL simulations in wind tunnel. In Proceedings of the MATEC Web of Conferences. EDP Sciences, Amsterdam, The Netherlands, 23–25 March 2016; Volume 45, p. 04012. [CrossRef]
117. Giberti, H.; La Mura, F.; Resmini, G.; Parmeggiani, M. Fully mechatronical design of an HIL system for floating devices. *Robotics* **2018**, *7*, 39. [CrossRef]
118. Arino, X.; de Wilde, J.; Russo, M.; Grytøyr, G.; Tognarelli, M. Forced oscillation model tests for determination of hydrodynamic coefficients of large subsea blowout preventers. In *Proceedings of the International Conference on Offshore Mechanics and Arctic Engineering*; American Society of Mechanical Engineers: New York, NY, USA, 2015; Volume 1: Offshore Technology; Offshore Geotechnics.
119. Symétrie. Motion Hexapods—Notus. Available online: https://symetrie.fr/en/hexapods/notus/ (accessed on 31 October 2024).
120. LUIS (Leibniz Universität IT Services). Hardware Specifications of Cluster Compute Nodes. Available online: https://docs.cluster.uni-hannover.de/doku.php?id=resources:computing_hardware (accessed on 31 October 2024).
121. Tartari Filho, S.C.; Cabral, E.L.L. Kinematics and workspace analysis of a parallel architecture robot: The Hexa. In Proceedings of the 18th International Congress of Mechanical Engineering, Orlando, FL, USA, 5–11 November 2005.
122. Gharahsofloo, A.; Rahmani, A. An efficient algorithm for workspace generation of Delta robot. In *Proceedings of the International Journal of Robotics*; K.N. Toosi University of Technology: Teheran, Iran, 2015.
123. Prause, I.; Charaf Eddine, S.; Corves, B. Comparison of parallel kinematic machines with three translational degrees of freedom and linear actuation. *Chin. J. Mech. Eng.* **2015**, *28*, 841–850. [CrossRef]
124. Heidenhein (Dr. Johannes Heidenhain GmbH). Angle Encoder Modules, MRP Series. Available online: https://www.heidenhain.com/products/angle-encoders/modules/mrp (accessed on 31 October 2024).

robotics

MDPI

Article

Three-Degree-of-Freedom Cable-Driven Parallel Manipulator with Self-Sensing Nitinol Actuators

Francesco Durante [1,*], Terenziano Raparelli [2] and Pierluigi Beomonte Zobel [1]

[1] Department of Industrial and Information Engineering and Economy (DIIIE), University of L'Aquila, P.le Pontieri 1, Località Monteluco, 67100 L'Aquila, Italy; pierluigi.zobel@univaq.it
[2] Department of Mechanical and Aerospace Engineering (DIMEAS), Politecnico di Torino, Corso Duca degli Abruzzi 24, 10129 Torino, Italy; terenziano.raparelli@polito.it
* Correspondence: francesco.durante@univaq.it

Abstract: This paper presents the design and analysis of a novel 3-degree-of-freedom (3-DOF) parallel manipulator equipped with self-sensing Ni-Ti (Nitinol) actuators. The manipulator's architecture and mechanical design are elucidated, emphasizing the integration of Nitinol actuators. The self-sensing technique implemented in a previous work was extended to a 20 mm actuator length, and the actuator was used to design the 3-DOF manipulator. Kinematic analyses were conducted to evaluate the manipulator's performance under various operating conditions. A dynamic model was implemented for the dynamic dimensioning of the actuators, which work synergistically with a bias spring. The manipulator was realized, and a control strategy was implemented. Experimental tests, although documenting some positioning accuracy issues, show the efficacy and potential applications of the proposed manipulator in robotics and automation systems, highlighting the advantages of self-sensing Nitinol actuators in small parallel manipulator designs.

Keywords: parallel manipulator; Ni-Ti actuators; kinematic analysis; dynamic analysis; robotics; cable robot

1. Introduction

Parallel manipulators have gained significant attention in robotics due to advantages such as high rigidity, accuracy, and payload capacity. In recent years, there has been a growing interest in integrating smart materials, particularly shape memory alloys (SMAs), into robotic systems to enhance their performance and capabilities. Among SMAs, nickel–titanium (Ni-Ti) alloys stand out for their unique properties, including high energy density, biocompatibility, and shape memory effect, making them ideal candidates as actuators [1–6] in many applications in different fields, such as in aerospace [7–11], biomedicine [12–17], wearables [18–21], micro-systems [22–25], and robotics [26–32]. There are not many works in the literature regarding manipulators with parallel architecture with SMA actuators. In [33], a parallel device for the orientation of the platform connected to the base via a flexible central rod in superelastic Nitinol is described. Three Nitinol wires arranged at 120° are used as actuators. The flexible structure reflects a system with two DOFs. The dimensions are contained in a cylinder of 65 mm in diameter and 75 mm in height. For the movement, a structural model allows position control through the knowledge of the forces applied by the actuators, which are detected using force sensors on each actuator. The experimental tests were carried out with the movement of just one actuator, and the authors also plan to use a position detection system. Ref. [34] describes a parallel manipulator that uses actuators made by a carriage on a slide connected to an elastic plate that is bent by the actuator and shortens. The wires are kept in an eccentric configuration concerning the plate to amplify the shortening. Control is implemented with on–off commands based on feedback using a signal from linear variable displacement transducers (LVDTs), one for each actuator. The system has three DOFs and good performance. Still, it is rather large, given that the actuators are bulky in order to achieve the amplification effect, i.e., they have a length of approximately 220 mm and a transversal dimension of roughly 40 mm × 50 mm. The device is contained in a cylinder with a diameter of 230 mm and a height greater than 220 mm. In [35], the device BAPAMAN is presented. It is a module comprising a base and a platform connected by three legs arranged at 120°. Each includes two articulated segments, and all the connections between the two segments of the leg and between the leg with the base and platform are made using flexible joints. There is a cylindrical hinge between the leg and the base and between the leg segments, while there is a universal joint between the leg and the platform. The platform has five DOFs with respect to the base: three translational and two rotational. The system is designed to be organized into multiple serial modules to constitute a snake robot instead of a parallel manipulator. As for the movements, they are implemented in binary mode. Each leg features two actuators, giving the module eight different configurations. The overall dimensions of the single module are contained in a cylinder with a diameter of 180 mm and a height of 200 mm. In [36], a parallel manipulator with a square base and a platform that implements four SMA springs arranged at the vertices as actuators, each with a steel antagonist spring, is presented. The movements are implemented in binary mode. It features three DOFs for roll pitch and one along the vertical axis. The system has small dimensions, 30 mm × 30 mm × 34 mm, and performs large movements: a rotation of the platform of 30° and a stroke in the vertical direction of 12 mm, thanks to the high deformation of the SMA springs. Given the structure, it is probably not very rigid, losing a peculiar characteristic of parallel architecture robots. The biggest limit, however, is the actuators' binary operation, which leads to a limited number of possible configurations, placing the end effector discreetly in the working volume. Reference [37] describes a system with a triangular base and platform connected with a universal joint and SMA springs at the vertices. It features two DOFs: pitch and roll. The dimensions are contained in a cylinder with a diameter of 92 mm and a height of 114 mm. A position control with an on–off type command in the closed loop is implemented using an IMU sensor. In [38], a 3-DOF parallel manipulator with SMA wire actuators is presented. It features a base and a platform. The base and platform are connected to the actuators, which are constantly tensioned by a bias spring between the base and the platform. The manipulator implements proportional position closed-loop control, using the signal generated by a potentiometer for each actuator. The project functions well, but the system is rather bulky, given the presence of position sensors. The dimensions are contained in a cylinder with a diameter of 82 mm and a height of approximately 210 mm.

This paper presents the design and analysis of a novel 3-DOF parallel manipulator utilizing self-sensing Ni-Ti actuators. Integrating self-sensing capabilities within the actuation system eliminates the need for external sensors, simplifies the overall design, and reduces the system's complexity, size, and cost. Also, using Ni-Ti actuators allows for smooth and silent actuation mechanisms. The proposed parallel manipulator is compact and lightweight. It suits various tasks requiring the positioning and manipulation of small objects, such as pick-and-place operations, assembly tasks, and biomedical applications. The paper describes the design methodology, i.e., kinematic modeling and analysis of the manipulator's workspace, singularity conditions, error analysis, and dynamic modeling for the actuation system dimensioning of the proposed 3-DOF parallel manipulator and the investigation of dynamic performance characteristics, including integrating self-sensing Ni-Ti actuators. Furthermore, the experimental validation of the realized manipulator prototype is conducted to assess its performance for positional accuracy. The results, except for a limited area of the working volume in which positioning errors outside the expected range occur, prove the effectiveness of the project by demonstrating that for the creation of cable-driven manipulators, it is possible to use self-sensing shape memory alloy wires exploiting their self-sensing characteristic, thus allowing compact dimensions to be obtained. Compared to the devices in the scientific literature, the proposed manipulator is similar to the one described last. A comparison with that manipulator has been conducted to prove the actual project's performance.

The contents of the paper are organized as described below. In Section 2, Materials and Methods, the mechanical design of the manipulator is reported (Section 2.1), it is addressed through the technical specification and functional design (Section 2.1.1), and then continues with the characterization of the Ni-Ti actuator (Section 2.1.2). Following these are the forward kinematic analysis (Section 2.1.3), the inverse kinematic analysis (Section 2.1.4), and the differential kinematic analysis (Section 2.1.5). The next section contains a dynamic analysis to verify the sizing of the actuator and spring system (Section 2.1.6). Then, the next subsection (Section 2.1.7) describes the detailed project. Subsequently, in Section 2.2, the control is described with regard to the hardware (Section 2.2.1) and the control strategy (Section 2.2.2). In Section 3, the results are presented. Section 3.1 presents the results relating to the estimation of the error on the working volume boundaries based on the errors detected on the single actuator. Subsequently, the prototype of the manipulator is presented (Section 3.2), and subsequently, the results of the positioning tests relate to the movement of the individual actuators and tests for trajectory tracking (Section 3.3). In Section 4, a discussion of the results is carried out, possible future developments are proposed, a comparison is made between the proposed system and a similar one, and conclusions are drawn.

2. Materials and Methods

2.1. Mechanical Design of the Manipulator

2.1.1. Manipulator's Technical Specification and Functional Design

The manipulator has to serve as a proof of concept for a small manipulator with parallel architecture, using shape memory alloy wires as actuators. Taking as a reference a Stewart platform, some preliminary analyses suggest that it is convenient to consider technical specifications for the design of the device, as follows:

- Number of DOFs: 3 without wrist and gripper;
- Overall dimensions, including actuators, contained in a cylinder with a diameter of 45 mm and a height of 140 mm;
- Diameter of the circle circumscribed to the plan projection of the working volume by height, in the range of 10 mm $\times$ 2 mm;
- Actuator's block control volume not more than 20,000 mm^3;
- Overall mass without the electronics for control of not more than 0.05 kg;
- The manipulator is intended to manipulate tiny mass objects, so the considered load capability is 0.05 N.

Following the specifications above, the kinematic architecture of Landsberger's robot is considered [39]. A system consisting of a base and a platform, with the dimensions of the platform reduced to zero, is considered according to the scheme in Figure 1 [40].

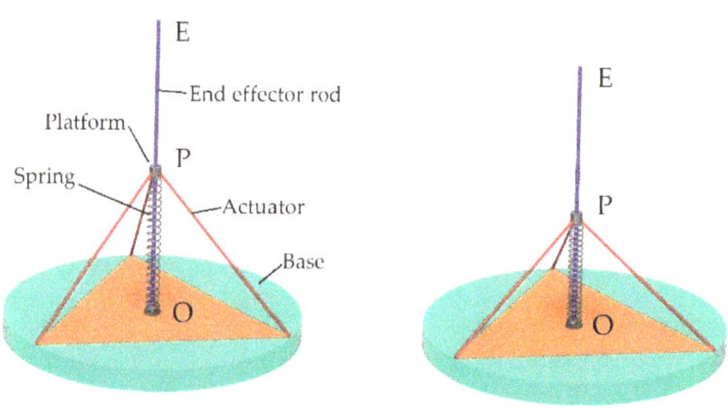

Figure 1. Concept of the manipulator.

The proposed idea is to implement Ni-Ti wires as actuators using the self-sensing effect. This way, it is possible to eliminate motors, drums, and external sensors, thus avoiding an increase in overall dimensions. The actuators are anchored to the base, each to the three vertices of an equilateral triangle, and on the other side to the platform, which is a tiny body. The actuators, which work only by traction, are constantly tensioned thanks to a compressed bias spring placed between the central point of the base, O, and the point P, representing the platform.

To amplify the movements, a small rod is used, which is coupled to the base via a joint consisting of a sphere articulated on the base and drilled to accommodate the end of the rod in a prismatic coupling. This joint allows the rod to rotate and slide along the direction of its axis relative to the base. At the other end, the rod is integral to the platform, P, and extends beyond it. The upper end of the rod, E, is at a fixed distance from the platform; it constitutes the end effector of the manipulator to which a tool can be applied.

2.1.2. Characterization of Actuators Using Self-Sensing Effect

Based on preliminary evaluations of the hypothesized architecture, a system with actuators approximately 20 mm long can satisfy the technical specification. The work [41] demonstrated that a Ni-Ti wire can work well as an actuator with a self-sensing effect. In that work, the effectiveness of a wire with a length of 200 mm was demonstrated. Here, the hypothesis is to work with a much shorter actuator. Therefore, before proceeding with the design of the manipulator, it is necessary to experimentally verify the effectiveness of the self-sensing process on a wire with a length of 20 mm.

The same Ni-Ti wire whose characteristics are detailed in Table 1 was considered, and the same experimental setup described in [41] was used but with a modification of the index of the graduated scale, which was brought to a greater length to amplify the movement generated by the actuator for better visualization. In this case, the actuator will have a total excursion equal to approximately 0.7 mm, considering a percentage shortening of 3.5%, so length variations of a hundredth of a mm must be detectable. Changes have been made to the electrical components of the control system to adapt them to the different electrical parameters of the shorter wire. The electrical measurement resistance was brought to 3.9 Ohm and the supply voltage to 1.1 V.

Table 1. Technical specification of the wire.

Flexinol 150HT
Wire diameter: 150 μm
Linear resistance: 50 Ω/m
Maximum recovery force: 10.4 N
Nominal recovery force: 3.2 N
Deformation force: 0.6 N
Recommended deformation ratio: 3–5%
Austenite start temperature: 88 °C
Austenite finish temperature: 98 °C
Martensite start temperature: 72 °C
Martensite finish temperature: 62 °C

Tests were conducted to characterize the wire's unitary deformation (ε) relationship with electrical resistance (R) to obtain the model for self-sensing implementation. In particular, the model was obtained for the actuator in an antagonistic configuration with a bias spring. The resulting model was represented as a linear one

$$\varepsilon = 26.374 - 17.419\,\mathrm{R} \tag{1}$$

whose relative graph is shown in Figure 2.

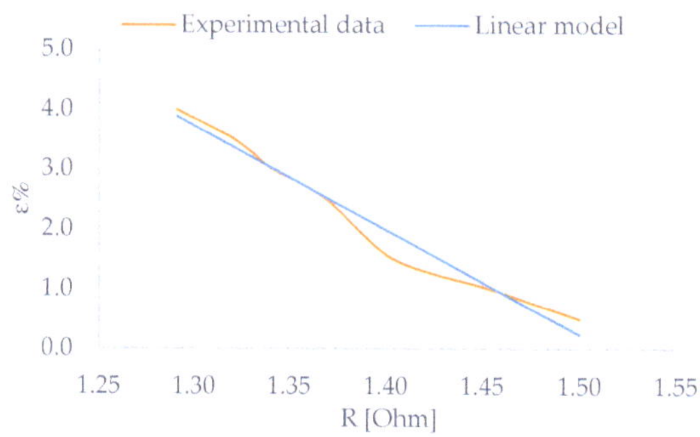

Figure 2. A linear mathematical model for the electrical resistance deformation for the NiTi alloy wires is adopted for the control.

Positioning tests were conducted using the model and the setup described. A precision potentiometer connected to the pulley [41] was added to acquire the actual position using a data acquisition board. It turns out that the actuator works well over the entire operating range (0.7 mm), except for the value close to the minimum length, i.e., for an elongation equal to 0.1 mm. Figure 3 shows the system responses for stepped input signals. There is an error due to oscillations of approximately 0.04 mm.

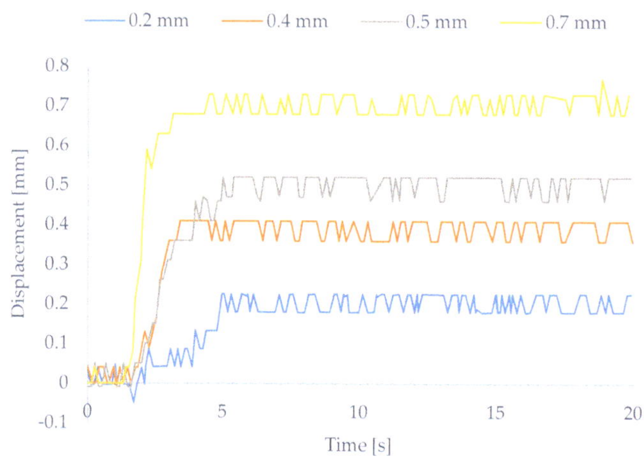

Figure 3. Responses of the 20 mm long actuator with the self-sensing process to step inputs of 0.2 mm, 0.4 mm, 0.5 mm, and 0.7 mm. The responses are quick with a precise mean value, but oscillations occur with an amplitude of less than 0.02 mm.

2.1.3. Direct Kinematic Model—Domain Analysis

For the working volume analysis, the direct kinematic model was considered. With reference to Figure 4, we name O_1, O_2, and O_3 as the three points at which the actuators are attached to the base, with each at a distance b from the center of the base, O. A fixed frame attached to the base is considered, such that the origin is located at the center of the base, with the x-axis pointing toward the point O_1, the z-axis pointing upward, and the y-axis pointing coherently with a right-handed reference. Each actuator is represented by a vector l_i (i = 1, 2, 3) from O_i to the point P, and all the actuators have the same initial length.

The first step in determining the direct kinematic model is to solve for the position of the point *P* given the values of the actuator's lengths, l_i. For each actuator, the loop closure equations can be written as

$$\mathrm{OP}\text{-}\mathrm{OO}_i = l_i \quad i = 1,2,3. \tag{2}$$

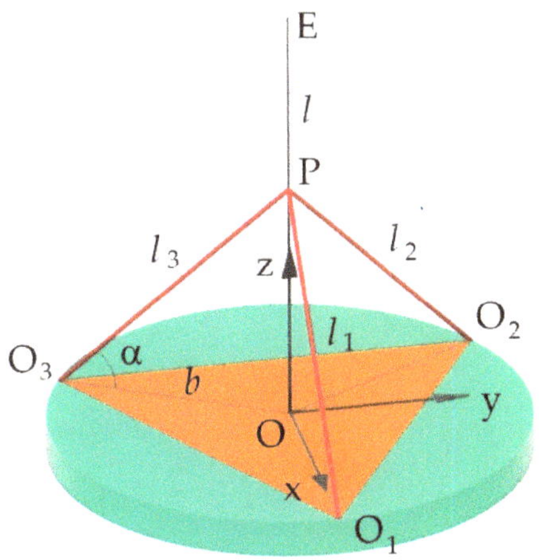

Figure 4. Parameters for the kinematic model of the manipulator.

The vector OP is as follows:

$$\mathrm{OP} = [x_P, y_P, z_P]^T \tag{3}$$

while OO_i have a fixed magnitude and are as

$$\mathrm{OO}_1 = [b,\ 0,\ 0]^T \tag{4}$$

$$\mathrm{OO}_2 = \left[-\frac{1}{2}b,\ \frac{\sqrt{3}}{2}b,\ 0\right]^T \tag{5}$$

$$\mathrm{OO}_3 = \left[-\frac{1}{2}b, -\frac{\sqrt{3}}{2}b,\ 0\right]^T \tag{6}$$

We can substitute Equations (3)–(6) into Equation (2), resulting in three vector equations with three unknowns, the coordinates of the point P, for a given set of l_i. Therefore, we have to solve the three equations with three unknowns simultaneously for the position of the platform (point P) to be determined.

Before solving Equation (2), the values of the manipulator's design parameters, namely the actuators' rest length and the base's dimension *b*, must be defined. To do that, we can choose a resting length for the actuators and then derive the dimensions of the base from it.

The choice of these parameters derives from an optimization process between conflicting needs. In fact, for a given length of the actuators, the excursion of the end effector in the vertical direction is more extensive, and consequently, the same is true for the working volume, the smaller the angle α. That involves a greater overall dimension of the basis and, therefore, a bigger value for *b*.

From a quantitative point of view, the distance OP, given the size of the base b and the actual length of the actuators, l_i, is

$$OP = l_i \sqrt{1 - \left(\frac{b}{l_i}\right)^2} \quad (7)$$

Once the value b has also been hypothesized, with relation (7), it is possible to calculate the distance OP, considering the variation in the actuator's length and the end effector's vertical excursion. Therefore, based on what has been said, we will proceed by hypothesizing different values for b, evaluating the best compromise solution between the excursion of the end effector, which we want to be large, and the overall size of the base, which we wish to be small. Table 2 shows the results for the cases relating to the actuator, with nominal length dimensions equal to 20 mm and different hypothesized values for dimension b. From the data observation, we chose to work with b = 16.5 mm as it constitutes a good compromise between the base's size and the end effector's excursion.

Table 2. Parameters' comparison for the choice of the characteristic dimension of the robot, b.

b (mm)	α_M (°)	α_m (°)	OP_M (mm)	OP_m (mm)	$\Delta(OP)$ (mm)
10.00	60.00	58.79	17.32	16.51	0.81
11.00	56.63	55.25	16.70	15.86	0.84
12.00	53.13	51.56	16.00	15.12	0.88
13.00	49.46	47.66	15.20	14.26	0.93
14.00	45.57	43.50	14.28	13.28	1.00
15.00	41.41	38.99	13.23	12.14	1.08
16.00	36.87	34.00	12.00	10.79	1.21
16.50	34.41	31.25	11.30	10.01	1.29
17.00	31.79	28.26	10.54	9.14	1.40
18.00	25.84	21.15	8.72	6.96	1.75
19.00	18.19	10.12	6.24	3.39	2.86

Once the geometric data characterizing the actuators were defined, the direct kinematic problem for determining the working volume was solved.

By substituting (3)–(6) into Equation (2) and then applying the norm and squaring, we can obtain the following three equations:

$$\begin{cases} (x_P - b)^2 + (y_P - 0)^2 + (z_P - 0)^2 = {l_1}^2 \\ \left(x_P + \frac{1}{2}b\right)^2 + \left(y_P - \frac{\sqrt{3}}{2}b\right)^2 + (z_P - 0)^2 = {l_2}^2 \\ \left(x_P + \frac{1}{2}b\right)^2 + \left(y_P + \frac{\sqrt{3}}{2}b\right)^2 + (z_P - 0)^2 = {l_3}^2 \end{cases} \quad (8)$$

These were solved with Mathematica code. An equation of the 16th order of x_P was obtained:

$$a_1 \cdot {x_P}^{16} + a_2 \cdot {x_P}^{14} + a_3 \cdot {x_P}^{12} + a_4 \cdot {x_P}^{10} + a_5 \cdot {x_P}^{8} + a_6 \cdot {x_P}^{6} + a_7 \cdot {x_P}^{4} + a_8 \cdot {x_P}^{2} + a_9 = 0 \quad (9)$$

where the expressions of the coefficients a_i are omitted because of their length.

This equation is in the 16th order of x_P and leads to 16 solutions. Since the power is always even, the solution is facilitated because it is possible to solve an 8th-degree equation. Six of these solutions are coincident, while the remaining two are symmetric with respect to the xy plane. By substituting x_P into Equation (8), it is possible to calculate y_P and z_P.

The second step concerns the calculation of the coordinates of the end-effector E (x_E, y_E, z_E). The distance OP is a function of the coordinates of the point P:

$$OP = \sqrt{x_P^2 + y_P^2 + z_P^2} \quad (10)$$

Considering the distance of the end effector E from P as l, we have

$$x_E = x_P \frac{\mathrm{OP} + l}{\mathrm{OP}} \quad (11)$$

$$y_E = y_P \frac{\mathrm{OP} + l}{\mathrm{OP}} \quad (12)$$

$$z_E = z_P \frac{\mathrm{OP} + l}{\mathrm{OP}} \quad (13)$$

Thus, the direct kinematic problem is solved.

A direct kinematic analysis was carried out using the model described to visualize the boundaries of the working volume. Using a Matlab code, the lengths of the actuators were made to vary systematically in sequence, and the coordinates of the end effector in physical space were calculated. The actuators' lengths varied between 19.3 mm and 20 mm, with variations of 0.035 mm for 1386 configurations. The size of the rod was considered equal to l = 120 mm. Figure 5 shows the working volume.

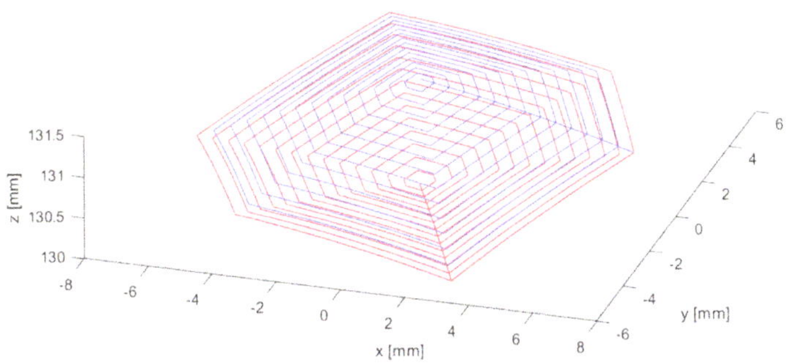

Figure 5. The boundaries of the working volume of the manipulator. Upper boundaries, blue; lower boundaries, red.

This resulted in a volume with a hexagonal shape in plan with an apothem equal to approximately 6 mm and with a maximum height equal to approximately 1.3 mm, as per the project.

2.1.4. Inverse Kinematic Model

The inverse kinematic problem is very straightforward. Starting from a position of the end effector in the cartesian space, E(x_E, y_E, z_E), it is possible to compute the coordinates of the point P using Equations (14)–(16):

$$x_P = x_E \frac{\mathrm{OE}}{\mathrm{OE} - l} \quad (14)$$

$$y_P = y_E \frac{\mathrm{OE}}{\mathrm{OE} - l} \quad (15)$$

$$z_P = z_E \frac{\mathrm{OE}}{\mathrm{OE} - l} \quad (16)$$

Then, by means of Equation (8), it is possible to solve for the coordinates in the joint space, l_1, l_2, l_3.

2.1.5. Jacobian and Singularity Analysis

Starting from Equation (2), the differentiation must be carried out with respect to the time of both members. It should be considered that the derivative with respect to the time of the terms l_i has two components, one represented by the variation in length and the other by the variation in orientation with respect to the reference system of the base. However, the derivative of the segments $\mathrm{OO_i}$ is equal to zero since their lengths and orientations are constant. The OP vector also has two components, one due to the change in length and another due to the change in orientation (length and rotation variations of the spring with respect to the base reference). Both of these components contribute to the determination of the velocity of point P.

We find

$$\frac{d}{dt}\mathrm{OP} - \frac{d}{dt}\mathrm{OO_i} = \frac{d}{dt}l_i \tag{17}$$

$$\mathbf{V}_\mathrm{P} = \dot{l}_i\mathbf{s}_i + \boldsymbol{\omega}_\mathrm{i} \times l_i\mathbf{s}_i \tag{18}$$

where $\mathbf{s}_i$ is the unit vector associated with the i-th actuator. To eliminate the angular velocities of actuators, we can dot-multiply both members of Equation (18) by $\mathbf{s}_i$. We find

$$\mathrm{s}_i \cdot \mathrm{V_P} = \dot{l}_i\mathrm{s}_i \cdot \mathrm{s}_i + \omega_\mathrm{i} \cdot (l_i\mathrm{s}_i \times \mathrm{s}_i) \tag{19}$$

According to [42], two Jacobians can be individuated in (19). The first one (type I Jacobian) is the coefficient of the velocity vector $\mathrm{V_P}$, and the other (type II Jacobian) is the coefficient of the scalar $\dot{l}_i$

$$J_q = I \tag{20}$$

$$J_x = \begin{bmatrix} s_{1x} & s_{1y} & s_{1z} \\ s_{2x} & s_{2y} & s_{2z} \\ s_{3x} & s_{3y} & s_{3z} \end{bmatrix} \tag{21}$$

The first Jacobian can give rise to the I type of singularities. After inversion, it can be useful for calculating joint speeds to obtain the assigned speeds of the end effector. It is, therefore, similar to the Jacobian that is determined for manipulators with serial architecture, and the singularities associated with it can be defined as inverse kinematic singularities. Since, in this case, it is equal to the identity matrix, there are no singularities in the inversion. Singularities, however, occur at the edge of the working volume when the spring is at its minimum and maximum lengths.

The second one can give rise to the II type of singularities. These can occur when the end effector is locally movable, even when all the actuated joints are locked. For this reason, we can refer to them as the forward kinematic singularities.

The singularities occur when

$$\det J_x = 0 \tag{22}$$

$$s_{1x}s_{2y}s_{3z} + s_{1y}s_{2z}s_{3x} + s_{1z}s_{2x}s_{3y} - \left(s_{1z}s_{2y}s_{3x} + s_{1y}s_{2x}s_{3z} + s_{1x}s_{2z}s_{3y}\right) = 0 \tag{23}$$

The equation should then be solved. We can analyze two straightforward cases. The first occurs when all the z components of the unit vectors associated with the actuators are zero. This happens when the actuators and points P and E all lie in the plane of the base. Although theoretically possible, this condition never occurs because, in the current case, points P and E are always far from the base plane. The second occurs when the x, y, or both components of the unit vectors associated with actuators are zero. Moreover, this is a required but not sufficient condition, since the actuators must be perpendicular to the upper base for the singularities to occur. This can be verified for an ideal architecture with a base and a platform of equal geometry. However, in the present case, where the platform is reduced to point P, the condition under the exam is never verified. The other type II singularities can also be excluded for the same reason.

2.1.6. Dynamic Analysis—Actuators' Dimensioning

The following applies regarding the dynamic behavior of the actuators and their interaction with the bias spring. The spring has the task of keeping the wires permanently taut. In other words, it must guarantee that the wires can recover their maximum length which requires the application of a deformation force for all the actuators simultaneously. In this sense, the most critical condition is the one in which the wire is close to its maximum length, both because the spring is in the condition of minimal compression and because the angle α is at its maximum value, which involves a component of the force of the spring, useful for lengthening the wires, being smaller. At the same time, the spring has to be sufficiently soft to be compressed by the actuators in the active work phase.

Concerning Figure 6, let F_i be the force developed by the i-th actuator and F the spring force that acts along the OP direction of the rod (end effector). The following relation is valid:

$$F = \sum_i F_i \sqrt{1 - \left(\frac{b}{l_i}\right)^2} \quad (24)$$

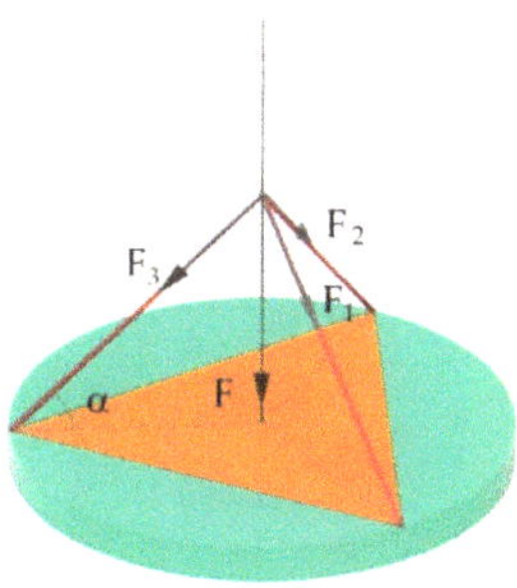

Figure 6. The force developed by the actuators F_i and their resultant F along the end effector rod axis. The spring has to apply a force opposite to F, as given by Equation (24).

The chosen wire for the actuators requires a low-temperature recovery force of 0.6 N and develops a high-temperature transformation force of 3.2 N.

Based on preliminary evaluations, a spring with a stiffness of k = 0.22 N/mm is compatible with the bias spring utilized to obtain the actuators' ε-R relationship (1). The free length of the spring is equal to 15.7 mm, and by choosing to fix its upper end on the rod below the point P at a distance equal to 3.3 mm, the force exerted by the spring, in a generic configuration, is equal to

$$F = k\,[15.7 - (OP - 3.3)] \quad (25)$$

A dynamic model is considered to verify the correct tensioning of the actuators by the spring over the entire working volume.

Given the actual position of point P and the force F exerted by the spring, to calculate the forces F_1, F_2, F_3 exerted by the three actuators, the system equation in (26) has to be solved:

$$F_1 \frac{O_{1i} - x_i}{\sqrt{\Sigma(O_{1i} - x_i)^2}} + F_2 \frac{O_{2i} - x_i}{\sqrt{\Sigma(O_{2i} - x_i)^2}} + F_3 \frac{O_{3i} - x_i}{\sqrt{\Sigma(O_{3i} - x_i)^2}} = x_i \frac{F}{OP} \qquad i = 1,2,3 \quad (26)$$

where x_i = cartesian coordinates of point P and O_{ji} = cartesian coordinates for attaching the points of the actuators to the base.

Using this model, a dynamic analysis was carried out on the entire boundary of the working volume to verify the spring. Figure 7 shows the results obtained with the dynamic model. Figure 7a shows the actuator's forces with the end effector lying on the upper working volume boundary, while Figure 7b shows the forces with the end effector lying on the lower working volume boundary, in contrast with the bias spring. The maximum spring force occurs on the lower boundary and is equal to 1.98 N, and the actuators' force has to be equal to 1.26 N for the equilibrium. Thus, since the actuators are capable of a force of 3.2 N, there is a margin to apply loads on the environment. On the upper boundary, the spring applies a minimum force of 1.69 N, and the actuators apply a minimum force of 1.22 N for the equilibrium. Since the actuators require a force of 0.6 N to be stretched at low temperatures, the spring is able to keep them taut and return them at maximum length, providing a margin to apply loads on the environment.

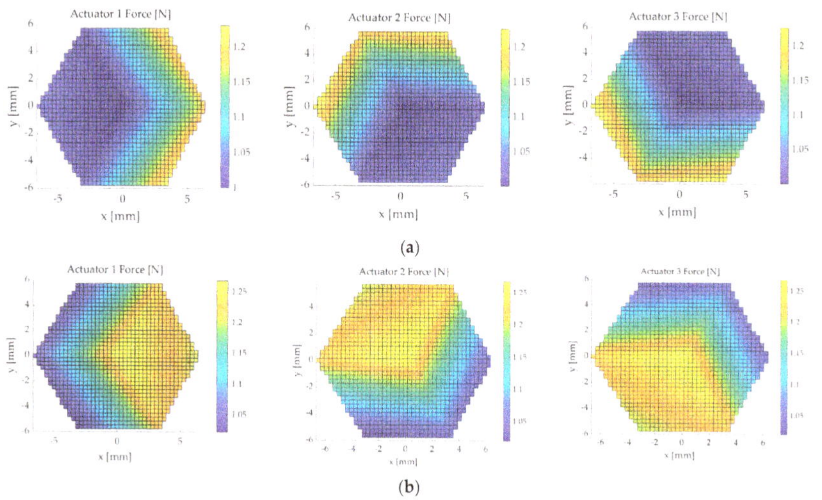

Figure 7. Forces applied by the actuators with the end effector lying on the upper working volume boundary (**a**) and lower working volume boundary (**b**) in contrast with the bias spring. The actuators and spring system are well-sized and succeed in the task of compressing the bias spring in every condition, while the actuators are always kept taut and stretched to their maximum length at low temperatures by the spring. In addition, excess forces result which are useful for applying loads to the environment.

2.1.7. Detailed Design

The components were designed in detail. The base is circular with a thickness of 6 mm and a diameter of 40.8 mm, and it was realized in nylon. In the center, a spherical groove is present for the articulation with the rod. The joint is realized by a spherical component with a diameter of 4.5 mm for the articulation to the base side, with a diametral hole for coupling with the rod. This one, with a diameter of 1 mm, can slide inside the hole to confer on the robot a third degree of freedom. The spherical component presents, in the upper part, a plane surface, which constitutes the base on which rests the lower extremity of the spring. The rod presents a lock for the upper extremity of the spring formed by an axisymmetric component. Above it, there is a cylindrical component for the lock of the upper extremity of the actuators. Finally, on the base, there are three holes disposed at 120° angular distance from each other, where the locks are coupled for the lower extremity of the actuators. Figure 8 shows the CAD model of the designed prototype.

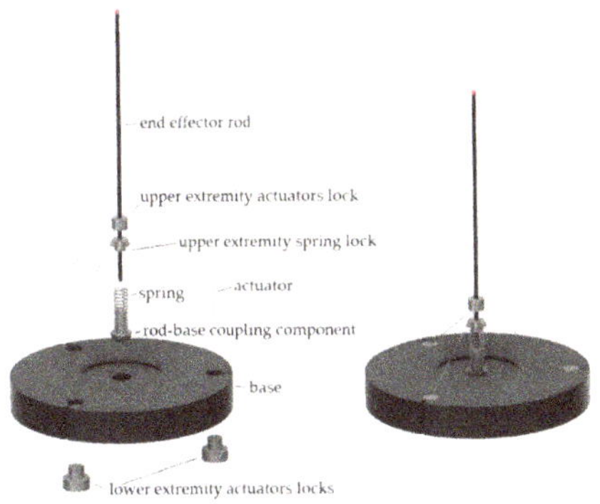

Figure 8. The CAD model of the manipulator.

2.2. Control System

In the next paragraphs, we will discuss the hardware and strategy for the control implemented.

2.2.1. Hardware

The hardware for the control is derived from the one used and described in [41]. The single actuator is driven by a homemade board with a drive circuit to drive the actuator and a data acquisition circuit to measure the electrical resistance to estimate its actual length. The board used for the one actuator was replicated three times on a single board to drive the three actuators to control the manipulator. Figure 9 shows the homemade drive board for the manipulator.

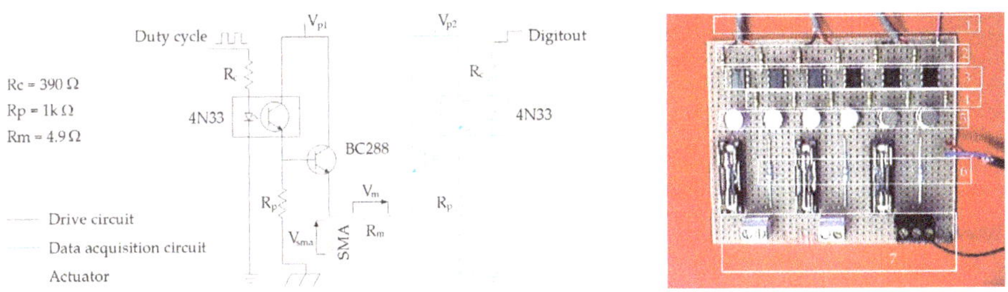

(a) (b)

Figure 9. The board used for the measurement of the electrical resistance and the driving of the actuator. (**a**) The electric scheme; (**b**) the physical homemade board in which the electric scheme was replicated three times for the three actuators: (1) controller connector, (2) R_c electrical resistances, (3) 4N33 optocoupler, (4) R_p electrical resistances, (5) BC288 transistors, (6) R_m measuring electrical resistance, and (7) Ni-Ti actuators connectors.

2.2.2. Control Strategy

As for control, we consider the command of the actuators in one direction and rely on the bias spring for movements in the opposite direction. The law adopted applies the maximum possible value of the signal when the actuator is far from the reference and becomes proportional (P) to the error near the reference according to the law shown in Figure 10.

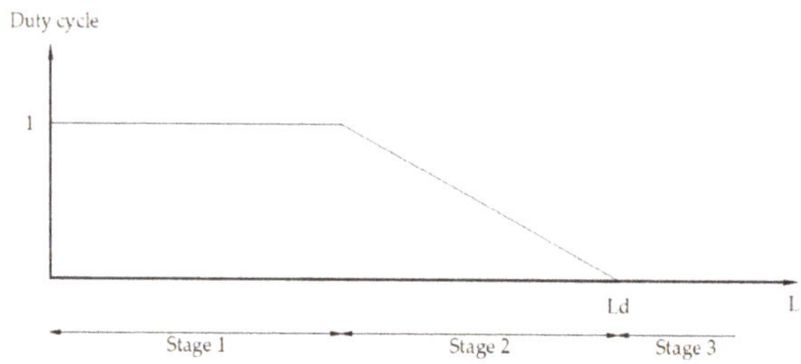

Figure 10. Control logic scheme: in stage 1, far from the target, the control supplies the maximum power. In stage 2, close to the target, the control becomes proportional. At stage 3, overcoming the target, the control cuts power off.

This control logic allows a faster approach phase than a purely proportional one.

The power and measurement circuits are powered alternately to estimate the actual deformation by the measurement circuit to control the actuator. This logic is adopted similarly for every one of the three actuators.

3. Results

3.1. Analysis of the Influence of Actuator Errors on Working Volume Errors

In Figure 3, it is shown that the single actuator positioning error was about 0.02 mm. A mapping of the errors in the working volume was derived through an estimation of the influence of the errors of the three actuators using the direct kinematic model. The calculation was carried out by commanding movements equivalent to the documented errors and combining them in the worst combination of the three actuators, which corresponds to double the amplitude of the error of a single actuator. In Figure 11, the derived map is shown. It can be seen that the error is relatively constant over the working volume. The maximum values occur at three vertices where two actuators are near the maximum contraction, and a higher movement amplification effect of the cable system occurs.

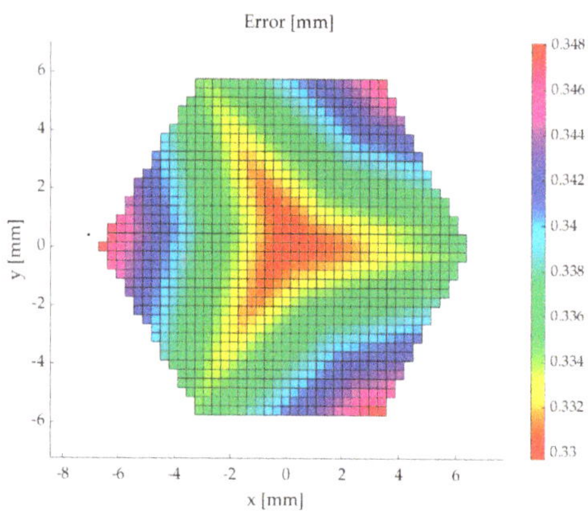

Figure 11. Sensitivity analysis of the error over the working volume domain due to the errors of the actuators (shown in Figure 3). It is possible to note a maximum absolute position error of 0.348 mm, corresponding to an error relative to the maximum transversal dimension of the working volume, equal to 2.7%.

3.2. Manipulator Prototype

The components were realized and assembled. Particular attention had to be paid to the realization of the actuators. They have to be of the same length: 20 mm. This goal was achieved using knots and mechanical end locks realized by machining. Figure 12 shows the realized prototype.

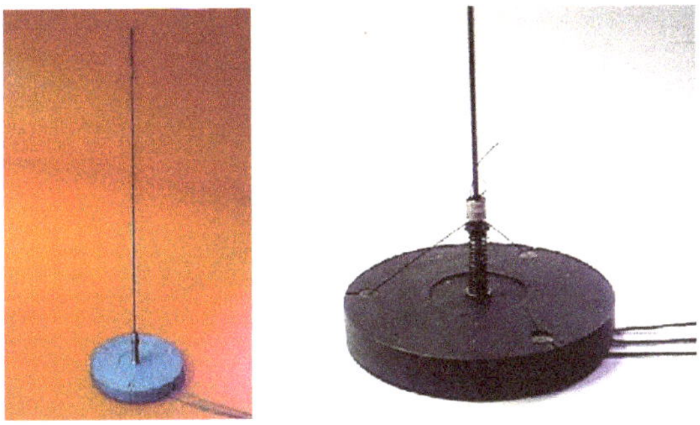

(a) (b)

Figure 12. The realized prototype. An overall view showing the base and the end effector rod (**a**). A zoom in on the base: it is possible to see the spherical joint between the end effector and the base, as well as the spring and the NiTi wires (**b**).

3.3. Experimental Tests

The first experimental tests were about movement, involving one actuator at a time. Using lateral photography at a distance of about 4 m, it was possible to document the movements of the end effector on a graduated scale. The possibility of positioning by a single actuator one at a time was verified on the overall range of motion. Figure 13 shows the manipulator executing a single actuator movement (a) and the movement to the position (2, 0, 131.3) as viewed by the camera with the optic axis intersecting the z-axis and pointing toward the positive direction of the y-axis (b).

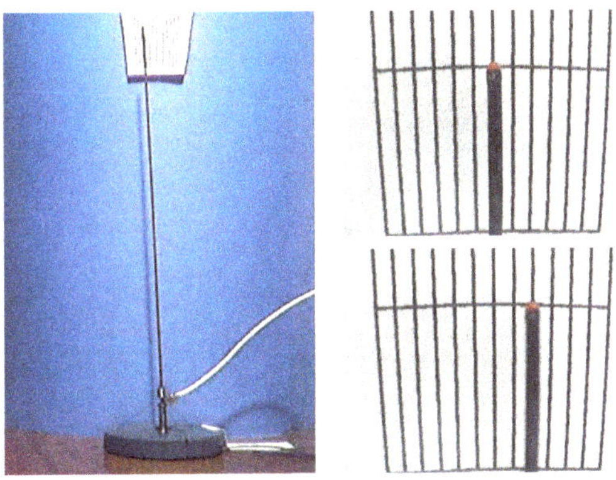

(a) (b)

Figure 13. The first experimental tests were related to the positioning of an actuator at the time. A calibrated index indicates the actual position on a graduated scale where every notch corresponds to 1 mm (**a**). The positioning from the resting position to 2 mm (**b**).

Once the functionality of the single actuators was demonstrated, some tests were conducted involving coordination between the actuators in trajectory control tasks. The first involved the end effector in describing a hexagonal trajectory centered on the working volume. Subsequently, trajectories arranged without particular symmetries with respect to the working volume were tested.

As for the measurement of the movements, a camera was placed above the robot, with its optical axis pointing downward and coincident with the manipulator's vertical axis. Video of the movement was recorded for every test. Subsequently, keyframes were extracted from the videos and analyzed in a CAD environment. To have a reference for the evaluations, a paper screen with a hexagonal hole, through which the robot's end effector can be seen, was placed and centered with the vertical axis of the system and in the horizontal plane in which the motion of point E is expected to lie. Using this setup, only trajectories in the horizontal planes can be analyzed. Therefore, for a more complete characterization, there is a need for a more sophisticated setup and procedure to detect trajectories in planes different from the horizontal or with evolutions in tridimensional space.

The first trajectory tracking tests were related to hexagon trajectories centered on the working volume. Then, trajectories with less symmetry with respect to the working volume were considered as references. These described triangles whose vertices were chosen without particular conditions.

Figure 14 compares the reference trajectories and the achieved movements relating to representative cases of the tests described. Figure 14a presents the test relative to a hexagon trajectory tracking. Starting from the resting position with the end effector in (0, 0, 131.3), the trajectory reference was commanded to reach the vertices through its own coordinates in the following sequence: (−2, −3.6, 130.7), (−4, 0, 130.7), (−2, 3.6, 130.7), (2, 3.6, 130.7), (4, 0, 130.7), (2, −3.6, 130.7), (−2, −3.6, 130.7); the comparison is made by 23 points of the actual trajectory. Figure 14b presents the test relative to triangular trajectory tracking. Starting from the resting position with the end effector in (0, 0, 131.3), the trajectory was commanded to reach the vertices through the own coordinates in the following sequence: (−0.3, −2, 130.7), (−4, 1.7, 130.7), (3.4, 1, 130.7); the comparison is made by 13 points of the actual trajectory.

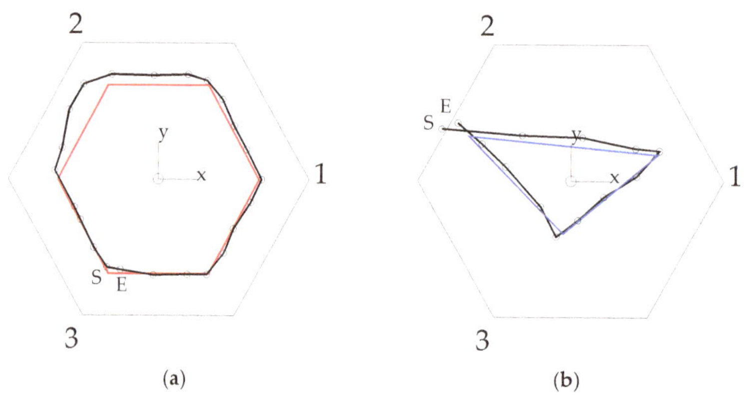

Figure 14. Trajectory tracking tests. The numbers 1, 2, and 3 represent the directions through which the action of the corresponding actuator moves. The letters S and E represent the trajectory's start and end positions. The circles represent the points of the actual trajectory. (**a**) Representative case of the first trajectory-tracking test: the reference (red line) is an almost regular hexagon with an edge of 4 mm. (**b**) Representative case of the second trajectory-tracking test: the reference is a general triangle (blue line).

Figure 15 shows the errors of the trajectory tracking tests computed by the distance between the achieved positions, i.e., the black circles, and the reference trajectories, with the red line, and the blue line for the relative tests documented in Figure 14.

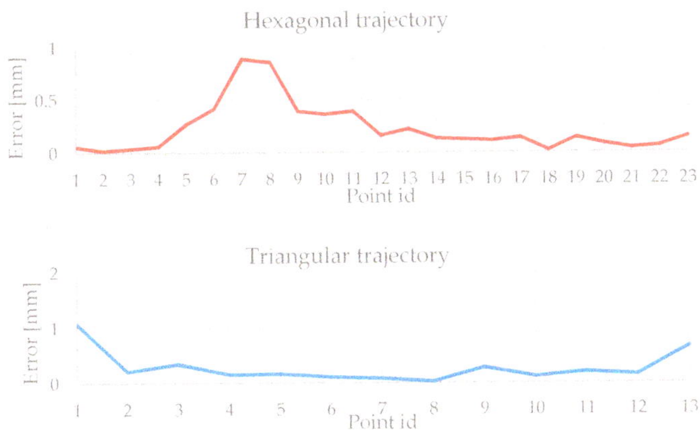

Figure 15. The hexagonal trajectory test shows a maximum error of almost 1 mm in a confined zone, while in the rest of the trajectory, the error is less than 0.2 mm. The triangular trajectory test shows similar maximum and minimum values for the errors.

4. Discussion and Conclusions

The manipulator responds well to positioning in a large part of the working volume, but there is a confined area where the error is considerably larger. In the area where the error is small, this is lower than the prediction based on the error detected for the single actuator. This is because the synergy between the three actuators ensures that the errors due to oscillations detected for the single actuator are dampened, reducing the error of the manipulator compared to what can be predicted based on a composition of the errors characterizing the individual actuators (Figure 11).

As for the area in which the error is most significant, it is clear that this derives from a discrepancy in the functioning of actuator number 2 compared to the others, which presents a more intense action than it should. It turns out that the positioning brings the end effector closer to the attack of actuator 2 than it should. This effect is noticeable in both trajectory tests.

Regarding the causes of this behavior, different lengths at rest can lead to this problem. Still, the authors believe that this problem has a limited effect, given that particular care was taken in constructing the actuators.

Another possible cause, and more probable than the first, is that actuator number 2 responds differently than the others due to the physical composition of the alloy and that it needs a slightly different electrical resistance deformation model from the one implemented for all three actuators. A possibility that should, therefore, be investigated in the future is the characterization of the actuators one by one, equipping each with a strain-electric resistance model to implement the sensor effect.

Another possibility could be to calibrate the system in the area where the errors are most significant and create a lookup table that can make corrections to increase precision.

To provide evidence of the advantages of the proposed device, a comparison was carried out with a similar device. The comparison was based on the number of DOFs, working volume, the accuracy in the positioning, the overall dimensions, the actuator's block control volume, and the mass. Figure 16 shows the considered device, and Table 3 compares the proposed device with the other under study.

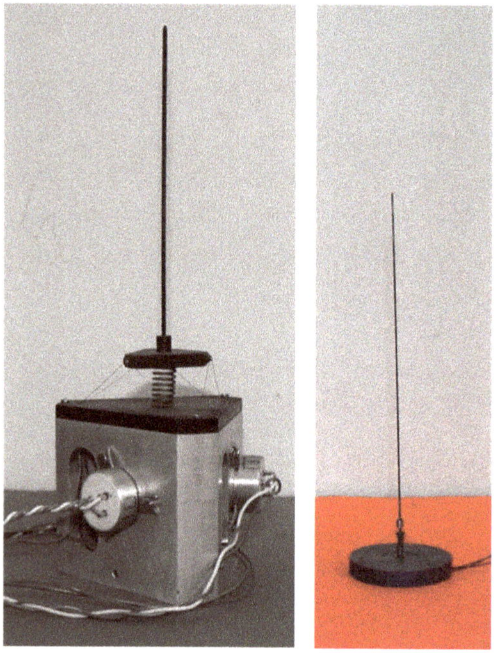

(a) (b)

Figure 16. The manipulator used for comparison with the proposed device: (**a**) parallel manipulator with three DOFs actuated by Ni-Ti wires and controlled by the use of potentiometers [38]; (**b**) the proposed manipulator.

Table 3. Characteristics of the devices under comparison. DOFs = degrees of freedom; WV = working volume dimensions as diameter of the circle circumscribed to the shape of the WV in plant by height [mm × mm]; PA = typical positioning accuracy [mm]; OD = overall dimensions intended as diameter by height of the smaller cylinder containing the device [mm × mm]; ABV = actuator's block control volume [mm^3]; M = mass of the device without electronics for control [kg]. Brackets (a)–(b) correspond to the devices shown in Figure 15; (c) represent the reference requirements.

Device	DOFs	WV	PA	OD	ABV	M
(a)	3	12 × 2	0.3	82 × 207	138,600	0.245
(b)	3	12 × 1.3	1/0.3	40.8 × 120	7844	0.012
(c)	3	10 × 2	0.1	45 × 140	20,000	0.050

From the comparison results, concerning the compared device, the proposed one has a smaller score just for the height of the working volume, with a result of about 66% of the compared one and a lower accuracy due to the problem of the malfunctioning of the actuator number 2, as discussed above. In all the other indices, the proposed device outperforms the compared one. From the comparison, it can be concluded that the proposed device has a better score for the requirements considered.

In [41], a study was carried out by the authors on an SMA wire actuator aimed at verifying the best controller with proportional (P), derivative (D), or integral (I) components or their combinations. In the future, the optimized controller will be implemented in the manipulator in question in order to verify possible improvements for the position accuracy.

Although the manipulator can move on spatial trajectories, the experimental trajectory tracking tests only involved flat trajectories in the horizontal plane. This is because having a setup for the survey and documentation of three-dimensional trajectories is not easy. The problem is difficult to solve given the small size of the working volume. A procedure that uses a small marker to be detected using a stereo camera is currently being studied. This study will constitute a future development of the present work.

Author Contributions: Conceptualization, F.D.; methodology, F.D., P.B.Z. and T.R.; validation, F.D.; formal analysis, F.D., P.B.Z. and T.R.; investigation, F.D., P.B.Z. and T.R.; data curation, F.D.; writing—original draft preparation, F.D.; writing—review and editing, F.D.; visualization, F.D., P.B.Z. and T.R.; supervision F.D., P.B.Z. and T.R. All authors have read and agreed to the published version of the manuscript.

Funding: This research received no external funding.

Data Availability Statement: Data are contained within the article.

Acknowledgments: The authors gratefully acknowledge the contribution of Emiliano Valentini in the experimental test reported in this paper.

Conflicts of Interest: The authors declare no conflicts of interest.

References

1. Kittinanthapanya, R.; Sugahara, Y.; Matsuura, D.; Takeda, Y. Development of a Novel SMA-Driven Compliant Rotary Actuator Based on a Double Helical Structure. *Robotics* **2019**, *8*, 12. [CrossRef]
2. Britz, R.; Motzki, P.; Seelecke, S. Scalable Bi-Directional SMA-Based Rotational Actuator. *Actuators* **2019**, *8*, 60. [CrossRef]
3. Copaci, D.; Blanco, D.; Moreno, L. Flexible Shape-Memory Alloy-Based Actuator: Mechanical Design Optimization According to Application. *Actuators* **2019**, *8*, 63. [CrossRef]
4. Schmelter, T.; Theren, B.; Fuchs, S.; Kuhlenkötter, B. Development of an Actuator for Translatory Movement by Means of a Detented Switching Shaft Based on a Shape Memory Alloy Wire for Repeatable Mechanical Positioning. *Crystals* **2021**, *11*, 163. [CrossRef]
5. Ballester, C.; Copaci, D.; Arias, J.; Moreno, L.; Blanco, D. Hoist-Based Shape Memory Alloy Actuator with Multiple Wires for High-Displacement Applications. *Actuators* **2023**, *12*, 159. [CrossRef]
6. Dauksher, R.; Patterson, Z.; Majidi, C. Characterization and Analysis of a Flexural Shape Memory Alloy Actuator. *Actuators* **2021**, *10*, 202. [CrossRef]
7. Ameduri, S.; Concilio, A.; Favaloro, N.; Pellone, L. A Shape Memory Alloy Application for Compact Unmanned Aerial Vehicles. *Aerospace* **2016**, *3*, 16. [CrossRef]
8. Bovesecchi, G.; Corasaniti, S.; Costanza, G.; Tata, M. A Novel Self-Deployable Solar Sail System Activated by Shape Memory Alloys. *Aerospace* **2019**, *6*, 78. [CrossRef]
9. Liu, M.; Wang, Z.; Ikeuchi, D.; Fu, J.; Wu, X. Design and Simulation of a Flexible Bending Actuator for Solar Sail Attitude Control. *Aerospace* **2021**, *8*, 372. [CrossRef]
10. Costanza, G.; Delle Monache, G.; Tata, M.; Filosi, S. Development of SMA Spring Linear Actuator for an Autonomous Lock and Release Mechanism: Application for the Gravity-Assisted Pointing System in Moon to Earth Alignment of Directional Devices. *Aerospace* **2022**, *9*, 735. [CrossRef]
11. Dimino, I.; Vendittozzi, C.; Reis Silva, W.; Ameduri, S.; Concilio, A. A Morphing Deployable Mechanism for Re-Entry Capsule Aeroshell. *Appl. Sci.* **2023**, *13*, 2783. [CrossRef]
12. Braun, D.; Weik, D.; Elsner, S.; Hunger, S.; Werner, M.; Drossel, W. Position Control and Force Estimation Method for Surgical Forceps Using SMA Actuators and Sensors. *Materials* **2021**, *14*, 5111. [CrossRef] [PubMed]
13. Kim, H.; Jang, S.; Do, P.; Lee, C.; Ahn, B.; Kwon, S.; Chang, H.; Kim, Y. Development of Wearable Finger Prosthesis with Pneumatic Actuator for Patients with Partial Amputations. *Actuators* **2023**, *12*, 434. [CrossRef]
14. Deng, E.; Tadesse, Y. A Soft 3D-Printed Robotic Hand Actuated by Coiled SMA. *Actuators* **2021**, *10*, 6. [CrossRef]
15. Kotb, Y.; Elgamal, I.; Serry, M. Shape Memory Alloy Capsule Micropump for Drug Delivery Applications. *Micromachines* **2021**, *12*, 520. [CrossRef] [PubMed]
16. Shen, J.; Chen, Y.; Sawada, H. A Wearable Assistive Device for Blind Pedestrians Using Real-Time Object Detection and Tactile Presentation. *Sensors* **2022**, *22*, 4537. [CrossRef] [PubMed]
17. Copaci, D.; Serrano, D.; Moreno, L.; Blanco, D. A High-Level Control Algorithm Based on sEMG Signalling for an Elbow Joint SMA Exoskeleton. *Sensors* **2018**, *18*, 2522. [CrossRef] [PubMed]
18. He, J.; Lu, Y.; Wang, L.; Ma, N. On the Improvement of Thermal Protection for Temperature-Responsive Protective Clothing Incorporated with Shape Memory Alloy. *Materials* **2018**, *11*, 1932. [CrossRef] [PubMed]
19. Srivastava, R.; Alsamhi, S.; Murray, N.; Devine, D. Shape Memory Alloy-Based Wearables: A Review, and Conceptual Frameworks on HCI and HRI in Industry 4.0. *Sensors* **2022**, *22*, 6802. [CrossRef]
20. Lee, J.; Han, M. Design and Evaluation of Smart Textile Actuator with Chain Structure. *Materials* **2023**, *16*, 5517. [CrossRef]
21. Helps, T.; Vivek, A.; Rossiter, J. Characterization and Lubrication of Tube-Guided Shape-Memory Alloy Actuators for Smart Textiles. *Robotics* **2019**, *8*, 94. [CrossRef]

22. Raparelli, T.; Zobel, P.B.; Durante, F. A proposed methodology for the development of microgrippers: An application to a silicon device actuated by shape memory alloy wires. *Int. J. Mech. Eng. Technol.* **2018**, *9*, 235–249.
23. Zainal, M.; Sahlan, S.; Ali, M. Micromachined Shape-Memory-Alloy Microactuators and Their Application in Biomedical Devices. *Micromachines* **2015**, *6*, 879–901. [CrossRef]
24. Garcés-Schröder, M.; Hecht, L.; Vierheller, A.; Leester-Schädel, M.; Böl, M.; Dietzel, A. Micro-Grippers with Femtosecond-Laser Machined In-Plane Agonist-Antagonist SMA Actuators Integrated on Wafer-Level by Galvanic Riveting. *Proceedings* **2017**, *1*, 385. [CrossRef]
25. Subendran, S.; Kang, C.; Chen, C. Comprehensive Hydrodynamic Investigation of Zebrafish Tail Beats in a Microfluidic Device with a Shape Memory Alloy. *Micromachines* **2021**, *12*, 68. [CrossRef]
26. Maffiodo, D.; Raparelli, T. Flexible Fingers Based on Shape Memory Alloy Actuated Modules. *Machines* **2019**, *7*, 40. [CrossRef]
27. Zeng, X.; Wu, Y.; Han, S.; Liu, Y.; Xiu, H.; Tian, F.; Ren, L. Theoretical and Experimental Investigations into a Crawling Robot Propelled by Piezoelectric Material. *Micromachines* **2021**, *12*, 1577. [CrossRef]
28. Perez-Sanchez, V.; Garcia-Rubiales, F.; Nekoo, S.; Arrue, B.; Ollero, A. Modeling and Application of an SMA-Actuated Lightweight Human-Inspired Gripper for Aerial Manipulation. *Machines* **2023**, *11*, 859. [CrossRef]
29. Raparelli, T.; Beomonte Zobel, P.; Durante, F. Mechanical design of a 3-dof parallel robot actuated by smart wires. In Proceedings of the EUCOMES 2008—The 2nd European Conference on Mechanism Science, Cassino, Italy, 17–20 September 2008; Springer: Dordrecht, The Netherlands, 2009; pp. 271–278. [CrossRef]
30. Shin, J.; Han, Y.; Lee, J.; Han, M. Shape Memory Alloys in Textile Platform: Smart Textile-Composite Actuator and Its Application to Soft Grippers. *Sensors* **2023**, *23*, 1518. [CrossRef]
31. Shi, L.; Guo, S.; Li, M.; Mao, S.; Xiao, N.; Gao, B.; Song, Z.; Asaka, K. A Novel Soft Biomimetic Microrobot with Two Motion Attitudes. *Sensors* **2012**, *12*, 16732–16758. [CrossRef]
32. Rajagopalan, R.; Petruska, A.; Howard, D. A Bi-State Shape Memory Material Composite Soft Actuator. *Actuators* **2022**, *11*, 86. [CrossRef]
33. Sreekumar, M.; Singaperumal, M.; Nagarajan, T.; Zoppi, M.; Molfino, R. A compliant miniature parallel manipulator with shape memory alloy actuators. In Proceedings of the IEEE International Conference on Industrial Technology, Mumbai, India, 15–17 December 2006; pp. 848–853. [CrossRef]
34. Elwaleed, A.K.; Mohamed, N.A.; Nor, M.J.M.; Mustafa, M.M. A new method for actuating parallel manipulators. *Sens. Actuators A Phys.* **2008**, *147*, 593–599. [CrossRef]
35. Giuseppe, C. Experimental characterization of a binary actuated parallel manipulator. *Chin. J. Mech. Eng.* **2016**, *29*, 445–453. [CrossRef]
36. AbuZaiter, A.; Ng, E.L.; Kazi, S.; Mohamed Ali, M.S. Development of Miniature Stewart Platform Using TiNiCu Shape-Memory-Alloy Actuators. *Adv. Mater. Sci. Eng.* **2015**, *2015*, 928139. [CrossRef]
37. Ranjith Pillai, R.; Ganesan, M. Mechatronics Design and Kinematic Analysis of SMA Spring Actuated Parallel Manipulator. *J. Phys. Conf. Ser.* **2021**, *1969*, 012011. [CrossRef]
38. Raparelli, T.; Zobel, P.B.; Durante, F. Design of a parallel robot actuated by shape memory alloy wires. *Mater. Trans.* **2002**, *43*, 1015–1022. [CrossRef]
39. Landsberger, S.E.; Sheridan, T.B. A minimal, minimal linkage: The tension-compression parallel link manipulator. In Proceedings of the IMACS/SICE International Symposium on Robotics, Mechatronics, and Manufacturing Systems, Kobe, Japan, 16–20 September 1992; pp. 493–500.
40. Raparelli, T.; Beomonte Zobel, P.; Durante, F. A robot actuated by shape memory alloy wires. In Proceedings of the Industrial Electronics, 2002. ISIE 2002, Proceedings of the 2002 IEEE International Symposium on, L'Aquila, Italy, 8–11 July 2002; IEEE: Piscataway, NJ, USA, 2002; Volume 2, pp. 420–423. [CrossRef]
41. Durante, F.; Raparelli, T.; Beomonte Zobel, P. Resistance Feedback of a Ni-Ti Alloy Actuator at Room Temperature in Still Air. *Micromachines* **2024**, *15*, 545. [CrossRef] [PubMed]
42. Gosselin, C.; Angeles, J. Singularity Analysis of Closed-Loop Kinematic Chains. *IEEE Trans. Robot. Autom.* **1990**, *6*, 281–290. [CrossRef]

robotics

MDPI

Article

Surgical Staplers in Laparoscopic Colectomy: A New Innovative Flexible Design Perspective †

Dhruva Khanzode [1,2,‡], Ranjan Jha [1,2,‡], Alexandra Thomieres [3,‡], Emilie Duchalais [3,‡] and Damien Chablat [4,5,*,‡]

1 CSIR-Central Scientific Instruments Organisation, Chandigarh 160030, India; dhruva.csio20a@acsir.res.in (D.K.); ranjan.jha@csio.res.in (R.J.)
2 Academy of Scientific and Innovative Research (AcSIR), Ghaziabad 201002, India
3 Centre Hospitalier Universitaire, F-44000 Nantes, France; alexandra.thomieres@chu-nantes.fr (A.T.); emilie.dassonneville@chu-nantes.fr (E.D.)
4 École Centrale Nantes, Nantes Université, CNRS, LS2N, UMR 6004, F-44000 Nantes, France
5 Research Center for Industrial Robots Simulation and Testing, Technical University of Cluj-Napoca, 400114 Cluj-Napoca, Romania
* Correspondence: damien.chablat@cnrs.fr
† This paper is an extended version of our paper published in Khanzode, D., Jha, R., Thomieres, A.; et al. Design of a Surgical Stapler for Laparoscopic Colectomy. In Proceedings of the New Trends in Medical and Service Robotics (MESROB 2023), Craiova, Romania, 7–10 June 2023; pp. 71–80.
‡ These authors contributed equally to this work.

Abstract: This article describes the development of a flexible surgical stapler mechanism, which serves as a fundamental tool for laparoscopic rectal cancer surgery, addressing the challenges posed by difficult types of accessibility using conventional instruments. The design of this mechanism involves the incorporation of a stacked tensegrity structure, in which a flexible beam serves as the central spine. To assess the stapler's range of operation, an analysis of the workspace was conducted by examining collaborative Computed Tomography (CT) scan data obtained from different perspectives (Axial, Coronal, and Sagittal planes) at various intervals. By synthesizing kinematic equations, Hooke's law was employed, taking into account rotational springs and bending moments. This allowed for precise control of the mechanism's movements during surgical procedures in the rectal region. Additionally, the study examined the singularities and simulations of the tensegrity mechanism, considering the influential eyelet friction parameter. Notably, the research revealed that this friction parameter can alter the mechanism's curvature, underscoring the importance of accurate analysis. To establish a correlation between the virtual and physical models, a preliminary design was presented, facilitating the identification of the friction parameter.

Keywords: surgical stapler; anastomosis; laparoscopy; tensegrity mechanism

1. Introduction

Over the last two decades, significant advancements have been achieved in medical science. Surgical procedures that were once difficult, risky, and painful, such as open-heart surgeries, now have safer alternatives. Endoscopies and catheterization procedures have become mainstream, replacing the need for extensive open surgeries. Minimally invasive surgical techniques have also gained popularity, becoming the preferred choice for many traditional surgeries. These procedures involve small incisions on the body, through which specialized slender instruments with miniature tools are inserted. These tiny instruments have undergone modernization and are presently operated through robotic systems. Robotics is gaining increasing prominence across various domains within medical science, encompassing tasks such as diagnosis, drug administration, and therapy.

In surgical procedures, a frequent obstacle is the constrained room for maneuvering surgical instruments within the designated surgical area. To overcome this challenge, the creation of innovative, compact surgical tools necessitates a thorough examination of the available workspace. Traditionally, computed tomography scans (CT scans) are employed for space assessment. CT scans utilize a blend of X-ray technology and computer processing to generate intricate visuals of the body's internal structures, encompassing bones, muscles, organs, and blood vessels. In comparison to conventional X-rays, CT scans offer a more comprehensive and precise dataset.

One common medical procedure is total mesorectal excision. It is used when there is rectal cancer or inflammation that requires removing part or all of the rectum. In this surgery, the upper part or the whole rectum is detached from the colon, and then the remaining parts (colon and lower rectum) are joined together. This connection of the colon and rectum is called colorectal anastomosis. A specialized surgical tool called a "surgical stapler" is used for colorectal anastomosis [1].

The emergence of the surgical stapler as a pivotal medical innovation can be traced back to 1908, when Victor Fischer and Hümér Hültl [2] introduced this groundbreaking device [3]. This instrumental apparatus emerged in response to the imperative of reducing the risk of wound contamination in abdominal surgeries by effectively containing gastrointestinal fluids. Initially referred to as a "mechanical stitching device" in academic literature [3], the Fischer–Hültl surgical stapler featured a design incorporating four rows of U-shaped staples constructed from steel wires, measuring 17 cm and 11 cm in length. Two enduring design features, the final B-shaped staple configuration [4], facilitating tissue perfusion, and the strategic staggered arrangement of staple pins [5], were integral aspects of this innovation.

The subsequent evolution of surgical staplers witnessed refinements and innovations throughout the 20th century, with researchers such as Aladár Petz and H. Friedrich making significant contributions, introducing innovations like the "L" shaped stapler. A pivotal milestone occurred in 1964 when Mark Ravitch, Leon Hirsch, and Felicien Steichen, under the United States Surgical Corporation (USSC), introduced the modern surgical stapler [3]. This innovation marked a paradigm shift with the introduction of disposable staple cartridges, fundamentally reshaping surgical stapling technology. USSC also pioneered the development of surgical staplers with circular stapling zones, leading to designations such as TA (Thoracoabdominal) and GIA (Gastrointestinal Anastomosis) becoming synonymous with surgical practice. Subsequent corporate developments had profound impacts on the trajectory of surgical stapling technology within the medical field [6].

In modern surgical practice, five main types of staplers are commonly used: TA (Thoraco-Abdominal), GIA (Gastro-Intestinal Anastomosis), Endo GIA (Endoscopic Gastro-Intestinal Anastomosis), EEA (End–End Anastomosis), and Skin Stapler. Notably, the TA stapler does not include an integrated cutting mechanism for tissues after stapling, which requires manual separation. On the other hand, the EEA stapler is known for its formation of circular staples [7], and the Skin Stapler is employed for closing surface wounds. The TA stapler is primarily utilized in veterinary surgical procedures [8]. In contrast, both GIA and Endo GIA staplers are extensively used in abdominal surgeries, with the latter specifically designed for minimally invasive surgical procedures [9,10].

At present, Endo GIA staplers come in three distinct configurations: Passive Articulated Wrist Type (PAW) [11], Active Articulated Wrist Type (AAW) [12], and Radial Reload Type Staplers (RR). The PAW version achieves wrist flexion by exerting pressure on the jaw against the abdominal wall. In contrast, the AAW model incorporates a lever mechanism to articulate the wrist into predetermined bending angles. The RR-type stapler, exclusively offered by Covidien Inc., Westmeath, Ireland , is highly esteemed for pulmonary surgeries [13]. Nonetheless, a notable drawback of the RR-type stapler is its need for a considerably larger incision for entry into the body, which may contradict the principles of "minimally invasive surgery" [14]. Consequently, there is a pressing need for a surgical stapler capable of accessing the surgical site through laparoscopic openings while preserving the functionality of an RR-type stapler within the body.

In response to the challenges associated with the positioning and maneuverability of endoscopic surgical staplers, this article delves into the design and development of a hyper-redundant flexible surgical stapler specifically tailored for laparoscopic procedures. This innovative stapler is meticulously engineered to provide adaptable orientation within intricate surgical sites, thereby enhancing the efficiency of tissue sealing and division when compared to conventional endoscopic surgical staplers. In summary, the realm of medical science has witnessed remarkable progress, resulting in safer and more effective surgical interventions. The adoption of minimally invasive techniques and the introduction of specialized surgical staplers have played pivotal roles in modern healthcare practices, contributing significantly to improved patient outcomes and expedited recovery periods.

This article explores the design and development of a highly adaptable, flexible surgical stapler tailored for laparoscopic procedures, as detailed in [15]. The primary objective is to address the challenges associated with positioning and manoeuvrability that conventional endoscopic surgical staplers, often encounter [16]. The envisioned stapler exhibits the capability to dynamically adjust its orientation within intricate surgical sites, consequently enhancing the efficiency of tissue sealing and cutting when compared to current endoscopic staplers.

The surgical stapler comprises four fundamental components: the upper jaw, lower jaw, stapling cartridge, and wrist, as depicted in Figure 1. For the flexible surgical stapler design discussed in this article, the upper and lower jaws are designed with a stacked tensegrity mechanism incorporated in the body structure, as shown in Figure 9 in section 4 of the article [17]. This design enables both jaws to flex within the same plane, with the bending of these jaws being controlled by the actuation of tendons integrated within the tensegrity mechanism.

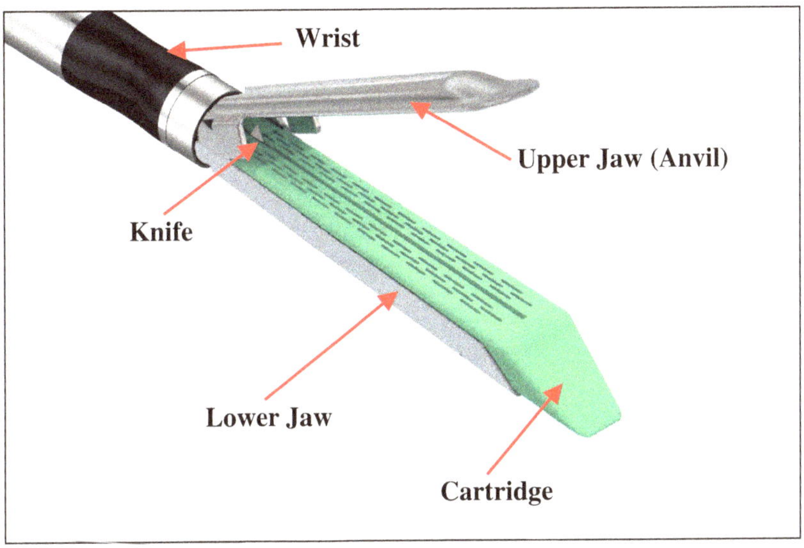

Figure 1. Various parts of a conventional endoscopic surgical stapler.

The stapling cartridge is intended to house surgical staple pins and a cutting knife, featuring a firing mechanism that engages the tissue and activates the knife simultaneously [18,19]. Additionally, the wrist, attached at the base of the jaw, is flexible and functions within the same plane as the upper and lower jaws, facilitated by the actuation of additional tendons. The mechanism responsible for articulating the upper and lower segments has been engineered as a decoupled system designed for application in the context of the jaw. The design is currently undergoing development to ensure the required force for securely closing the jaws while preventing potential tissue trauma.

To execute the full range of operations for the flexible surgical stapler, four distinct actuators come into play: one for bending the wrist, another for activating the staple pins and knife, a third for opening and closing the jaws, and finally, a unified actuation mechanism for synchronously bending the upper and lower jaws.

In [20], a simple kinematic model and its singular configurations were introduced for three segments in the simulation only. Extending from that, this article aims to discuss the design process of a multi-segmented mechanism utilizing CAD modeling, which integrates the results of mathematical modeling, kinematic simulations, and singularity analysis presented in the associated research for n-segments. The article also covers the prototyping of this design through SLA 3D printing and the subsequent dynamic experimentation conducted on the prototype.

2. Workspace Analysis

In the field of design and modeling, the first crucial step is to carefully examine the given problem statement and discern the inherent limitations and constraints that will shape the design process. When it comes to designing a flexible surgical stapler, the paramount constraint revolves around the available workspace. To formulate a practical design, it becomes imperative to delve into the stapler's potential maneuverability within the surgical environment. Frequently, this available space proves to be quite confined, presenting significant challenges for the efficient navigation of surgical instruments. Therefore, when crafting a novel, compact surgical tool, it becomes paramount to conduct an exhaustive analysis of the actual spatial allowances for the tool's mobility.

In this context, the standard approach involves the utilization of computed tomography (CT) scans to evaluate the accessible space. To conduct our analysis, we compiled data from a cohort of patients who had undergone colon surgery. The volume of the abdominal workspace is notably influenced by various factors, including the patient's age, gender, weight, and physiological attributes. To exemplify our research, we opted to focus on data obtained from a specific male patient, aged 65, with a height of 1.66 m and a weight of 55 kg. This particular patient was selected due to his unique physiological characteristics, which yielded the most confined workspace volume for the flexible surgical stapler

We then proceeded to acquire CT scan data for the general surgical site, encompassing three anatomical planes: Axial, Coronal, and Sagittal. These data points were spaced at 3.33-millimeter intervals for each plane. A skilled surgeon delineated the dimensional parameters of the surgical site on the data, denoted by red and blue markings in Figure 2, respectively, employing Vue PACS software, (version 12.1.6.1005, CARESTREAM Health, New York, NY, USA). Initially, these data underwent analysis to approximate the maximum workspace volume available for the flexible stapler. Subsequently, we pinpointed the most confined cross-sectional area within the surgical site, which came out to be 2613 mm^2 in the axial plane, 8036 mm^2 in the coronal plane and 6875 mm^2 in the sagittal plane.

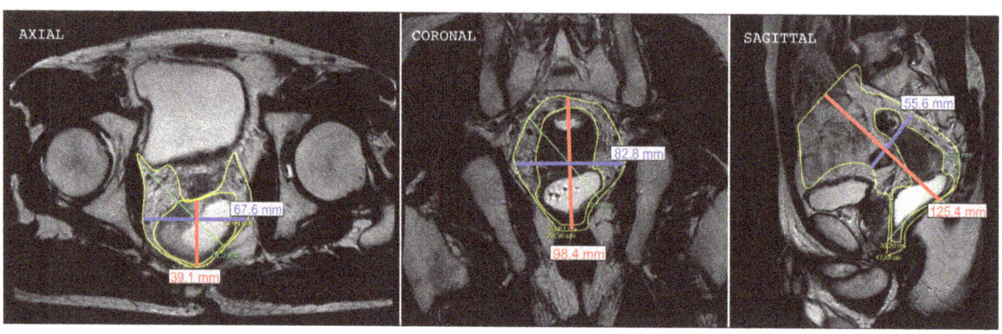

Figure 2. Sample CT scan data for workspace evaluation in Axial, Coronal, and Sagittal planes. The red and blue markings represent the available dimension (mm) in each plane.

Using this dataset, a three-dimensional model was crafted to simulate the placement of the stapler and its laparoscopic insertion. This simulation enables us to ascertain the optimal size of the tool that can be effectively utilized within the most restricted available space. By taking these factors into account, we can develop a surgical stapler capable of adeptly and precisely navigating challenging anatomical spaces during minimally invasive procedures.

3. Mechanism Design of Surgical Stapler

When dealing with a flexible mechanism, deformations or significant deflections often display nonlinear behavior. To streamline the process of kinematic modeling, the widespread practice is to employ the constant curvature technique. Constant curvature mechanisms consist of a finite number of curved segments, each characterized by a set of arc parameters that can be analytically transformed. The piece-wise constant curvature approximation (PCCA), as previously elucidated in research [21], provides a method for determining the direct kinematics of the mechanism. This method takes cable tension as input and yields the corresponding posture as output. Notably, the PCCA remains applicable even when a flexible mechanism possesses a singular flexible spine spanning all its segments [22]. This technique has proven its efficacy in a variety of flexible mechanisms, including those utilized in the realms of medical science and surgery.

In this article, our attention is directed toward a specific multi-segmented mechanism, which is arranged in a vertical stack as visually represented in Figure 3. This configuration embodies a novel formulation of the tensegrity mechanism detailed in [23]. A continuous flexible beam, allowing for in-plane deformation while preserving significant stiffness within that same plane (depicted in Figure 3), serves as the link connecting the base to the mobile platform of each segment. The continuous beam, denoted as $(A_1^0 A_n^0)$, stretches from the mechanism's base to its tip. To manipulate the shape of the mechanism, two cables are employed, extending from A_1^i to A_n^i and from A_1^j to A_n^j. Additionally, a set of springs is positioned along the cables to represent the elastic stress generated by the outer protective skin.

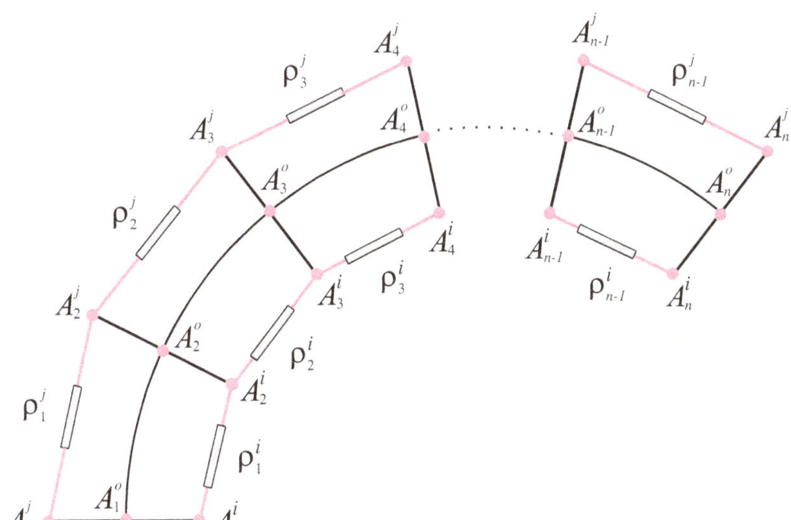

Figure 3. The tensegrity mechanism understudy with a stacked multi-segmented model with a flexible beam.

For analysis, our focus centres on a specific segment, namely, $A_1^o A_2^o$, and we utilize the piece-wise constant curvature approximation (PCCA). In this single segment, the base platform and the moving platform are interconnected by two cables or tendons, specifically designated as ρ_1^j and ρ_1^i, which are positioned on either side of the central spine. Additionally, two springs, each with stiffness coefficients k_1^j and k_1^i", are placed between $(A_1^j\ A_2^j)$ and $(A_1^i\ A_2^i)$, respectively. For subsequent segments $A_2^o A_3^o$, $A_3^o A_4^o, \ldots, A_{n-1}^o A_n^o$, a similar definition will be applicable, with sides remaining the same for each stage, respectively.

The act of contracting the cables leads to angular displacement within the flexible beam of the central spine. This effect is observed in all three segments, consequently causing the mechanism to bend. The cable's length is quantified as $\rho_n^{i/j}$.

3.1. Kinematics Equations of One Segment

In order to determine the locations of the points on the mobile platform, it becomes imperative to investigate the deformation of a beam under torsion. We assume that the distance between A_1^o and A_2^o is h_1 when the mechanism is in its neutral position. Also, A_1^o is the centre point of $A_1^i A_1^j$, where, $A_1^i A_1^o = A_1^o A_1^j = l_1$.

The coordinates of the fixed points are

$$\mathbf{a}_1^j = \begin{bmatrix} -l_1 & 0 \end{bmatrix}^T \quad \mathbf{a}_1^i = \begin{bmatrix} l_1 & 0 \end{bmatrix}^T \tag{1}$$

and in the moving reference frame

$$\mathbf{a}_2^j = \begin{bmatrix} -l_1 & 0 \end{bmatrix}^T \quad \mathbf{a}_2^i = \begin{bmatrix} l_1 & 0 \end{bmatrix}^T \tag{2}$$

where, $\mathbf{a}_n^{i/j}$ is the vector representation of graphical point $A_n^{i/j}$.

During the formulation of the multi-segmented structure, the mechanism's width remained consistent with that shown in Equations (1) and (2). This determination was made after examining the stability between a mechanism with a consistent width and one with a tapering width, as depicted in [24].

To correlate cable tensions and spring forces in one segment with its radius of curvature, we employ a lumped parameter model. In the context of a beam, the springs and cables exert forces on its ends, leading to the generation of a torsional torque:

$$\mathcal{T}_1 = \mathcal{T}_{f_1^j} + \mathcal{T}_{f_1^i} + k_1^j(\rho_1^j - d_1^j)l_2 - k_1^i(\rho_1^i - d_1^i)l_2 \tag{3}$$

where $\mathcal{T}_{f_1^j} = -f_1^j l_2$ and $\mathcal{T}_{f_1^i} = f_1^i l_2$. Equation (4) represents Hooke's law with a rotational spring and the bending moment. In this equation, K_θ stands for the effective bending stiffness (measured in N·m/rad), h_1 signifies the arc length between A_1^o and A_2^o, f_1 and f_2 denote the forces applied by the two cables, and r_1 represents the radius of curvature (as illustrated in Figure 4). In the context of a single section, it can be expressed as follows:

$$h_1/r_1 = \mathcal{T}_1/K_\theta \tag{4}$$

The characterization of the beam can be denoted by its bending radius r_1 and the angle θ_1.

$$r_1 = h_1/\theta_1 \quad \text{and} \quad \theta_1 = \tau_1/K_\theta \tag{5}$$

This definition holds true when we bend in one direction, whether to the left or right, and it enables us to determine the position of A_2^o.

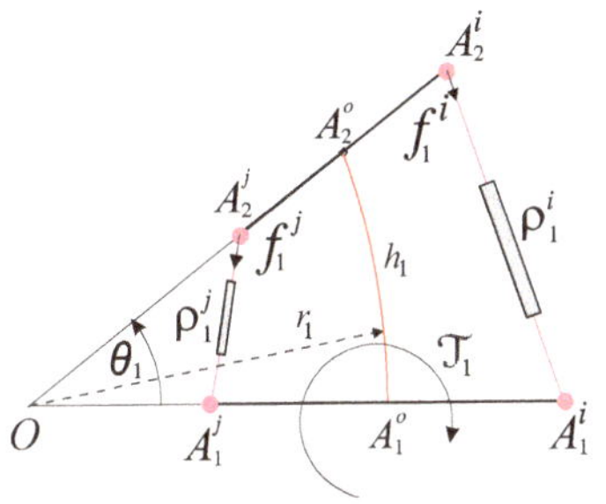

Figure 4. Bending moment equilibrium [22] when two forces f_1 and f_2 are applied to generate the torque $\mathcal{T}_1$.

The segment $(A_2^j A_2^i)$ is normal to the beam at point A_2^o and the coordinates of these points are

$$\mathbf{a}_2^o = \begin{bmatrix} r_1 - r_1 \cos(\theta_1) \\ r_1 \sin(\theta 1) \end{bmatrix} \quad (6)$$

$$\mathbf{a}_2^j = \begin{bmatrix} r_1 - r_1 \cos(\theta 1) - l_2 \cos(\theta 1) \\ R \sin(\theta 1) + l_2 \sin(\theta 1) \end{bmatrix} \quad (7)$$

$$\mathbf{a}_2^i = \begin{bmatrix} r_1 - r_1 \cos(\theta 1) + l_2 \cos(\theta 1) \\ R \sin(\theta 1) - l_2 \sin(\theta 1) \end{bmatrix} \quad (8)$$

The inverse kinematic model for a mechanism featuring a flexible beam can be presented as follows:

$$||\mathbf{a}_1^j - \mathbf{a}_2^j|| = \rho_1^j, \quad ||\mathbf{a}_1^i - \mathbf{a}_2^i|| = \rho_1^i \quad (9)$$

These equations can likewise be represented as functions of the spring lengths; hence,

$$(-l_1 - r_1 + r_1 \cos(\theta 1) - l_2 \cos(\theta_1))^2 + (-r_1 \sin(\theta_1) + l_2 \sin(\theta_1))^2 = (\rho_1^j)^2 \quad (10)$$

$$(l_1 - r_1 + r_1 \cos(\theta_1) + l_2 \cos(\theta_1))^2 + (-r_1 \sin(\theta_1) - l_2 \sin(\theta_1))^2 = (\rho_1^i)^2 \quad (11)$$

Similar to previous assumptions, for other sections, we can write the equations in the same way for the positions of the joints. The inverse kinematic model of the sections is given by

$$||\mathbf{a}_{n-1}^j - \mathbf{a}_n^j|| = \rho_{n-1}^j, \quad ||\mathbf{a}_{n-1}^i - \mathbf{a}_n^i|| = \rho_{n-1}^i, \quad (12)$$

3.2. Force Transmissions between Segments

To determine the position of the mobile platform as a consequence of the application of force in the tendons or cables, it becomes imperative to compute the radius of curvature and the angle of inclination that align with the kinematic model.

The optimization problem $\mathcal{S}$ for a single segment is written using Equation (9), as follows:

$$\mathcal{S} = \text{Minimize}\Big((||\mathbf{a}_1^j - \mathbf{a}_2^j|| - \rho_1^j)^2 + (||\mathbf{a}_1^i - \mathbf{a}_2^i|| - \rho_1^i)^2\Big) \quad (13)$$

with $\rho_1^i \geq 0$, $\rho_1^j \geq 0$, $\rho_1^i \leq 3$, $\rho_1^j \leq 3$. This function is generalized for n-segments by summing the squares of the equations associated with each segment.

For each segment, it is necessary to assess the bending moment and its transfer to the preceding one. When $f_n^i > f_n^j$, the beam exhibits a rightward bend. The cables pass through the eyelet, introducing friction denoted as μ. Figure 5 visually illustrates the transmission of forces before and after the cable passes through the eyelets on the mobile platform. The examination of segment stability, while considering cable friction in the eyelets, allows for the derivation of force transmission coefficients [22].

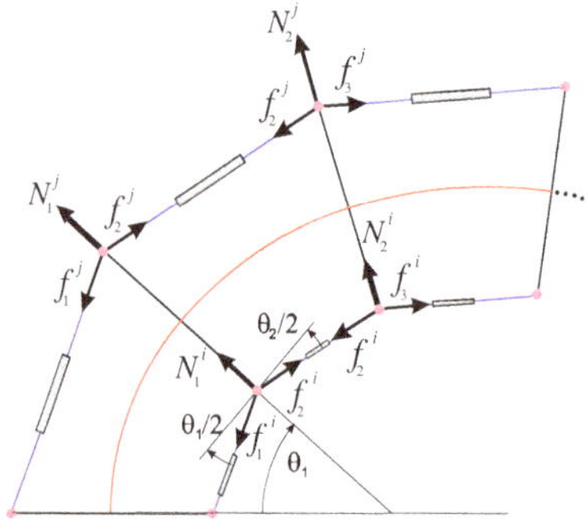

Figure 5. Static equilibrium of three segments.

For actuators on the right side of the central axis, the tensions in the cables reduce from the base to the end as

$$\frac{f^i_{n+1}}{f^i_n} = \left(\frac{1-\mu\sin(\theta_1/2)}{1+\mu\sin(\theta_1/2)}\right) \tag{14}$$

For actuators on the left side of the central axis, the tensions in the cables increase from the base to the end as

$$\frac{f^j_{n+1}}{f^j_n} = \left(\frac{1+\mu\sin(\theta_1/2)}{1-\mu\sin(\theta_1/2)}\right) \tag{15}$$

3.3. Motion Simulations

In this section, we suggest conducting a simulation study using the SIROPA library [25] within the MAPLE software (Version 2023, Maplesoft, Waterloo, ON, Canada). The primary aim of this investigation is to examine the behavioural characteristics of the mechanism. It is crucial to emphasize that, at this juncture, specific model parameters are undefined, requiring a thorough selection of materials and a comprehensive comprehension of the system's workspace.

To visually illustrate the bending motion of the mechanism in response to different force magnitudes, we refer to Figure 6, where red curve depict the central spine and blue lines depict the tendons/cable. This figure provides a visual representation of the bending motion, as it responds to force magnitudes ranging from $f_1 - f_2 = 0 + \epsilon$ to 17 N. The range set between 0 and 17 N was meticulously in accordance with the study's constraints. This range encompasses the utmost extent of bending achievable for the mechanism outlined in the article under the considered limitations. The inclusion of a small offset symbolized as ϵ, holds great importance in preventing a situation in which the radius of curvature tends to infinity, as in practicality, the equality of both forces is impossible.

Nevertheless, the introduction of length restrictions on the springs, each featuring a spring constant of 100 N/m, gives rise to a distinct set of challenges related to achieving the intended range of motion. The system reaches an extreme configuration in various scenarios: when one of the springs reaches its minimum length, experiences full compression, or when the cable aligns tangentially with the curved configuration.

To tackle these challenges, we resolve this matter by imposing constraints on the spring lengths, restricting them to only 40% of their initial length in the home pose. This particular modification results in a maximum bending angle of around 1.15 radians (≈130°). This angular parameter serves as a crucial metric in assessing the performance of the mechanism under these specific constraints, providing valuable insights into its operational capabilities.

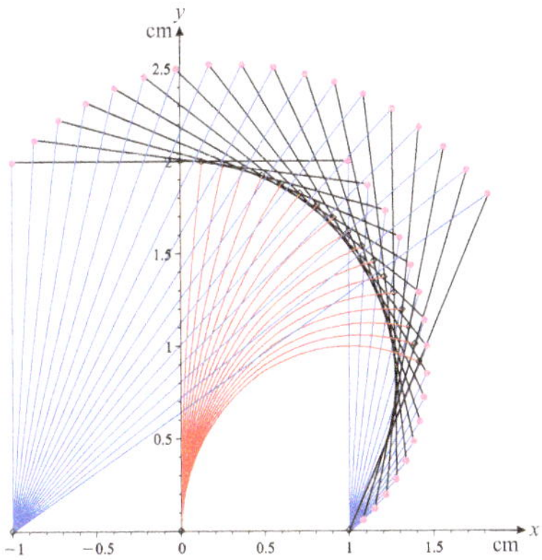

Figure 6. Simulation of the bending of one segment with $l_1 = l_2 = 1$ cm and $h_1 = 2$ cm.

In our simulation, we leverage the power of inverse kinematics to solve for the parameters outlined in Equations (9) and (12), employing the same optimization algorithm as used for a single segment. Notably, in the absence of friction, our simulation of the mechanism's flexion yields a seamless and continuous curve across each segment, as vividly demonstrated in Figure 7 (left). To carry out this optimization and determine the beam's curvature, we harnessed the capabilities of the MAPLE software (Version 2023, Maplesoft, Waterloo, ON, Canada) optimization package, utilizing cable forces and spring tensions as crucial inputs.

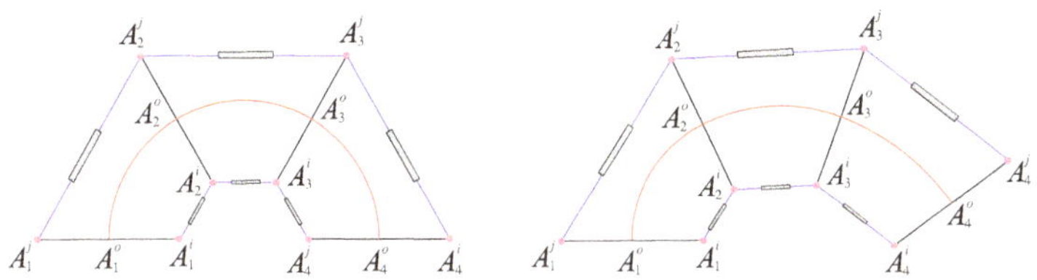

Figure 7. Simulation of bending for three segments with $\mu = 0$ (**left**) and $\mu = 0.2$ (**right**).

When we introduce friction into the system, a notable change occurs in the tension levels of the two cables within each segment. This alteration has a direct impact on the curvature, which is visually evident in Figure 7 (right). It is essential to acknowledge that our model, while informative, maintains a simplification in which it assumes a consistent curvature angle across all segments.

Implementing a tensegrity mechanism for the creation of a flexible stapler in the confined and intricate workspace of a patient's body brings about a reduction in the cross-sectional dimensions of each individual segment. However, this design presents a significant challenge. The slender cross-section width causes the folding beam to intersect with the actuating cables, even at slight deviation angles, resulting in an enlarged radius of curvature, as illustrated in Figure 8.

To ensure unimpeded access to the patient's workspace, it becomes imperative to incorporate an auxiliary joint, which resembles a wrist, before reaching the stapler section. Alternatively, one may address this issue by increasing the number of segments. However, it is crucial to consider that augmenting the segment count may potentially introduce heightened friction between the cables and the eyelets through which they pass. Careful engineering consideration is necessary to strike a balance between these design factors.

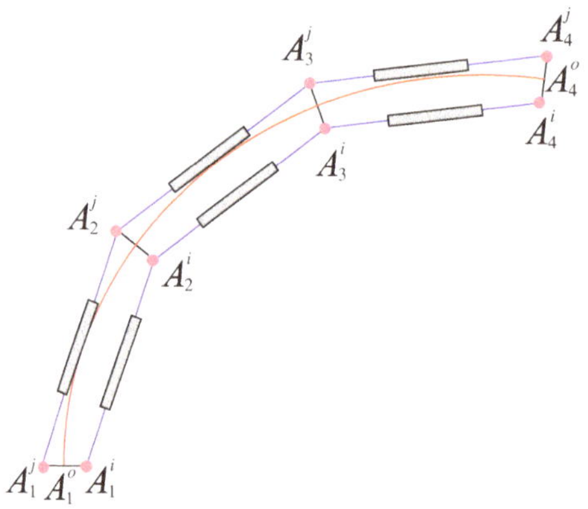

Figure 8. Bending singularity occurring due to reduced cross-section

4. Stapler Design and Prototype

In this section, we are interested in the intricate process of designing and bringing to life the kinematic equations that were introduced in the preceding segment. Building upon a robust mathematical model, extensive simulations, and a thorough examination of potential singularities, we have meticulously crafted a sophisticated multi-segmented mechanism with a total of ten segments (referred to as $n = 10$). This innovative creation, depicted in Figure 9, takes the form of a slotted beam-type flexible jaw mechanism, which is uniquely endowed with the ability to bend within a single plane, aligning perfectly with the plane in which staple insertion occurs.

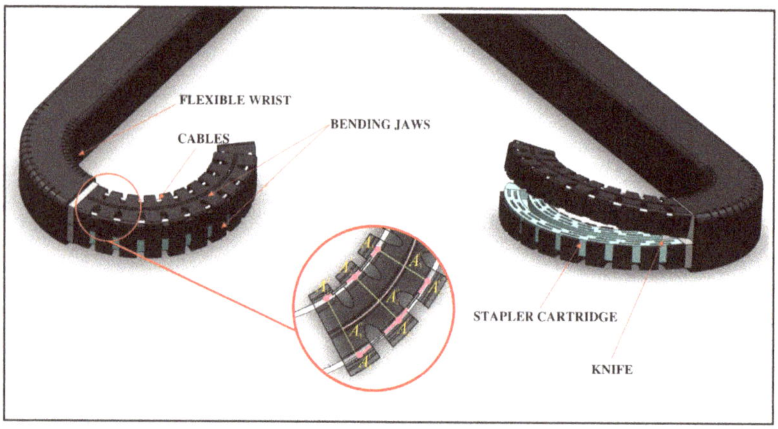

Figure 9. FLexible surgical stapler design

To further elevate its maneuverability and versatility, our design incorporates a compound flexible wrist, seamlessly working in conjunction with the flexible jaws. As an added feature, we've also integrated a knife into the stapler cartridge, strategically positioned to expedite tissue separation once the staple pins have been successfully deployed. This comprehensive design approach represents a significant step forward in the field, optimizing the performance and functionality of this medical tool for improved patient outcomes.

After completing the CAD modeling and rigorous simulations, we moved on to the next phase, in which we brought our vision to life through the creation of a scaled-up prototype. This prototype, enlarged by 200% along the centroid, represents the stapler jaws and was fabricated using cutting-edge 3D printing technology (see Figure 10). Specifically, we employed a Stereo-lithography Apparatus (SLA)-type 3D printer, which utilizes a liquid polymer resin as its foundational material and employs UV radiation to solidify the intended pattern. For this particular prototype, we selected "MONOCURE 3D FLEX 100" resin due to its high flexibility and structural weaknesses. To enhance the overall rigidity of the 3D printed part, we added a small amount of "ANYCUBIC 3D printing UV sensitive (Aqua)" resin, known for producing hard and rigid 3D-printed components. The mixing ratio for these resins was 80–20%, with MONOCURE flexible resin making up 80% and ANYCUBIC (Aqua) accounting for the remaining 20%. The curing process took approximately 120 min, resulting in a composite material with distinctive properties, as outlined in Table 1.

Table 1. Material properties of 3D-printed parts.

Properties	Values
Density (ρ)	1290.57 kg/m^3
Ultimate Tensile Strength (UTS)	1.6416 MPa
Poisson's ratio (ν)	0.35
Young's Modulus (E)	4.83 MPa

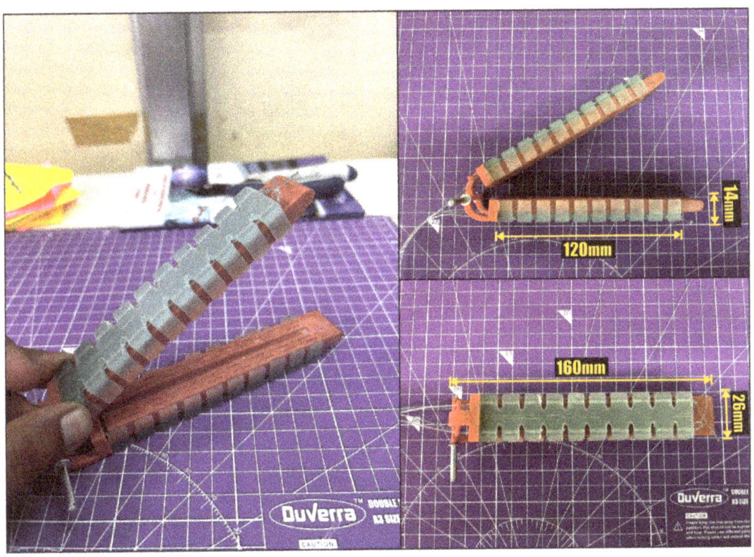

Figure 10. The 3D-printed prototype of flexible surgical stapler.

Following the fabrication of the prototype, we conducted a series of rigorous bending tests to evaluate its mechanical performance under varying force levels. The testing apparatus employed a pulley system, with weights attached to one of the cables incorporated into the 3D-printed prototype. To measure the prototype's flexibility and resilience, we marked its centerline tip and observed only the tip's movement during bending. The resulting bending motion was meticulously traced on a sheet of plotting paper to map the trajectory of the tip. Various weights, ranging from 100 g to 1500 g, were incrementally applied as depicted in Figure 11, and the resultant displacement of the prototype's tip in both the X and Y directions, along with its angular deflection (α) relative to the central axis at the neutral position, were meticulously recorded and detailed in Table 2 and depicted in Figure 12. These tests provide critical insights into the mechanical behavior and performance of our prototype.

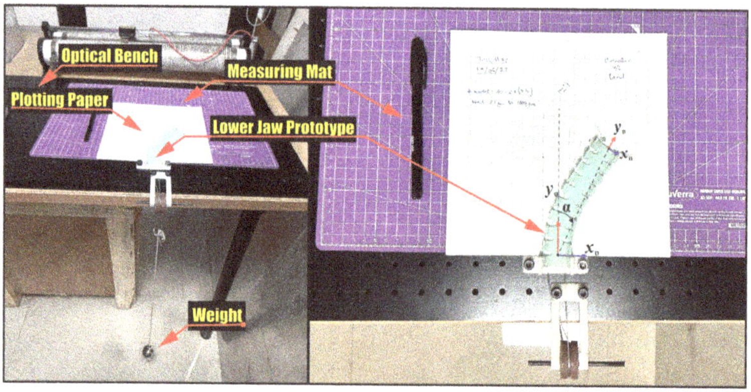

Figure 11. Testing of 3D-printed prototype of lower Jaw.

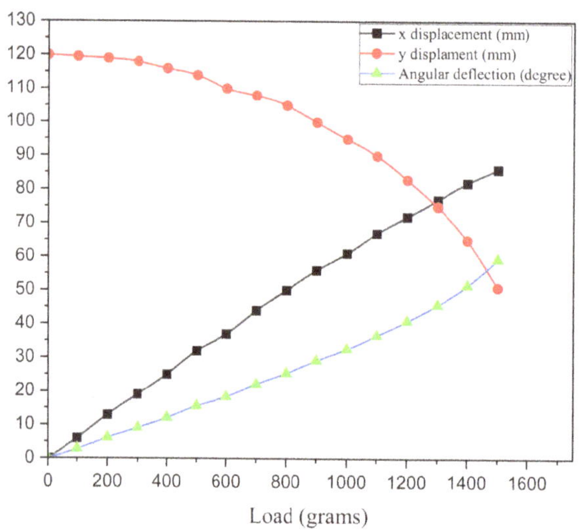

Figure 12. Plot of x displacement, y displacement and angular deflection of prototype tip (α) vs. load applied, as presented in Table 2.

Table 2. Table for load v/s deflection of the tip of the 3D-printed lower jaw prototype.

Sl. No.	Load (g)	X Displacement (mm)	Y Displacement (mm)	Angular Deflection (α) (deg)
1	0	0	120	0
2	100	6	119.5	2.86
3	200	13	119	6.23
4	300	19	118	9.14
5	400	25	116	12.16
6	500	32	114	15.67
7	600	37	110	18.43
8	700	44	108	22.16
9	800	50	105	25.46
10	900	56	100	29.24
11	1000	61	95	32.70
12	1100	67	90	36.66

Table 2. *Cont.*

Sl. No.	Load (g)	X Displacement (mm)	Y Displacement (mm)	Angular Deflection (α) (deg)
13	1200	72	83	40.94
14	1300	77	75	45.75
15	1400	82	65	51.59
16	1500	86	51	59.33

5. Conclusions and Future Work

In this article, we have investigated a flexible mechanism designed for application within a stapler during laparoscopic total mesorectal excision procedures for rectal cancer and other colorectal interventions. These procedures introduce challenges for traditional tools due to constrained accessibility and the limited space available for manoeuvring such stapler tools. The mechanism comprises three stacked tensegrity structures, which consist of rigid bodies, linear springs, actuation cables, and a flexible beam. The dimensions of the stapler are determined based on various factors, including patient anatomy and the constraints imposed by laparoscopy. Currently, patient scans are under analysis to define the surgical workspace.

We employed a lumped parameter model to analyze the mechanism, factoring in the presence or absence of friction in each rigid segment's eyelet. Our findings underscore the impact of friction on curvature radius, necessitating precise analysis. Ongoing research focuses on understanding the friction between the mobile platform eyelets and cables as we add more sections, anticipating increased overall friction that may potentially affect curvature beyond expectations.

The materials used are under evaluation, especially the sheathing covering the mechanism. The stapler's outer skin, which is made of elastomer, doubles as a spring for sealing and medical equipment suitability. This research signifies a vital step in advancing surgical stapler technology for complex procedures, enabling more effective and precise surgeries in challenging anatomical regions.

Our future goal is to gather more data and enhance data analysis, constructing a three-dimensional workspace representation. This will facilitate better analysis of the surgical stapler design and identification of engineering and anatomical constraints for improved problem-solving and design optimization.

Author Contributions: Conceptualization, D.K., R.J., A.T., E.D. and D.C.; methodology, all authors; Maple software, D.K. and D.C.; experimental validation, D.K.; formal analysis, D.K.; investigation, D.K.; resources, D.K., A.T. and E.D.; data curation, A.T.; writing—original draft preparation, D.K.; writing—review and editing, R.J. and D.C.; visualization, D.K.; supervision, E.D., R.J. and D.C.; project administration, E.D., R.J. and D.C.; funding acquisition, E.D., R.J. and D.C. All authors have read and agreed to the published version of the manuscript.

Funding: This research is funded by the Indian Council of Medical Research, New Delhi, India under the "Senior Research Fellowship" awarded to Mr. Dhruva Rajesh Khanzode (File no. 5/3/8/46/ITR-F/2022). This research was supported by the project New smart and adaptive robotics solutions for personalized minimally invasive surgery in cancer treatment—ATHENA, funded by European Union—NextGenerationEU and Romanian Government, under National Recovery and Resilience Plan for Romania, contract no. 760072/23.05.2023, code CF 16/15.11.2022, through the Romanian Ministry of Research, Innovation and Digitalization, within Component 9, investment I8.

Data Availability Statement: Data are contained within the article.

Acknowledgments: This research was supported by CSIR-Central Scientific Instruments Organization, Chandigarh, India, Centre Hospitalier Universitaire de Nantes, France and Centrale Nantes, France.

Conflicts of Interest: The authors declare no conflict of interest.

References

1. Gaidry, A.D.; Tremblay, L.; Nakayama, D.; Ignacio, R.C. The History of Surgical Staplers: A Combination of Hungarian, Russian, and American Innovation. *Am. Surg.* **2019**, *85*, 563–566. [CrossRef]
2. Robicsek, F.; Konstantinov, I. Hümer Hültl: The father of the surgical stapler. *J. Med Biogr.* **2001**, *9*, 16–19. [CrossRef]
3. Akopov, A.; Artioukh, D.Y.; Molnar, T.F. Surgical Staplers: The History of Conception and Adoption. *Ann. Thorac. Surg.* **2021**, *112*, 1716–1721. [CrossRef]
4. Demertzis, S.; Beslac, O.; Mettler, D.; Zalokar, D.; Spangler, T.; Hausen, B.; Swanstrom, L. Beyond the B: A new concept of the surgical staple enabling miniature staplers. *Surg. Endosc.* **2015**, *29*, 3674–3684. [CrossRef]
5. Kostrzewski, S.; Aranyi, E.; Scirica, P. In-Situ Loaded Stapler. US Patent 9364217B2, 14 June 2016.
6. McGuire, J.; Wright, I.; Leverment, J. Surgical staplers: A review. *J. R. Coll. Surg. Edinb.* **1997**, *42*, 1–9.
7. Adams, R.D.; Main, L.O.; Swaffar, S.E.; Pugsley, C.H., Jr.; Gordon, D.P. Endoscopc Stapler. US Patent 6302311, 16 October 2001.
8. Tobias, K.M. Surgical Stapling Devices in Veterinary Medicine: A Review. *Vet. Surg.* **2007**, *36*, 341–349. [CrossRef]
9. Schemmer, P.; Friess, H.; Dervenis, C.; Schmidt, J.; Weitz, J.; Uhl, W.; Büchler, M.W. The Use of Endo-GIA Vascular Staplers in Liver Surgery and Their Potential Benefit: A Review. *Dig. Surg.* **2007**, *24*, 300–305. [CrossRef]
10. Holzmacher, J.L.; Luka, S.; Aziz, M.; Amdur, R.L.; Agarwal, S.; Obias, V. The use of robotic and laparoscopic surgical stapling devices during minimally invasive colon and rectal surgery: A comparison. *J. Laparoendosc. Adv. Surg. Tech.* **2017**, *27*, 151–155. [CrossRef]
11. Bolanos, H.; Norwalk, E.; Sherts, C.R.; Southport; Pelletier, A.T.; Wallingford. Endoscopic Stapler. US Patent 690269, 5 November 1997.
12. Milliman, K.L.; Bethel; Viola, F.J.; Hook, S.; Orban, J.; Norwalk; Lehn, R.F.; Stratford. Surgical Stapling Apparatus. US Patent 586531, 2 February 1999.
13. Ema, T. The experience of using Endo GIA Radial Reload with Tri-Staple Technology for various lung surgery. *J. Thorac. Dis.* **2014**, *6*, 1482–1484. [CrossRef]
14. Rivadeneira, D.E.; Verdeja, J.C.; Sonoda, T. Improved access and visibility during stapling of the ultra-low rectum: A comparative human cadaver study between two curved staplers. *Ann. Surg. Innov. Res.* **2012**, *6*, 11. [CrossRef]
15. Ruffo, G.; Sartori, A.; Crippa, S.; Partelli, S.; Barugola, G.; Manzoni, A.; Steinasserer, M.; Minelli, L.; Falconi, M. Laparoscopic rectal resection for severe endometriosis of the mid and low rectum: Technique and operative results. *Surg. Endosc.* **2012**, *26*, 1035–1040. [CrossRef]
16. de Calan, L.; Gayet, B.; Bourlier, P.; Perniceni, T. Chirurgie du cancer du rectum par laparotomie et par laparoscopie. *EMC Chir.* **2004**, *1*, 231–274. [CrossRef]
17. Furet, M.; Lettl, M.; Wenger, P. Kinematic Analysis of Planar Tensegrity 2-X Manipulators. In Proceedings of the International Symposium on Advances in Robot Kinematics, Bologna, Italy, 1–5 July 2018; Springer: Berlin/Heidelberg, Germany, 2018; Volume 8, pp. 153–160. [CrossRef]
18. Chekan, E.; Whelan, R. Surgical stapling device-tissue interactions: What surgeons need to know to improve patient outcomes. *Med Devices Evid. Res.* **2014**, *7*, 305. [CrossRef]
19. Ghosh, S.; More, N.; Kapusetti, G. Surgical staples: Current state-of-the-art and future prospective. *Med. Nov. Technol. Devices* **2022**, *16*, 100166. [CrossRef]
20. Khanzode, D.; Jha, R.; Thomieres, A.; Duchalais, E.; Chablat, D. Design of a surgical stapler for laparoscopic colectomy. In Proceedings of the International Workshop on Medical and Service Robots, Craiova, Romania, 7–10 June 2023; Springer: Berlin/Heidelberg, Germany, 2023; pp. 71–80. [CrossRef]
21. Gravagne, I.; Rahn, C.; Walker, I. Large deflection dynamics and control for planar continuum robots. *IEEE/ASME Trans. Mechatron.* **2003**, *8*, 299–307. [CrossRef]
22. Kato, T.; Okumura, I.; Song, S.E.; Golby, A.J.; Hata, N. Tendon-Driven Continuum Robot for Endoscopic Surgery: Preclinical Development and Validation of a Tension Propagation Model. *IEEE/ASME Trans. Mechatron.* **2015**, *20*, 2252–2263. [CrossRef]
23. Khanzode, D.; Jha, R.; Chablat, D.; Duchalais, E. Stapler design with stacked tensegrity mechanisms for surgical procedures. In Proceedings of the International Design Engineering Technical Conferences and Computers and Information in Engineering Conference, St. Louis, MI, USA, 14–17 August 2022; American Society of Mechanical Engineers: New York, NY, USA, 2022; Volume 86281, p. V007T07A037. [CrossRef]
24. Khanzode, D.; Jha, R.; Duchalais, E.; Chablat, D. Stacked Tensegrity Mechanism for Medical Application. In Proceedings of the International Symposium on Advances in Robot Kinematics, Bilbao, Spain, 26–30 June 2022; Springer: Berlin/Heidelberg, Germany, 2022, pp. 139–148.
25. Jha, R.; Chablat, D.; Baron, L.; Rouillier, F.; Moroz, G. Workspace, joint space and singularities of a family of delta-like robot. *Mech. Mach. Theory* **2018**, *127*, 73–95. [CrossRef]

robotics

MDPI

Article

Neural Network Mapping of Industrial Robots' Task Times for Real-Time Process Optimization

Paolo Righettini *, Roberto Strada and Filippo Cortinovis

Department of Engineering and Applied Sciences, University of Bergamo, 24044 Dalmine (BG), Italy; roberto.strada@unibg.it (R.S.); filippo.cortinovis@unibg.it (F.C.)
* Correspondence: paolo.righettini@unibg.it

Abstract: The ability to predict the maximal performance of an industrial robot executing non-deterministic tasks can improve process productivity through time-based planning and scheduling strategies. These strategies require the configuration and the comparison of a large number of tasks in real time for making a decision; therefore, an efficient task execution time estimation method is required. In this work, we propose the use of neural network models to approximate the task time function of a generic multi-DOF robot; the models are trained using data obtained from sophisticated motion planning algorithms that optimize the shape of the trajectory and the executed motion law, taking into account the kinematic and dynamic model of the robot. For scheduling purposes, we propose to evaluate only the neural network models, thus confining the online use of the motion planning software to the full definition of the actually scheduled task. The proposed neural network model presents a uniform interface and an implementation procedure that is easily adaptable to generic robots and tasks. The paper's results show that the models are accurate and more efficient than the full planning pipeline, having evaluation times compatible with real-time process optimization.

Keywords: robot process optimization; task scheduling; real-time management; task time mapping; AI in robotics

1. Introduction

The throughput of an industrial application using one or more robots with a fixed base depends on the cycle time of the movement executed by the robot to perform the task. Therefore, the choice of movements with the shortest cycle time increases the number of operations in the reference time, enhancing the robot application's productivity.

The cycle time of pick and place operations depends on several aspects, such as the kinematic structure of the robot, its inertial characteristics, and the properties of its actuation and transmission systems, which all concur to limit the maximal achievable velocities and accelerations; an important role is played also by the trajectory's shape, by the motion profile on it, and by its position inside the robot's working volume.

The choice of the robot's movement with the shortest cycle time is straightforward for fixed pick and place positions. In this case, an offline procedure based on the robot's software environment can be used. In more complex cases, however, the pick and place positions are not known in advance, and moreover might be non-stationary; for such cases, then, three types of pick and place tasks may be distinguished: (a) tasks on fixed pick and place targets, which are known in advance and do not change in time; (b) stationary pick and place tasks, in which the targets do not move while the robot is performing its operations but which are dynamically reconfigured and a priori unknown; (c) on-the-fly pick and place tasks, in which the targets are not known in advance and also move while the robot operates on them.

Tasks of type (b) and (c) occur, e.g., in modular multi-robot work cells where a series of robots operates on a transportation system that intermittently or continuously conveys the items to be processed. Each robot, which, as exemplified in Figure 1, must sort, manipulate, and rearrange items randomly loaded onto the transportation system, operates therefore under time constraints in a time-varying and non-deterministic environment.

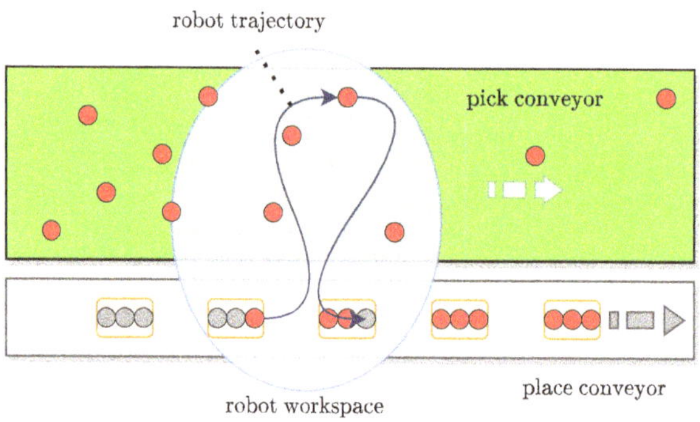

Figure 1. Schematic representation of a non-deterministic and time-constrained on-the-fly manipulation task.

In this kind of application, the definition of the robot's movement with the shortest cycle time is a complex process, requiring an optimization strategy based on the changing scenario that the robot is presented with and on the robot's characteristics. For stationary pick and place positions, the process optimization problem concerns the sequencing of the tasks. For on-the-fly manipulation, an additional optimization problem, i.e., the computation of the best position at which the robot intercepts its moving target, must be preliminarily solved.

Additional problems that must be solved after a task has been scheduled are first the geometric motion planning, which defines the shape of the trajectory to be executed by the robot, and then the generation of the motion law along the geometric path.

The load balancing and task assignment problems for assembly lines and multi-robot cells have been treated in the current literature from an operational research perspective. In [1–5], the actual capabilities of the robot are indeed expressed in terms of its cycle times, which are considered not as a function of the properties of the robot and of the tasks but either as a random variable [1], as an average value to which other independent contributions (such as waiting times) might added [2,3], or as a single fixed value [4,5].

The intercept problem both for manipulators and for mobile robots, in which the precise estimation of the task times becomes a focal point, emerges in a variety of applications and has been treated according to several approaches. In [6], a Particle Swarm Optimization method is proposed for a coal gangue processing plant served by a conveyor and by Cartesian robots; [7] discusses a heuristic algorithm for the real-time solution of the rendezvous problem between unmanned aerial vehicles operating in a leader-follower configuration. In [8], the intercept problem for objects moving on a conveyor belt was solved online within a bisection search algorithm, which requires the evaluation of the task times of a robot moving with trapezoidal velocity profiles. A similar concept was described earlier in [9], in which the traveling time curve of the robot is iteratively approximated online; the possibility of recomputing the intercept position in case of variations of the moving target's trajectory is also included. In [10], target tracking and collision avoidance with moving obstacles are jointly treated using a Model Predictive Control framework in which neural networks are used to implement the models of the robot and of the obstacle dynamics.

These works are characterized by the online definition of the task time curves of the robot, or, as in [10], by a reactive approach that does not require the evaluation of the task time curve. Alternative approaches are instead based on the pre-evaluation of the robot task times. This general idea is discussed in [11], in which the execution times are obtained from the Reflexxes Motion Library [12]; an early example of an analogous concept can be found in [13], in which experimental data collected from the actual use of the robot are arranged in a simple task time lookup table. More recently, also [14] solved the optimal intercept problem for pick and place tasks on a rotating table using a two-dimensional task time table adapted to the radial geometry of the problem; the motion planning of each trajectory is performed using the commercial software that accompanies the robot controller. In a similar vein, [15] constructs a task time table needed for pick and place operations on a moving conveyor. In that work, the dependence of the cycle time function on the manipulator's specific properties is highlighted. The procedure for the determination of the cycle time data is not described, as reliance is again placed on the Reflexxes Motion Library. The authors assume that the velocity of the conveyor is fixed, and only a rectangular portion of the workspace of the robot is considered. In [16], the concept of a task-dependent map emerges in relation not to the execution times but rather to the energy consumption associated with each operation. The resulting energy map is three-dimensional and expressed in a cylindrical system.

To maximize the throughput of the robot, it is necessary to minimize the execution times of its tasks; hence, the minimization of the travel time along a given geometric path under constraints determined by the physical properties of the robot has received considerable attention and has been treated successfully according to a variety of methods. The algorithm described by Bobrow et al. [17] constructs time-optimal trajectories characterized by C^0 velocities using a numerical integration approach. Further refinements were proposed by [18]. A similar result is obtained through an algorithm based on reachability analysis developed in [19], while other commonly applied approaches are based on convex optimization [20]. These methods differ in terms of ease of implementation and computational cost; a further significant difference lies in their ability to include more or less straightforwardly third-order constraints whose main purpose is the limitation of the motion jerk. Such constraints were included in an exact but cumbersome way within the numerical integration approach [21], whereas methods based on convex optimization were shown to be able to consider motion jerk constraints rather straightforwardly, though not exactly [22,23]. A commonality of these works is that the geometric path is considered a given; the problem of generating suitable geometric trajectories remains therefore out of their scope. The geometric path planning, indeed, is more application specific and thus less amenable to be treated in a standard way. Different strategies and ideas, a review of which can be found in [24], can be specialized for use in diverse fields such as race driver modeling [25], routing of mobile robots for agricultural applications [26], CNC milling [27,28], and fixed-based robots operating in cluttered environments [29].

The current literature reveals the following gaps:

- the scheduling, the optimal intercept computation, and the trajectory execution time minimization are treated in isolation, without considering the interdependencies between them;
- the scheduling algorithms often do not consider accurately that different tasks are associated with different execution times;
- the cycle or task time maps are developed without a focus on the optimal (as opposed to feasible) times;
- the cycle time maps or tables cover only a portion of the useful workspace of the manipulator and appear tailored to a specific application;
- the methods employed for the generation of the cycle time maps are known to scale poorly to problems featuring a high number of degrees of freedom.

To address these gaps, this work proposes a mapping of the robot's task time using neural networks. The use of a neural network model results in a task time map covering the entire workspace of the robot and presenting a uniform software interface.

For the collection of the data needed to train the map, the authors take into account all the aspects that determine the task time, namely the kinematics and dynamics of the manipulator, the geometric motion planning, and the task time optimization. This paper illustrates therefore the geometric planning and task time minimization procedures that enable the generation of the task time data directly from simulations. Since these software modules are based on the mechanical model of the robot, its properties and limitations are reflected in the generated training dataset and therefore also in the trained task time map. The proposed approach remains applicable also in cases in which the software modules for the generation of appropriate datasets are not available, since the neural network model can also be trained with experimental data, whose collection is however more expensive and time consuming.

The geometric planning and task time minimization are relatively demanding procedures, which should be invoked sparingly during real-time computations. The evaluation of the task time map, on the contrary, is orders of magnitude faster, making the map itself an essential component of the real-time optimizing scheduler that solves the intercept problem and defines the robot's tasks. In particular, since the pick and place targets are moving, the task has to be defined in a useful time window, before the targets move past the robot's reach. The task time map is therefore used to speed up these computations.

The main contributions of the article are:

- the development of a systematic architecture for the representation of the task time maps as neural networks;
- the illustration of a computational procedure for the generation of the dataset needed to train the task time models;
- the demonstration of the use of the task time maps to solve the optimal intercept problem in an industrial application case.

The paper is structured as follows. An outline of the problem and an overview of the proposed solution are first presented. The proposed methodologies and the necessary algorithms for the creation of the task time map are then illustrated; their implementation in relation to a full-featured case study is then shown. The results concerning two applications of the task time map are finally discussed: one related to pick and place operations on items carried by a conveyor that advances intermittently, the other related to on-the-fly manipulation of objects moving on a conveyor that advances continuously and with slowly adjustable speed.

2. Methods

2.1. Problem Statement and Solution Outline

The paper denotes as a task or job a configurable elementary operation that can be performed by the robot. Chaining multiple elementary tasks results in a composite task; a task that is executed in a repeating pattern constitutes a work cycle. The scheduler is a proper software algorithm that selects and configures the next job assigned to the robot among all the possibilities. Several scheduling strategies that aim at the maximization of the throughput of the system, such as the Shortest Job Next or the Highest Response Ratio Next rules, require the knowledge of the execution times associated with each of the possible tasks. To implement this kind of scheduler, a map that associates the jobs that can be executed inside the workspace of the robot to their minimal execution time is then needed. If required by the application, the map can be used by the scheduler also for solving the optimal intercept problem, which likewise is governed by the task time function.

An elementary task is defined by a geometric path and by the motion profile on it. The geometric path $\gamma : [u_{lb}, u_{ub}] \rightarrow \Omega \subset \mathbb{R}^{n_f}$ is a function mapping a real number belonging to the interval $[u_{lb}, u_{ub}]$ into the n_f-dimensional robot's task space Ω. A point in Ω is denoted as $\boldsymbol{p}$; $\boldsymbol{v}$ and $\boldsymbol{a}$ denote likewise velocities and accelerations.

The motion profile on the geometric path is representable as a function $u : [t_s, t_f] \rightarrow [u_{lb}, u_{ub}]$, where $[t_s, t_f]$ is the time interval in which the motion is executed. Therefore, to define a task, two issues arise, namely the definition of the function γ and of the function u. Once these two functions have been defined, the task execution time $T_{task} = t_f - t_s$ is also defined. In robotic applications, the definition of the geometric path and of the motion profile is typically decoupled. The geometric path is commonly defined by the Geometric Trajectory Planner (GTP); the motion profile can be on the other hand defined by a separate software module, the Task Time Optimizer (TTO). The GTP and the TTO can be thought of as two higher-order functions, denoted each as G and H, that accept a certain number of inputs and have as output, respectively, functions γ and u . Concerning the GTP module, it can be said that its implementation is highly application-specific. In general, however, the geometric definition of a task requires a set of position, velocity, and acceleration boundary conditions plus a set of further configuration parameters; the output is a geometric path. Denoting as $\boldsymbol{c}_G$ the further configuration parameters, we can write the following:

$$\gamma \leftarrow G(\boldsymbol{c}_G, \boldsymbol{p}_s, \boldsymbol{v}_s, \boldsymbol{a}_s, \boldsymbol{p}_f, \boldsymbol{v}_f, \boldsymbol{a}_f) . \tag{1}$$

As already stated in the introduction, standard algorithms for the TTO exist; irrespectively of the implementation details, these algorithms can be assumed to take the form of a higher-order function:

$$u \leftarrow H(\boldsymbol{c}_H, \gamma, \boldsymbol{p}_s, \boldsymbol{v}_s, \boldsymbol{a}_s, \boldsymbol{p}_f, \boldsymbol{v}_f, \boldsymbol{a}_f) , \tag{2}$$

where $\boldsymbol{c}_H$ are configuration parameters of the task time optimizer. The output of H, usually determined through an optimization procedure that takes into consideration the properties of the robot, is the motion profile u.

The functions G and H must be such that the following equalities hold:

$$\gamma(u(t_s)) = \boldsymbol{p}_s \tag{3}$$
$$\dot{\gamma}(u(t_s)) = \boldsymbol{v}_s \tag{4}$$
$$\ddot{\gamma}(u(t_s)) = \boldsymbol{a}_s \tag{5}$$
$$\gamma(u(t_f)) = \boldsymbol{p}_f \tag{6}$$
$$\dot{\gamma}(u(t_f)) = \boldsymbol{v}_f \tag{7}$$
$$\ddot{\gamma}(u(t_f)) = \boldsymbol{a}_f . \tag{8}$$

In light of assignments (1) and (2), it can be said that each elementary task is ultimately defined by the boundary conditions $\boldsymbol{p}_s$, $\boldsymbol{v}_s$, $\boldsymbol{a}_s$, $\boldsymbol{p}_f$, $\boldsymbol{v}_f$, $\boldsymbol{a}_f$ and by the parameters $\boldsymbol{c}_G$, $\boldsymbol{c}_H$. Therefore, the task execution time T_{task} is also a function of these same parameters. It is clear that the evaluation of T_{task} is an expensive procedure that requires the application of G and H.

In general terms, we define as the task time map $\hat{T}_{task}$ a function of the form:

$$\hat{T}_{task} = \hat{T}_{task}(\boldsymbol{c}_G, \boldsymbol{c}_H, \boldsymbol{p}_s, \boldsymbol{v}_s, \boldsymbol{a}_s, \boldsymbol{p}_f, \boldsymbol{v}_f, \boldsymbol{a}_f) \tag{9}$$

whose output nominally matches T_{task}.

Without loss of generality, this work considers applications where the configuration parameters are embedded into the GTP and TTO implementations. Additionally, since for motion control purposes the enforcement of acceleration boundary conditions is not strictly required, in the following discussion we do not consider them as input parameters. As a result of these assumptions, which still allow us to deal with most industrial cases of any practical interest, the task time map will assume the form: $\hat{T}_{task} = \hat{T}_{task}(\boldsymbol{p}_s, \boldsymbol{v}_s, \boldsymbol{p}_f, \boldsymbol{v}_f)$.

The paper presents the development of the task time map as a neural network model and the definition of a strategy for generating suitable training datasets. Moreover, the paper presents how the map is helpful for efficiently solving the optimal intercept problem. Further uses, discussed conceptually in this article, are related to implementing efficient scheduling algorithms usable in real-time software applications.

To precisely define the optimal intercept problem, the authors assume that the motion of the target (i.e., of the object to be intercepted by the robot) is a known curve in Ω of the form $\gamma_{tgt}(t)$. In addition, the time derivatives of γ_{tgt} are known. The solution of the following equation formally defines the optimal intercept problem:

$$T_{task}(\boldsymbol{p}_s, \boldsymbol{v}_s, \gamma_{tgt}(t), \dot{\gamma}_{tgt}(t)) = t\,. \tag{10}$$

The quantities $\boldsymbol{p}_s$ and $\boldsymbol{v}_s$ are determined by the current state of the robot. Equation (10) is therefore univariate; its solution yields the time $t_{intercept}$ and position $\gamma_{tgt}(t_{intercept})$ at which the robot can intercept the target. As the solution is in general not unique, the one characterized by the minimal $t_{intercept}$ should be selected.

In order to solve the optimal intercept Equation (10), numerical solving is required, which can involve a large number of iterations. However, substituting T_{task} with $\hat{T}_{task}$ can significantly reduce the computation time. This approach meets the soft real-time constraints, which are necessary because the environment in which the robot operates is not stationary.

Within a complete software stack for a practical robotic application, the GTP and TTO modules should be reasonably efficient, as they are used once for each scheduled task to fully define it; the task time map $\hat{T}_{task}$ on the other hand needs to be not only accurate but also much faster than the direct determination of T_{task} because both the solution of the optimal intercept and the task scheduling require its frequent evaluation.

The scheduling and planning operations are represented graphically in Figure 2 according to the above discussion, though with an eye towards the industrial case of interest, which is a pick and place line with conveyors moving with quasi-statically adjustable velocity.

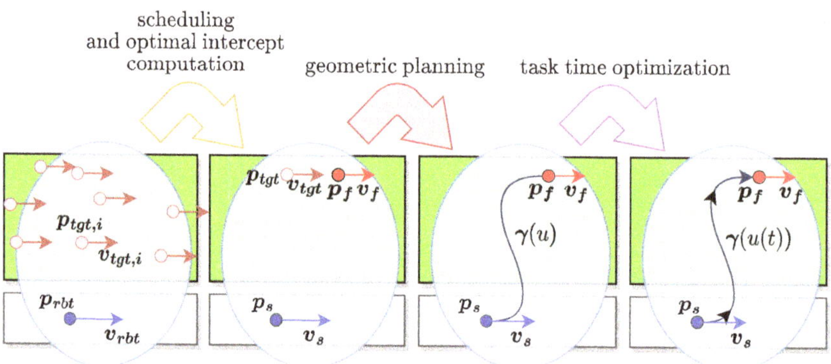

Figure 2. Pictorial representation of the scheduling, geometric motion planning, and task time optimization pipeline.

The scheduler requires as inputs the current position p_{rbt} and velocity v_{rbt} of the manipulator's end effector and a list of current positions $p_{tgt,1} \dots p_{tgt,n}$ and velocities $v_{tgt,1} \dots v_{tgt,n}$ of the pick and place targets, which are detected, e.g., by a vision system.

The scheduler performs two operations according to the information provided by the task time map. First, if the targets are moving, it computes the positions at which the robot should intercept them. Subsequently, it identifies the target that should be processed next among all the possibilities.

The outputs of these operations are the position and velocity boundary conditions of the trajectory needed to perform the pick or place task: the scheduler sets the starting position and velocity p_s and v_s equal to the current position and velocity of the robot; the final position and velocity p_f and v_f are instead defined according to the solution of the intercept problem of the chosen target. Once p_s, v_s, p_f, and v_f have been defined, the geometric planning can be performed, yielding a parametric curve $\gamma(u)$. This curve defines the geometry of the trajectory, but not the motion law on it, whose generation is demanded instead to the task optimization routine, which outputs the time reparametrization $\gamma(u(t))$ and in suborder the actual task execution time T_{task}.

The geometric trajectory planning is strongly tied to the type of application and can be developed only in relation to it. This work proposes a geometric planning approach suitable for pick and place applications where the manipulator can either perform pick and place operations on stationary targets or also follow non-stationary picking and placing positions, which move along a given direction. This problem arises when the manipulator is used to pick and place the payload from and onto moving conveyors. We note that the implementation of the Geometric Trajectory Planner defines classes of parametrized tasks, which, in our case, there are two: stationary and on-the-fly pick and place tasks.

In Figure 3, the input–output relationships between the environment, the scheduler, the GTP, and the TTO are illustrated in accordance with the description above.

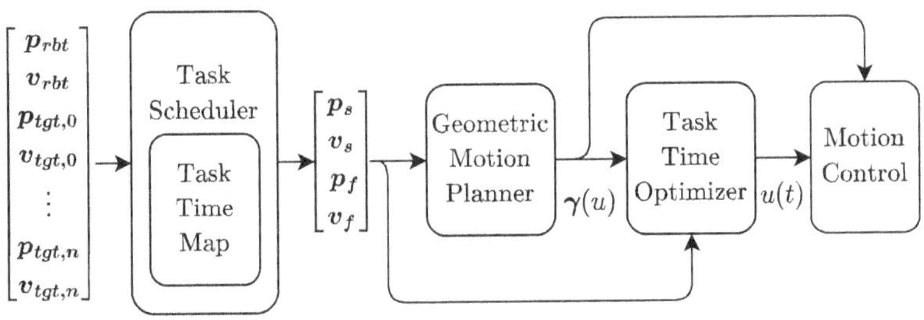

Figure 3. Input–output representation of the scheduling, geometric motion planning, and task time optimization pipeline.

Here, it can be seen clearly that the task time estimation needed by the scheduler is provided by the task time map, which ideally should coincide with the function that associates the tasks with their execution times. The exact evaluation of the task time function using the GTP and the TTO is however rather expensive; therefore, for scheduling purposes, the task time map is used in their stead to efficiently determine the task times.

To generate the map, a representative sample of the task times must be collected. The task time data can be obtained from the robot's tasks offline simulation, performed using the GTP and the TTO. The training data can also be obtained by experimental tests using the robot for the given payload, for the given boundary conditions, and for a set of trajectories whose starting and ending points thoroughly cover the working space. In this case, the cost and the time for collecting the data can be high, but the need for the development of the GTP and TTO is removed, as reliance can be placed on the robot's accompanying software. In addition, in this case, the proposed approach based on the creation of the task time map as a deep learning model remains still valid. Within this work, the authors focus on the computational approach, notwithstanding the fact that the experimental generation of the needed dataset remains a possibility.

The proposed approach is useful for applications that involve robots and movements with several degrees of freedom, where each task can be configured according to a large number of parameters, i.e., the initial and final positions and velocities of the robot. In these cases, the simpler table-based lookup approaches do not exhibit either a uniform interface or a general implementation procedure and are not suitable for the description of the task time map, not only because of the irregular geometry of the robot's workspace but also because of the large dimensionality of the problem. The representation of the task time map through the use of artificial neural networks overcomes the difficulties associated with more traditional table-based lookup structures: in addition to being inherently mesh-free and adaptable to the workspace geometry, it also offers the advantages both of a standard query interface and of a generation methodology that remains independent from the specific characteristics of the robot and of the application.

The paper also address the fact that an accurate and reliable model can be generated only from task time data that reflect:

- the kinematic and dynamic properties of the robot;
- the parameters of the actuators and transmission systems;
- the behavior of the Geometric Planning Module;
- the behavior of the Task Time Optimizer.

This work proposes therefore a systematic procedure that includes the following steps:

- the parametrization of the tasks to be executed by the robot using the GTP;
- the offline generation of a dataset of optimal execution times using the TTO;
- the generation, using deep learning models, of an accurate and efficiently queryable map of the optimal task times over the entire workspace of the robot.

2.2. Geometric Trajectory Planning

The geometric trajectory planner accepts as inputs the configuration parameters of the path's geometry. These are the initial and final configurations of the manipulator, expressed as position and velocity boundary conditions.

The geometric planner should generate as its output a parametric curve $\gamma(u)$, with parameter u belonging to an interval $[u_{lb}, u_{ub}]$. In this paper, curves having C^2 continuity are considered in order to avoid points along the trajectory necessarily characterized by unbounded motion jerk. Clearly, the curves should also belong entirely to the workspace of the robot and should not cross any singular configuration.

How to efficiently achieve these goals depends on the properties of the robot, which determine the shape both of the workspace and of the singular loci. Additional requirements are obstacle avoidance and specific shape characteristics, which furthermore depend on the robot's environment and on the type of task. Therefore, we postpone the detailed description of our Geometric Trajectory Planner implementation to the application case presented in Section 3.

2.3. Task Time Optimization

As already stated in the introduction, several methodologies are well described in the literature for the implementation of the Task Time Optimizer (TTO) as a constrained optimization process that accepts as inputs a geometric description of the trajectory and higher-order boundary conditions. The constraints are computed according to:

- the actual performances of the actuation drives and of the transmissions;
- the model of the mechanical dynamics of the manipulator;
- the properties of the payload.

The main output of the TTO is the motion along the geometric path that minimizes the travel time while remaining compatible with the performances of the robot. The minimal execution time can also be obtained as a further output of the TTO.

Let q be the parameters describing the robot pose in the joint space, and let the dynamics of the robot be expressed in the form:

$$M(q)\ddot{q} + C(q,\dot{q})\dot{q} + \nabla U(q) = \tau\,, \tag{11}$$

where M is the mass matrix, C is the Coriolis matrix, U is the potential energy, and τ are the torques exerted at the joints.

The task time optimization typically uses a robot's dynamics formulation that neglects the compliance of the transmission units and of the links; this simplification can be considered acceptable whenever these aspects have been appropriately considered during the design stage of the robot, as shown, e.g., in [30,31].

Using Lagrange equations, it can be straightforwardly shown that $C(q,\dot{q})$ is homogeneous with respect to $\dot{q}$:

$$C(q,\alpha\dot{q}) = \alpha C(q,\dot{q})\,. \tag{12}$$

Even though $\gamma(u)$ is typically formulated in the task space, the robot's inverse kinematic relationships can be applied to express the same motion in the joint space. The path-projected dynamic equations can be then expressed in the form:

$$m_i(u)\ddot{u} + c_i(u)\dot{u}^2 + g_i(u) = \tau_i\,, \tag{13}$$

where m_i, c_i, g_i are each the ith elements of the vectors:

$$\begin{aligned} \boldsymbol{m}(u) &= \boldsymbol{M}(\boldsymbol{q}(u))\boldsymbol{q}'(u) & (14)\\ \boldsymbol{c}(u) &= \boldsymbol{M}(\boldsymbol{q}(u))\boldsymbol{q}''(u) + \boldsymbol{C}(\boldsymbol{q}(u),\boldsymbol{q}'(u))\boldsymbol{q}'(u) & (15)\\ \boldsymbol{g}(u) &= \nabla U(\boldsymbol{q}(u))\,. & (16) \end{aligned}$$

Each power drive system (i.e., the driver-motor unit) actuating the robot is characterized by a maximum admissible torque $\tau_{pds,max,i}$, by a rated torque $\tau_{pds,rated,i}$, by a maximum velocity $\omega_{pds,max,i}$, and by an inertia $J_{m,i}$, all expressed at the motor shaft. Additionally, the transmission system is characterized by a reduction ratio $i_{t,i}$, by an efficiency $\eta_{t,i}$, and by an inertia $J_{t,i}$, typically expressed with respect to the input shaft. All these parameters are fundamental in determining the minimal execution time of a given trajectory; although they are considered as a given as far as the TTO is concerned, they must be carefully selected during the design phase of the manipulator. A procedure for the proper configuration of the power drive systems and transmissions is illustrated in [32].

The inertial properties of motor and transmission are easily projected on the output shaft and thus accounted for within $\boldsymbol{M}$.

Assuming a continuative use, the following constraints can be conservatively enforced:

$$\left| m_i(u)\ddot{u}(u) + c_i(u)\dot{u}^2(u) + g_i(u) \right| \leq \eta_{t,i} i_{t,i} \tau_{pds,rated,i}\,. \tag{17}$$

In this way, it can be ensured that the constraints on the RMS torque (and therefore also on the peak torque) are satisfied. The constraints on the maximum rotational speed of the actuators can moreover be written as:

$$-\frac{\omega_{pds,max,i}}{i_{t,i}} \leq q_i'(u)\dot{u} \leq \frac{\omega_{pds,max,i}}{i_{t,i}}\,. \tag{18}$$

As suggested in [20], the following change of variables is introduced: let

$$x(u) = \dot{u}^2(t(u))\,; \tag{19}$$

then:

$$x'(u) = 2\ddot{u}(t)\,. \tag{20}$$

Constraints equivalent to those of Inequality (18) can be written as:

$$q_i'^2(u)x(u) \leq \left(\frac{\omega_{pds,max,i}}{i_{t,i}}\right)^2 \tag{21}$$

while Inequality (17) can be rewritten as

$$\left| \frac{m_i(u)}{2}x'(u) + c_i(u)x(u) + g_i(u) \right| \leq \eta_{t,i} i_{t,i} \tau_{pds,rated,i}\,. \tag{22}$$

The optimal function $\dot{u}_{opt}(u) = \sqrt{x_{opt}(u)}$ is commonly intended as the one minimizing the total traveling or task time T_{task}, itself expressible as:

$$T_{task} = \int_{u_{lb}}^{u_{ub}} \frac{\mathrm{d}u}{\sqrt{x(u)}}\,. \tag{23}$$

The optimization problem described above, in which T_{task} is the cost function, needs to be discretized in order to be solved. The constraints are therefore enforced not on the entire interval of definition of the parameter u but only on a finite number of values u_j, with $j = 0, 1, \ldots, m-1$. To obtain a finite number of decision variables $\boldsymbol{x}$, the function $x(u)$ is assumed to be of the form:

$$x(u) = \boldsymbol{N}(u)^T\boldsymbol{x} \tag{24}$$

$$x'(u) = \boldsymbol{N}'(u)^T\boldsymbol{x}, \tag{25}$$

where $\boldsymbol{N}(u)$ is a suitable vector-valued shape function. The shape function can be isolated, leading to the reformulation of the jth constraint inequalities as:

$$\left(\frac{m_i(u_j)}{2}\boldsymbol{N}'(u_j) + c_i(u_j)\boldsymbol{N}(u_j)\right)^{\top}\boldsymbol{x} \leq \eta_{t,i} i_{t,i} \tau_{pds,rated,i} - g_i(u_j) \tag{26}$$

$$\left(\frac{m_i(u_j)}{2}\boldsymbol{N}'(u_j) + c_i(u_j)\boldsymbol{N}(u_j)\right)^{\top}\boldsymbol{x} \geq -\eta_{t,i} i_{t,i} \tau_{pds,rated,i} - g_i(u_j) \tag{27}$$

$$(q_i{}'(u_j))^2\boldsymbol{N}(u_j)^{\top}\boldsymbol{x} \leq \omega^2_{pds,max,i}\,, \tag{28}$$

which are linear with respect to the decision variables $\boldsymbol{x}$. The cost function T_{task} can be approximated as follows:

$$\widetilde{T}_{task} = \sum_{k=1}^{n} \frac{\Delta u}{\sqrt{\boldsymbol{N}(u_k)^{\top}\boldsymbol{x}}}, \tag{29}$$

leading to the optimization problem:

$$\widetilde{T}_{task,min} = \min_{x} \widetilde{T}_{task} \text{ under constraints (26)–(28)}\,. \tag{30}$$

As $\widetilde{T}_{task}$ is convex with respect to $\boldsymbol{x}$, the solution of the time reparametrization problem for a given trajectory can be found using appropriate convex optimization solvers. On the other hand, the structure of the problem and of the constraints suggests the setup of a different optimization problem, namely:

$$\Phi(\boldsymbol{x}) = [1, 1, \ldots, 1]\boldsymbol{x} \tag{31}$$

$$\Phi_{max} = \max_{x} \Phi(\boldsymbol{x}) \text{ under constraints (26)–(28)}\,. \tag{32}$$

Objective function $\Phi(\boldsymbol{x})$ is linear with respect to the decision variables; its optimization problem is therefore a linear program (LP). Maximizing $\Phi(\boldsymbol{x})$ results in the approximate maximization of the integral average of the squared velocity over the geometric path. While this does not lead to strictly minimal trajectory execution times, intuition suggests that the result should well approximate the optimal one. To confirm this expectation, the authors generated a set of random trajectories; each trajectory was optimized according to both the convex minimization problem (30) and the linear program (32), yielding, respectively, the trajectory execution times $T_{task,CM}$ and $T_{task,LP}$.

Figure 4 reports the comparison between the two methods. The histogram shows that the solution obtained from the linear programming approach is slightly worse than the one determined from the convex minimization. However, the gap between the two is typically less than 1%. Qualitative inspection of the obtained solutions also revealed that not only the task times but the overall motion profiles obtained with either method are very similar to each other.

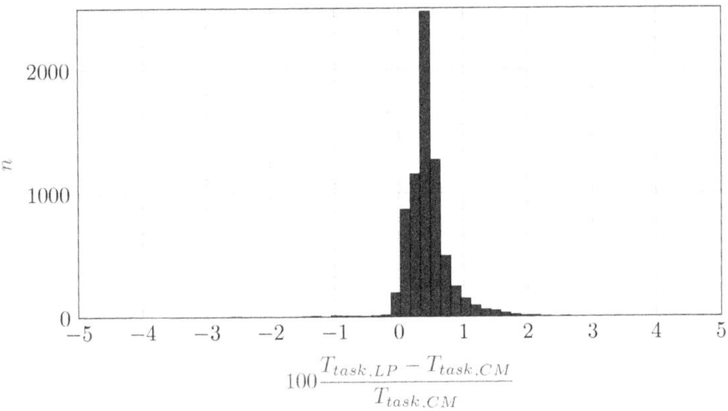

Figure 4. Statistical comparison between the trajectory execution times obtained through convex optimization and linear programming.

Given the greater computational efficiency of LP solvers, our final implementation of the TTO is reliant on the solution of Equation (32). Once a solution is available, the trajectory execution time T_{task} can still be determined from Equation (23). As shown in [23], once a first solution to the time optimal reparametrization has been found, it can be used to approximately compute additional constraints that are added to the problem described above in order to constrain not only the velocities and torques but also the time derivative of the torques. The resulting parametrization is then characterized by limited motion jerk.

2.4. Task Times Map Based on Neural Networks

The Geometric Trajectory Planner and the Task Time Optimizer can be jointly used to generate during an offline phase the data needed to construct the task time map, which itself will be used during the online phase by the task scheduler.

Consider a robot having an n_f-dimensional task space that performs motions that start and terminate with given positions and velocities. The procedure for the dataset generation is the following:

- sample the position and velocity boundary conditions $\boldsymbol{p}_s$, $\boldsymbol{p}_f$, $\boldsymbol{v}_s$, and $\boldsymbol{v}_f$;
- for each set of boundary conditions, generate a geometric path using the GTP;
- for each motion path, compute the optimal execution time T_{task} using the TTO and store explicitly the association between the inputs $(\boldsymbol{p}_s, \boldsymbol{p}_f, \boldsymbol{v}_s, \boldsymbol{v}_f)$ and T_{task}.

The dataset can then be used to train the task time model. Figure 5 represents the conceptual scheme of the training process that involves the data coming from the GTP, TTO, and boundary conditions; in this representation, the generation of the data and the training are simultaneous, even though in practice the two phases can be decoupled.

Once the functionality of the GTP is fixed, it is clear that the number of input variables it accepts increases with the number n_f of task space DOFs (degrees of freedom) of the manipulator.

Velocity boundary conditions that are a priori undefined can also occur in some applications, such as on-the-fly pick and place operations over one or multiple conveyor belts moving at slowly variable speeds, leading to a further increase in input variables. Considering, for example, a 6-DOF manipulator, a general trajectory is parametrized in terms of 24 position and velocity boundary conditions, to which a further 12 acceleration boundary conditions could be added if required by the application. Even considering the common special case of null terminal velocities, 12 position boundary conditions should be specified in order to completely define the motion path. On the other hand, lower DOF manipulators such as 3-DOF Cartesian robots or 4-DOF SCARA manipulators are also commonly encountered.

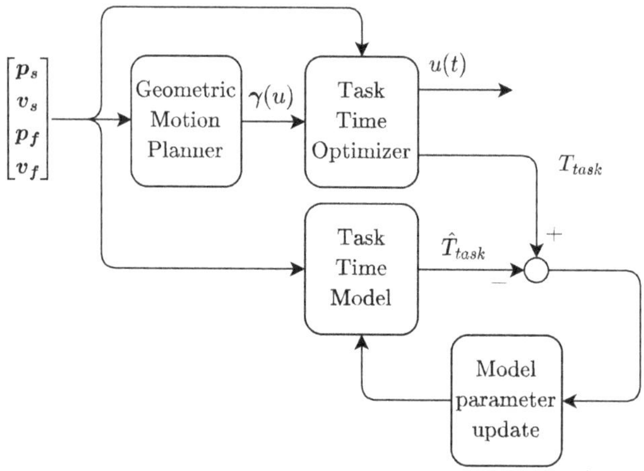

Figure 5. Conceptual diagram of the training process of the task time model; p_s, p_f, v_s, and v_f represent the position and velocity boundary conditions of the trajectory, while T_{task} and $\hat{T}_{task}$ denote each the actual task time of the trajectory and its estimation.

It can be concluded that the estimation architecture for the optimal task time map should be adaptable with minimal variations to the number of input parameters as well as powerful enough to deal with a potentially large input space.

These considerations tend already to rule out lookup tables; these could work well for inputs having lower dimensionality but quickly become unwieldy when high numbers of inputs are considered. Structured grids are also a poor match for manipulators having irregular working volumes. These problems can be mitigated but not completely solved using unstructured or adaptive lookup structures.

On the other hand, a number of mesh-free approximation methods exist, among which are radial basis function interpolation and Gaussian process regression; both these methods present attractive properties, such as being adaptable to irregular input spaces and having a low number of tuneable parameters. Their scalability to large datasets is however poor, as their training and evaluation time are each cubic and quadratic with respect to the number of data points. This might become a point of failure as the dimensions of the problem increase, since it can be expected that to achieve sufficient predictive capabilities, the amount of training data must increase.

This reasoning has led us to the use of a feedforward neural network as our approximation architecture, which has all the advantages of mesh-free methods and, compared to more traditional machine learning methods, can scale to handle larger datasets and problems having high dimensionality. In its most basic form, the neural network accepts the trajectory boundary conditions as inputs and has the task time estimate $\hat{T}_{task}$ as its single output. Auxiliary outputs could concern, e.g., the feasibility of the trajectory.

The architectures of the hidden layers and the training hyperparameters are, on the other hand, largely unconstrained and can be adapted to the specific dataset. Despite the internal differences, the training procedure remains largely the same; the neural network model is thus applicable with minimal variations to different robots performing various tasks.

In summary, with respect to the use of alternative methods, the following advantages can be expected from the use of a neural network:

- ability to sample the task space in an unstructured way;

- lower persistent memory usage (which is used not for the T_{task} samples directly but rather for the parameters of the neural network);
- more straightforward adaptability to different kinds of manipulators;
- tuneability of the internal architecture for achieving the desired performances;
- application-independent software interface and generation procedure.

3. Case Study Description

A meaningful case study for the proposed approach is constituted by a 4-DOF 5R manipulator designed for fast pick and place applications. Similarly to more commonly used 4-DOF SCARA manipulators, the 5R uses a screw-spline mechanism for the roto-translation of the end-effector along and around the vertical axis. As depicted in Figure 6, the in-plane motion of the 5R 's end-effector is obtained thanks to a planar mechanism that features two parallel kinematic chains actuated by two electric motors fixed to the frame of the robot.

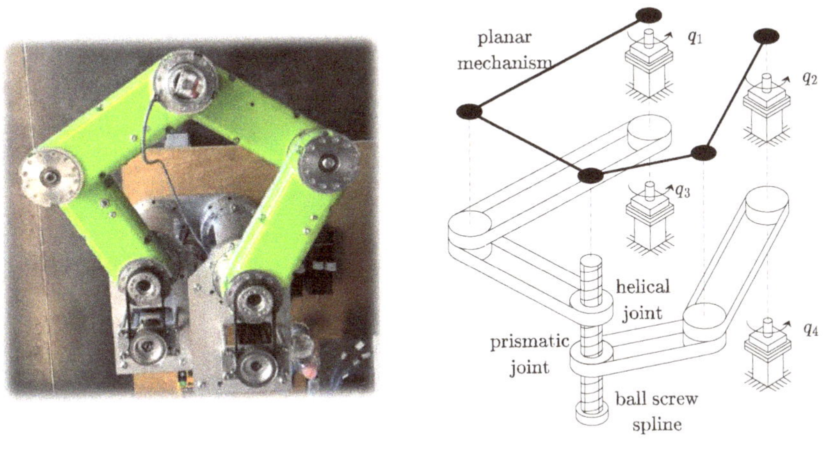

(a) The 5R robot. (b) Schematic representation of the 5R.

Figure 6. 5R manipulator at the University of Bergamo and schematic representation of its actuation and transmission systems.

In Figure 7, the notable points and angles needed to describe the planar linkage are shown. It should be noted that the actuated angles q_1 and q_2 are constrained to rotate within bounds enforced by mechanical endstops. In particular:

$$q_{1,min} = \ 60^\circ \leq q_1 \leq q_{1,max} = 210^\circ \tag{33}$$

$$q_{2,min} = -30^\circ \leq q_2 \leq q_{2,max} = 150^\circ\,. \tag{34}$$

Since the mechanical endstops are not symmetrical, the actual workspace will also be asymmetrical, despite the symmetry of the idealized kinematic scheme.

One peculiarity of this manipulator is constituted by the timing belt transmission systems housed inside the links, which allow to actuate the screw-spline using two additional brushless motors also attached to the robot's fixed frame. Interestingly, this mechanical arrangement leads not only to the reduction of the moving mass but also to the complete kinematic and dynamic decoupling between the motions of the screw-spline and of the planar arms.

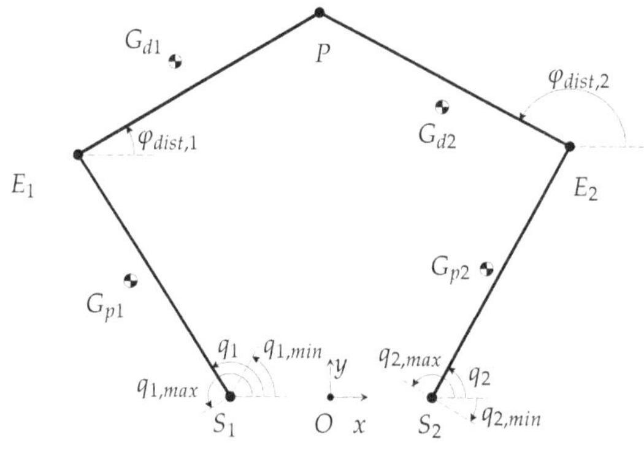

Figure 7. Notable points and angular quantities of the 5R planar linkage.

This is due to the unitary transmission ratio of the belt transmission systems and to the negligible friction torques of the ball bearings used to constrain the pulleys and the links. Under these conditions, it can be shown that the motion of the planar linkage and of the ball-screw-spline do not display a mutual influence; the former is exclusively determined by generalized coordinates q_1 and q_2 and motor torques τ_1 and τ_2, whereas the latter depends only on coordinates q_3 and q_4 and motor torques τ_3 and τ_4. Without introducing further assumptions, then, these two sub-systems can be analysed separately. In particular, the kinematics and dynamics of the screw-spline mechanism are invariant with respect to the joint configuration and are thus fully described by a constant Jacobian matrix. Let z_{ee} and ϕ_{ee} be the vertical position and the rotation of the end effector; let q_3 and q_4 be the joint coordinates associated, respectively, to the helical and to the prismatic joints constraining the screw-spline. The following kinematic relationships hold:

$$\begin{bmatrix} z_{ee} \\ \phi_{ee} \end{bmatrix} = \frac{1}{2\pi}\begin{bmatrix} -p_{ss} & p_{ss} \\ 0 & 2\pi \end{bmatrix} = \begin{bmatrix} q_3 \\ q_4 \end{bmatrix}. \tag{35}$$

The kinetic and potential energies of the screw-spline can therefore be expressed with respect to the joint variables as follows:

$$T_{ss} = \frac{1}{2}\left((J_3 + \frac{m_{ee}p_{ss}^2}{4\pi^2})\dot{q}_3^2 + (J_4 + \frac{m_{ee}p_{ss}^2}{4\pi^2} + J_{ee})\dot{q}_4^2 - \frac{m_{ee}p_{ss}^2}{2\pi^2}\dot{q}_3\dot{q}_4 \right) \tag{36}$$

$$U_{ss} = m_{ee}g\frac{p_{ss}}{2\pi}(q_4 - q_3). \tag{37}$$

The kinematics of the planar mechanism were instead obtained from the in-plane closure equation:

$$\overrightarrow{OS_1} + \overrightarrow{S_1E_1} + \overrightarrow{E_1P} = \overrightarrow{OS_2} + \overrightarrow{S_2E_2} + \overrightarrow{E_2P}. \tag{38}$$

The Jacobian matrices relating the velocities $\dot{E}_1$ and $\dot{E}_2$ to the joint velocities $\dot{q}_1$ and $\dot{q}_2$ can be trivially written as:

$$\boldsymbol{D}_{E,1} = l_{prox}\begin{bmatrix} -\sin(q_1) & 0 \\ \cos(q_1) & 0 \end{bmatrix} \tag{39}$$

$$\boldsymbol{D}_{E,2} = l_{prox}\begin{bmatrix} 0 & -\sin(q_2) \\ 0 & \cos(q_2) \end{bmatrix}. \tag{40}$$

The Jacobian analysis of the 5R end-effector kinematics is performed in such a way as to assign a clear geometric interpretation to the semi-Jacobian matrices, which relate the velocity at the joints to the velocity at the end-effector. These are:

$$\begin{bmatrix}(P-E_1)^\top \\ (P-E_2)^\top\end{bmatrix}\dot{P} = \begin{bmatrix}(P-E_1)^\top \boldsymbol{D}_{E,1} \\ (P-E_2)^\top \boldsymbol{D}_{E,2}\end{bmatrix}\begin{bmatrix}\dot{q}_1 \\ \dot{q}_2\end{bmatrix}. \tag{41}$$

From the analysis of these two matrices, it emerges that the singular configurations for the 5R are those for which two consecutive links are collinear. As will be shown, this characterization makes it straightforward to geometrically detect configurations that are approaching a singularity and to isolate a singularity-free portion of the workspace. A more in-depth discussion of the singularities for the five bar linkage can be found in [33].

Multiplying both sides of Equation (41) by the inverse of the matrix appearing on the left-hand-side, it is possible to highlight the Jacobian matrix relating the joint velocities q_1 and q_2 to the planar velocity at the end-effector:

$$\dot{P} = \boldsymbol{D}_P\begin{bmatrix}\dot{q}_1 \\ \dot{q}_2\end{bmatrix}. \tag{42}$$

The distal angles $\varphi_{dist,1}$ and $\varphi_{dist,2}$ can be expressed as:

$$\varphi_{dist,j} = \arctan\left(\frac{y_P - y_{E_j}}{x_P - x_{E_j}}\right). \tag{43}$$

Since the Jacobian matrices $\boldsymbol{D}_P$, $\boldsymbol{D}_{E,1}$, and $\boldsymbol{D}_{E,2}$ have already been made explicit, it is straightforward to find the Jacobian matrices $\boldsymbol{D}_{\varphi,dist,j}$ relating $\dot{\varphi}_{dist,j}$ to the joint velocities $\dot{q}_1$ and $\dot{q}_2$. Similarly, the Jacobian matrices of the centers of mass G_{p1}, G_{p2}, G_{d1}, G_{d2} of each link can be easily deduced from $\boldsymbol{D}_{E,1}$, $\boldsymbol{D}_{E,2}$, $\boldsymbol{D}_{\varphi,dist,1}$, $\boldsymbol{D}_{\varphi,dist,2}$, and $\boldsymbol{D}_P$.

Assuming that the planar mechanism operates in the horizontal plane, the potential energy U_{5R} of the 5R is constant; the kinetic energy of the planar system, on the other hand, can be written as:

$$\begin{aligned} T_{5R} = \frac{1}{2}\begin{bmatrix}\dot{q}_1 & \dot{q}_2\end{bmatrix}\Big(& m_{prox,1}\boldsymbol{D}_{G,p1}^\top \boldsymbol{D}_{G,p1} + \\ & m_{prox,2}\boldsymbol{D}_{G,p2}^\top \boldsymbol{D}_{G,p2} + \\ & m_{dist,1}\boldsymbol{D}_{G,d1}^\top \boldsymbol{D}_{G,d1} + \\ & m_{dist,2}\boldsymbol{D}_{G,d2}^\top \boldsymbol{D}_{G,d2} + \\ & J_{dist,1}\boldsymbol{D}_{\varphi,dist,1}^\top \boldsymbol{D}_{\varphi,dist,1} + \\ & J_{dist,2}\boldsymbol{D}_{\varphi,dist,2}^\top \boldsymbol{D}_{\varphi,dist,2} + \boldsymbol{J}_{prox}\Big)\begin{bmatrix}\dot{q}_1 \\ \dot{q}_2\end{bmatrix}, \end{aligned} \tag{44}$$

with:

$$\boldsymbol{J}_{prox} = \begin{bmatrix}J_{prox,1} & 0 \\ 0 & J_{prox,2}\end{bmatrix}. \tag{45}$$

The Lagrangian function of the entire system is:

$$\mathcal{L}(\boldsymbol{q}, \dot{\boldsymbol{q}}) = T_{5R} + T_{ss} - U_{5R} - U_{ss}. \tag{46}$$

Consequently, the dynamics of the system were obtained using Lagrange equations:

$$\frac{\mathrm{d}}{\mathrm{d}t}\frac{\partial \mathcal{L}}{\partial \dot{q}_k} - \frac{\partial \mathcal{L}}{\partial q_k} = \tau_k \tag{47}$$

and afterwards were rearranged to obtain relationships formally identical to Equation (11). A more detailed discussion of the dynamics of the 5R robot can be found in [34].

The selection of the transmission and of the power drive systems was performed according to the procedure detailed in [32]. All the parameters related to the actuation and transmission systems are thus fully defined and reported in Table 1. Accordingly, the contributions associated with the motors and with the transmissions are included as appropriate within the overall dynamics of the system.

Table 1. Main parameters of the four servoaxes.

	Servoaxis 1	**Servoaxis 2**	**Servoaxis 3**	**Servoaxis 4**
$\tau_{pds,rated}$, [N m]	0.7	0.7	0.36	0.36
$\tau_{pds,max}$, [N m]	1.4	1.4	0.72	0.72
$\omega_{pds,max}$, [rad s^{-1}]	500	500	500	500
J_m, [kg m^2]	1.7×10^{-5}	1.7×10^{-5}	2.4×10^{-6}	2.4×10^{-6}
η_t	~ 1	~ 1	~ 1	~ 1
J_t, [kg m^2]	2.485×10^{-5}	2.485×10^{-5}	2.05×10^{-5}	2.05×10^{-5}
i_t	64	64	10	10

During the motion, the manipulator should not cross any singular configuration, change its assembly, or violate the physical constraints introduced by the mechanical endstops. The actual workspace is consequently reduced to the configurations that satisfy the following conditions:

$$q_1 \geq q_{1,min} \tag{48}$$
$$q_1 \leq q_{1,max} \tag{49}$$
$$q_2 \geq q_{2,min} \tag{50}$$
$$q_2 \leq q_{2,max} \tag{51}$$
$$\overrightarrow{E_1P} \cdot \boldsymbol{j}(\overrightarrow{S_1E_1}) \leq 0 \tag{52}$$
$$\overrightarrow{E_2P} \cdot \boldsymbol{j}(\overrightarrow{S_2E_2}) \geq 0 \tag{53}$$
$$\overrightarrow{E_{avg}P} \cdot \boldsymbol{j}(\overrightarrow{E_1E_2}) \geq 0, \tag{54}$$

with

$$E_{avg} = \frac{1}{2}(E_1 + E_2) \tag{55}$$
$$\boldsymbol{i}(\overrightarrow{AB}) = \overrightarrow{AB}/||\overrightarrow{AB}|| \tag{56}$$
$$\boldsymbol{j}(\overrightarrow{AB}) = \left[-i_2(\overrightarrow{AB}),\ i_1(\overrightarrow{AB})\right]^\top. \tag{57}$$

In particular, the enforcement of inequalities (48)–(51) and conditions (52)–(54) (which were determined from the analysis of Equation (41)) prevents the crossing of singular configurations and the change of assembly or operating mode. In Figure 8, the useful workspace resulting from inequalities (48)–(54) is shown, both in the taskspace and in the jointspace. The boundaries of the filled area either are determined by the mechanical endstops or correspond to singularities. Inside the admissible area, the reciprocal condition number:

$$r_\sigma = \sqrt{\frac{\sigma_{min}}{\sigma_{max}}}, \tag{58}$$

with σ_{min} and σ_{max} being the minimum and maximum singular values of the matrix $\boldsymbol{D}_P\boldsymbol{D}_P^\top$, is represented. A further condition fixing a minimum value for the reciprocal condition number can be defined in order to stay sufficiently clear of singular configurations:

$$r_\sigma \geq \delta_\sigma\,. \tag{59}$$

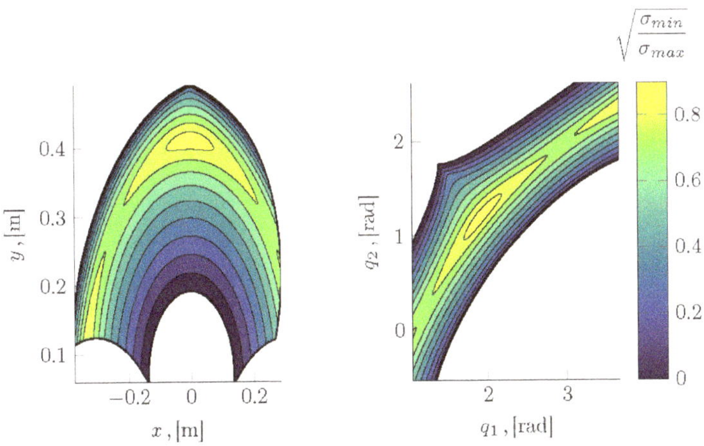

(**a**) Taskspace representation (**b**) Jointspace representation

Figure 8. Reciprocal condition number represented in the admissible portions of the workspace.

For the first numerical example detailed below, the parameter δ_σ has been set to 0.1; for the second application, instead, we selected $\delta_\sigma = 0.2$.

The main properties of the robot are:

- length of the proximal links $l_{prox} = 250\,\mathrm{mm}$;
- length of the distal links $l_{dist} = 250\,\mathrm{mm}$;
- frame length $l_f = 180\,\mathrm{mm}$;
- mass of the proximal links $m_{prox,1} = m_{prox,2} = 2.9\,\mathrm{kg}$;
- mass of the distal links $m_{dist,1} = m_{dist,2} = 2.9\,\mathrm{kg}$;
- barycentric inertia of the proximal links $J_{prox,1} = J_{prox,2} = 5.22 \times 10^{-2}\,\mathrm{kg\,m^2}$;
- barycentric inertia of the distal links $J_{dist,1} = J_{dist,2} = 5.22 \times 10^{-2}\,\mathrm{kg\,m^2}$;
- mass of the screw-spline and of the end-effector $m_{ee} = 0.36\,\mathrm{kg}$;
- screw-spline pitch $p_{ss} = 2\,\mathrm{mm}$;
- rotational inertia of the end-effector $J_{ee} = 6.40 \times 10^{-6}\,\mathrm{kg\,m^2}$;
- rotational inertia of the transmission system actuating the screw-spline helical joint $J_3 = 1.20 \times 10^{-6}\,\mathrm{kg\,m^2}$;
- rotational inertia of the transmission system actuating the screw-spline prismatic joint $J_4 = 1.20 \times 10^{-6}\,\mathrm{kg\,m^2}$.

The mass parameters of the links also take into account the masses of the pulleys and belts.

Geometric Planning for the 5R Robot

The GTP here proposed is configured for the parametrization of two different types of motions:

- stationary pick and place tasks, where the initial and final velocities are null;
- on-the-fly pick and place tasks, such as those represented in Figure 1, where the initial and final velocities are non-null and aligned to the x-axis.

For stationary pick and place motions, a simple portal-like trajectory is sufficient. On the contrary, for on-the-fly manipulation tasks, more complex composite trajectories, such as the one represented qualitatively in Figure 9, are needed.

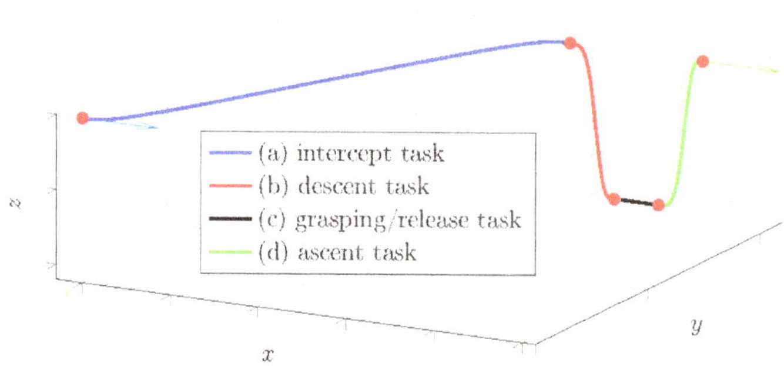

Figure 9. Examples of a composite trajectory for on-the-fly pick and place.

In that figure, it can be seen that the on-the-fly manipulation trajectory is constituted by four portions:

- an intercept motion (a), which happens in a slightly elevated plane and which terminates with an x-y position, a rotation and a velocity that match those of the pick and place target; its terminal position can be found, as will be shown, by solving the optimal intercept problem;
- a descent motion (b) through which the gripper approaches the target from above;
- a grasping/release motion (c) during which the gripping device operates to grasp or release the item;
- an ascent motion (d) through which the gripper attains the proper vertical clearance.

The motions (b) (c) and (d) constitute the tracking phase proper, since all have a velocity along the x-axis that matches the one of the pick and place target. The duration of motion (c) is determined by the operating time of the gripping device and therefore is not a target for optimization. The motions (b) and (d) are usually quite short, as the vertical excursion is typically very limited in pick and place applications. For the 5R robot considered in the example, moreover, the optimal time needed to perform a given vertical motion is a function only of the vertical displacement itself, due to the screw-spline being a linear time-invariant system decoupled from the planar five bar mechanism. Once the vertical displacement is fixed, the task times of (b) and (d) are known constants. In particular, considering a vertical displacement of the end-effector $\Delta z_{ee} = 20$ mm, we obtained the following task times:

$$T_{task(b)} = 0.076\,82\,\text{s} \tag{60}$$

$$T_{task(d)} = 0.076\,82\,\text{s}\,. \tag{61}$$

The fact that the $T_{task(b)}$ is equal to $T_{task(d)}$ is not surprising, given the symmetry of the problem, in which gravity:

- assists the acceleration phase of the descent motion (b);
- opposes the deceleration phase of the descent motion (b);
- opposes the acceleration phase of the ascent motion (d);
- assists the deceleration phase of the ascent motion (d).

In cases featuring different robots, it is entirely possible to construct time maps for motions (b) and (d) using the general approach here proposed. It should be remarked that, in general, the execution times of tasks (b) and (d) are a function only of the starting position and velocity; the dimensionality of the problem is therefore lower compared to the one associated with motion (a). Finally, it can be noted that the map associated with motion (a) needs to be evaluated more frequently for the solution of the optimal intercept problem, whereas maps for motions (b) and (d) are not invoked for this purpose. It can be easily observed in Figure 8 that the useful workspace is not convex, neither in the task space nor in the joint space. The geometric planning, even for simple point-to-point trajectories, is therefore not entirely straightforward. A variety of solutions might be adopted; the one here proposed features minimum-length trajectories planned in the task space.

This choice is motivated by the fact that, already, for $\delta_\sigma = 0.1$, the useful workspace is a normal domain, clearly having two function-like lower and upper boundaries, which are denoted in the following as $y_{lb}(x_{ee})$, $y_{ub}(x_{ee})$ and which can be numerically evaluated, e.g., using a zero-finding algorithm on the boolean function obtained from the simultaneous enforcement of Inequalities (48)–(54) and (59). This, in turn, allowed us to set up the geometric planning of a point-to-point minimum length motion as a convex quadratic program (CQP) that yields a control polyline, which can be easily postprocessed in order to fully define the trajectory.

Consider at first only the planar motion. Fixing the feasible initial and final positions $\boldsymbol{p}_s$ and $\boldsymbol{p}_f$ of the manipulator, the coordinates $x_{ee,s}$, $y_{ee,s}$, $x_{ee,f}$, and $y_{ee,f}$ are fixed. Let $\boldsymbol{x}_{sol}$ and $\boldsymbol{x}_{constr}$ be equally spaced arrays of n_{cqp} and m_{cqp} elements each:

$$\boldsymbol{x}_{sol} = \left[x_{sol,1} = x_{ee,s}, \quad x_{sol,2}, \quad \dots, \quad x_{sol,n_{cqp}} = x_{ee,f}\right] \tag{62}$$

$$\boldsymbol{x}_{constr} = \left[x_{constr,1} = x_{ee,s}, \quad x_{constr,2}, \quad \dots, \quad x_{constr,m_{cqp}} = x_{ee,f}\right]. \tag{63}$$

Let $\boldsymbol{y} = \left[y_{ee,1}, \quad y_{ee,2}, \quad \dots, \quad y_{ee,n_{cqp}}\right]$ be n_{cqp} decision variables.

Finally, let $\boldsymbol{N}_{\boldsymbol{x}_{sol}}(x_{ee})$ be the shape functions yielding the linear interpolation of $\boldsymbol{y}$ over $\boldsymbol{x}_{sol}$. The constraints of the problem can then be written in linear form as:

$$y_{ee,1} = y_{ee,s} \tag{64}$$

$$y_{ee,n_{cqp}} = y_{ee,f} \tag{65}$$

$$\boldsymbol{N}_{\boldsymbol{x}_{sol}}^{\top}(x_{constr,k})\boldsymbol{y} \leq y_{ub}(x_{constr,k}) \tag{66}$$

$$\boldsymbol{N}_{\boldsymbol{x}_{sol}}^{\top}(x_{constr,k})\boldsymbol{y} \geq y_{lb}(x_{constr,k}) \tag{67}$$

for $k = 1, \dots, m_{cqp}$. Finally, as the quadratic cost function (which has the same minimum point of the squared-length function):

$$L^2 = \sum_{i=2}^{n_{cqp}} (y_{ee,i} - y_{ee,i-1})^2 \tag{68}$$

admits a positive definite Hessian matrix, the minimum length polyline joining the initial and final configurations can be found as the result of a convex quadratic program.

The minimum length polyline so obtained constitutes the basic solution from which a variety of motions can be easily constructed, as exemplified in Figure 10. For example, it is entirely straightforward to add to each point a desired out-of-plane displacement, e.g., to obtain a motion suitable for stationary pick and place actions; the rotational coordinate of the end effector can also be added as dictated by the requisites for the specific application. It is then possible to join two consecutive line segments with circular arcs in order to obtain a C^1 continuous path whose arc parameter is easily computed. Furthermore, a smoothing operation can be applied to yield a C^2 path; given the fact that the geometric primitives we used are integrable in closed form, the filter we developed is constituted by an analytical integral average over a window, which slides along the arc parameter of the curve and whose size is calculated in relation to the length of each constitutive segment or arc. An analogous idea can be found in [28], in which, however, a discrete time FIR filter is used to obtain continuous geometric accelerations.

In addition, the changes of directions needed for on-the-fly pick and place operations can be prepended or appended to the trajectory as required to obtain the proper alignment with the tracking direction. These trajectories, in particular, are constructed assuming that the items are moving along one or more conveyor belts aligned to the x_{ee}-axis and such that the end-effector moves in plane in order to reach the position above the target with the velocity needed to initiate the subsequent descent, grasping/release, and ascent motions.

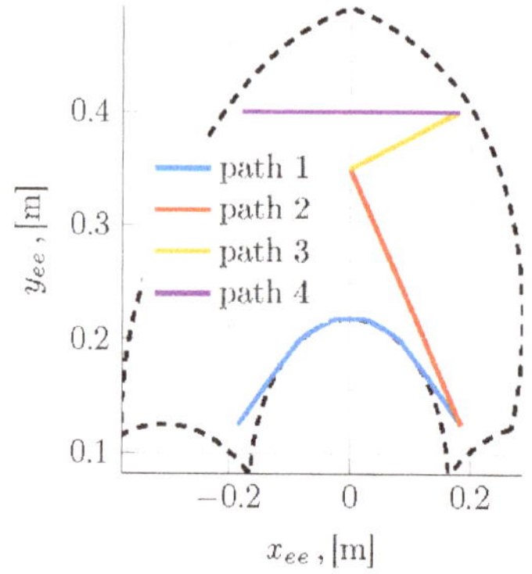

(**a**) Point-to-point paths.

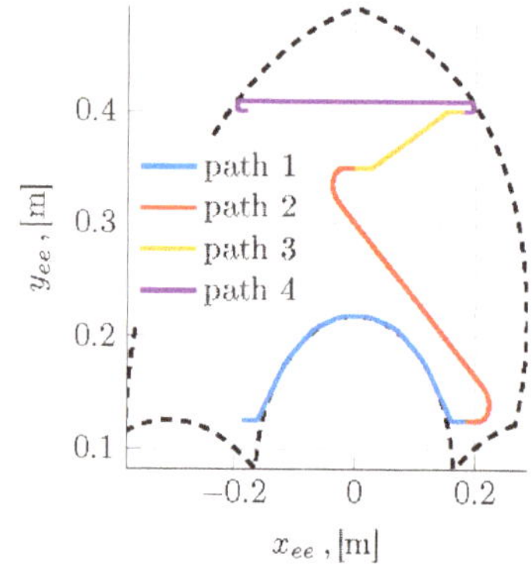

(**b**) Tracking paths.

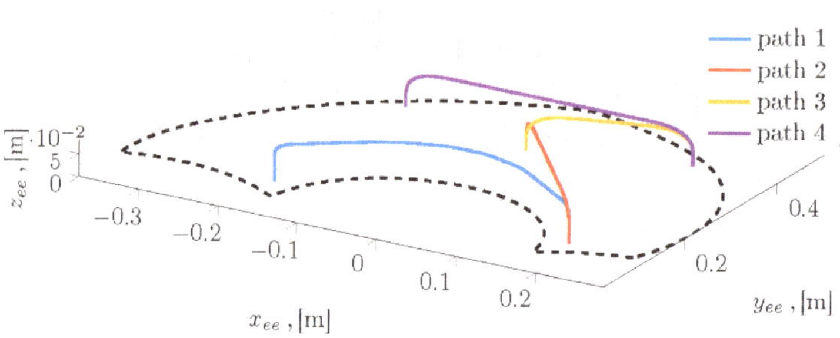

(**c**) Static pick and place paths.

Figure 10. Examples of point-to-point paths (**a**), tracking paths (**b**), and paths for stationary pick and place tasks (**c**).

Since both stationary and on-the-fly pick and place tasks have been considered at the geometric level, two families of parametric tasks are to be optimized in terms of execution times. Of these families, the former is characterized by null initial velocities, while the velocity boundary conditions of the latter are free parameters appearing alongside the initial and final positions. The task time models for these two kinds of motions accept therefore a different number of input parameters, and, as detailed in the following discussion, they are characterized by different internal architectures. Despite these differences, they offer a very similar query interface and are trained and deployed in an almost identical manner, showcasing the adaptability of the proposed method to different task typologies.

4. Computational Results and Discussion

4.1. Software and Hardware Setup

The task time datasets were generated using an in-house object-oriented C++ codebase. The main modules of our software implement:

- an abstract class representing a generic n-DOF manipulator, from which specific implementations (among which the one relative to the 5R robot) are derived;
- an abstract class representing a generic TTO, from which specific implementations are derived according to the different optimization methods;
- an abstract class representing a generic n-DOF geometric path, with its derived implementations;
- a set of routines implementing the GTP for the 5R robot;
- a utility class dedicated to the multithreaded generation of the dataset.

To solve the linear program (32), we use the Gnu Linear Programming Kit (GLPK); we treat the quadratic optimization problem arising during the geometric trajectory planning using the OSQP solver [35]. To accurately compute the task time T_{task} according to Equation (23), we implemented the algorithm proposed by [36]. This method was selected due to its ability to treat integrable singularities at finite points, which, in our case, occur when the motion has null velocities at the extremities; in those cases, we allowed small initial and final accelerations to mathematically ensure integrability.

Each model was trained on a dataset of 10^6 trajectories, whose initial and final poses were sampled according to a uniform random distribution in the task space, considering:

- a rotation range for the end-effector equal to $\Delta\phi_{ee} \in [-\pi, \pi]$;
- a range for the initial and final velocities equal to $v \in [0.05\,\mathrm{m\,s^{-1}}, 0.5\,\mathrm{m\,s^{-1}}]$;
- initial and final x-y positions that satisfy inequalities (48)–(54) and (59).

The neural network architectures were defined and trained using the PyTorch library. The main hyperparameters characterizing the training process are:

- loss function: MSE (Mean Squared Error)
- training algorithm: SGD (Stochastic Gradient Descent)
- batch size: 2048 data points.

The determination of the hyperparameters has not been automated; instead, they have been experimentally tuned in order to obtain a compromise between training speed and accuracy.

For both models, the accuracy and precision of the prediction are evaluated through the distributions of the prediction error and of the relative prediction error. The prediction error is defined as:

$$\varepsilon_{T_{task}} = T_{task} - \hat{T}_{task}, \tag{69}$$

where $\hat{T}_{task}$ is the task time estimate obtained from the trained model. On the other hand, the relative prediction error is defined as:

$$\varepsilon_{T_{task},\%} = 100\frac{T_{task} - \hat{T}_{task}}{T_{task}}. \tag{70}$$

Both error distributions were computed on the test dataset.

Three additional quantities that are considered to evaluate the soundness of the proposed methods are:

- $\overline{T}_{plan}$, the average execution time of the entire motion planning and optimization pipeline for a single trajectory;
- $\overline{T}_{eval,cpu}$, the average evaluation time of the task time map on a single CPU core;
- $\overline{T}_{eval,gpu}$, the average evaluation time of the task time map on the GPU.

These values were obtained by performing all the required computations on a machine running a Linux operating system and having the specifications detailed below:

- CPU: Intel® Core™(Santa Clara, CA, USA) i7-8750H processor 2.2 GHz (9 MB cache, up to 4.1 GHz, 6 processors);
- GPU: Nvidia® (Santa Clara, CA, USA) GeForce® (Santa Clara, CA, USA) GTX 1050, 4 GB, GDDR5;
- RAM: 16 GB DDR4 SO-DIMM, 2400 MHz.

Clearly, the usefulness of the task time map is predicated not only on the achievement of sufficient accuracy but also on evaluation times (either $\overline{T}_{eval,cpu}$ or $\overline{T}_{eval,gpu}$) much lower than the planning time $\overline{T}_{plan}$. Within the scope of this work, the average evaluation time of the models is conservatively determined through the execution of the model itself within the Python3 interpreter, whereas $\overline{T}_{plan}$ is determined using the C++ implementations of the GTP and TTO. These quantities are summarized for both the stationary and the on-the-fly pick and place case studies in Table 2. A histogram of the planning times for a representative sample of trajectories is reported in Figure 11.

Table 2. Comparison of the evaluation times of the neural network models and of the GTP-TTO algorithms.

Task Type	$\overline{T}_{eval,cpu}$, [µs]	$\overline{T}_{eval,gpu}$, [µs]	$\overline{T}_{plan}$, [µs]	$\overline{T}_{plan}/\overline{T}_{eval,cpu}$	$\overline{T}_{plan}/\overline{T}_{eval,gpu}$
Stationary Pick and Place	81	6	49,419	610.11	8236.5
On-the-fly Pick and Place	2310	73	56,985	24.7	780.6

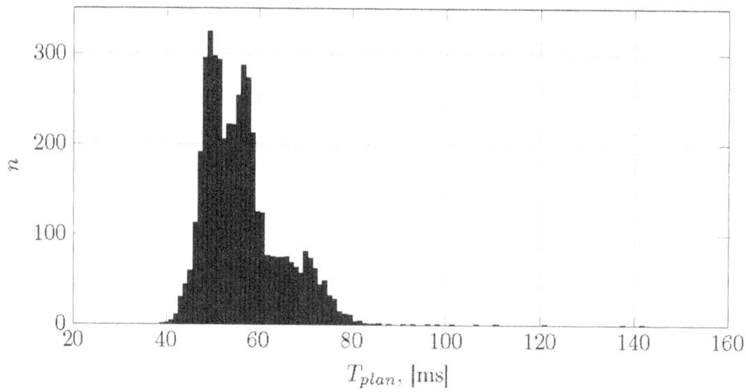

Figure 11. Histogram of measured planning times.

4.2. *Stationary Pick and Place Task Time Map*

A feedforward neural network having three hidden layers was selected; the main parameters of the network are summarized in Table 3.

Table 3. Main parameters of the neural network architecture for the stationary pick and place task time map.

	Layer 1	Layer 2	Layer 3
Type	Fully connected	Fully connected	Fully connected
Activation	ReLU	ReLU	Linear
Dropout probability	0.20	0.20	0.20
n_{in}	5	2048	2048
n_{out}	2048	2048	1

In Figure 12, the top panel shows the histogram of the prediction error $\varepsilon_{T_{task}}$, while the bottom one depicts the histogram of the relative prediction error $\varepsilon_{T_{task},\%}$. These errors are small both in absolute and relative terms; therefore, the map output cannot reliably discern two tasks whose duration difference is less than the error introduced. From the time duration point of view, these tasks are practically equivalent due to the limited percentual error. Regarding the effective behavior of the robot, we can discuss two main scenarios.

In the first, the command to the robot uses the boundary conditions and the map task time $\hat{T}_{task}$. Therefore, the robot executes the task in the given time. If $\hat{T}_{task} > T_{task}$, the task is executed with speeds and accelerations lower than those generated by the TTO. On the contrary, if $\hat{T}_{task} < T_{task}$, the task will be executed with higher speeds and accelerations, provided that the robot can reach these. In general, the TTO should optimize the task execution times considering conservative estimates of the robot's capabilities so that the small adjustments described here remain compatible with the actual peak performances of the manipulator.

In the second, the command to the robot considers only the boundary conditions. In this case, the task execution time is T_{task} due to the robot's TTO, and the robot can realize the task. The subsequent task can be executed after the completion of the previous one.

To verify that the use of the resulting model is compatible with the time requirements of an online scheduling algorithm, we evaluated the average time needed on the one hand to evaluate the neural network and on the other to execute the GTP and the TTO processes. The average evaluation times are:

- $\overline{T}_{eval,cpu} = 81.38\,\mu s$ on a single CPU core, using the Python-3 interpreter;
- $\overline{T}_{eval,gpu} = 5.96\,\mu s$ on the GPU, using the Python-3 interpreter;
- $\overline{T}_{plan} = 49.4\,ms$ on the CPU, using the C++ implementation of the GTP and TTO.

Even without the GPU, then, it can be observed that the evaluation of the model to yield $\hat{T}_{task}$ is two orders of magnitude faster than the direct evaluation of the task time T_{task} using the GTP and the TTO; further and significant speedups can moreover be obtained using the GPU.

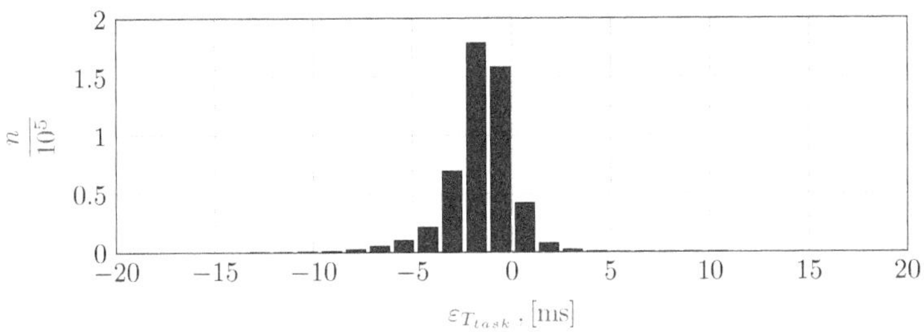

(a) Prediction error

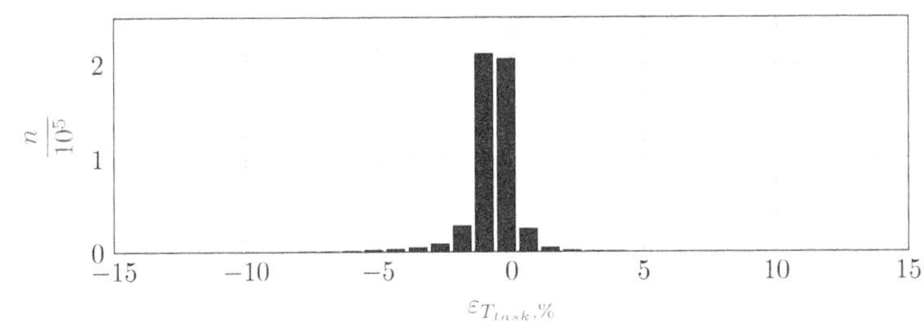

(b) Relative prediction error

Figure 12. Stationary pick and place task time prediction errors calculated on the test dataset.

A qualitative inspection of the predicted task times $\hat{T}_{task}$ was also performed as shown in Figures 13–16. In particular, by fixing the initial pose and the total rotation of the end-effector, it is possible to visualize the task time estimate as a function of the final Cartesian coordinates of the gripper. The initial configuration is shown in the figures with a red marker. It can be seen that $\hat{T}_{task}$ is positively correlated with the total displacement achieved at the end-effector. The isolines of $\hat{T}_{task}$ are not, however, merely radial, as they are a more general function not only of the non-linear kinematics and dynamics of the mechanism but also of the actuation systems, whose main properties and non-linearities (namely the torque and velocity saturations) are accounted for within the overall motion planning strategy and are thus reflected in the outputs of the task time model.

In this example, the rotational displacement has a rather small effect on the total execution time of the trajectory. It can be seen, however, that for small displacements, the execution of the end-effector rotation can constitute the bottleneck of the overall motion. This fact becomes apparent only when closely focusing on the size of the region enclosed in the isoline closest to the initial configuration considered at different magnitudes of the rotational displacement.

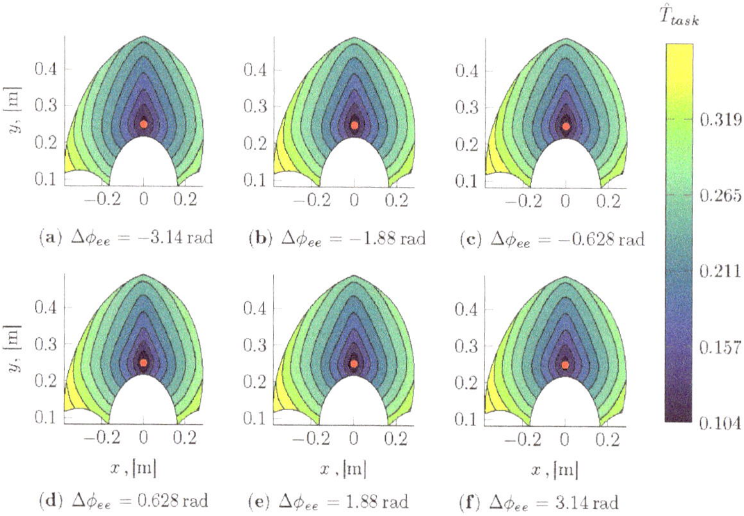

Figure 13. Stationary pick and place task time model evaluated at different end-effector rotational displacements $\Delta\phi_{ee}$; the initial configuration of the robot, shown as a red marker, has been fixed at $x_{ee} = 0.0$ m, $y_{ee} = 0.25$ m.

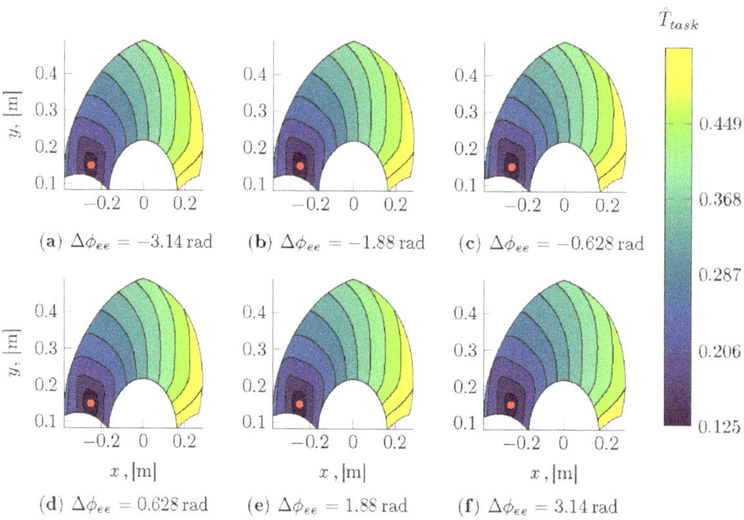

Figure 14. Stationary pick and place task time model evaluated at different end-effector rotational displacements $\Delta\phi_{ee}$; the initial configuration of the robot, shown as a red marker, has been fixed at $x_{ee} = -0.25$ m, $y_{ee} = 0.15$ m.

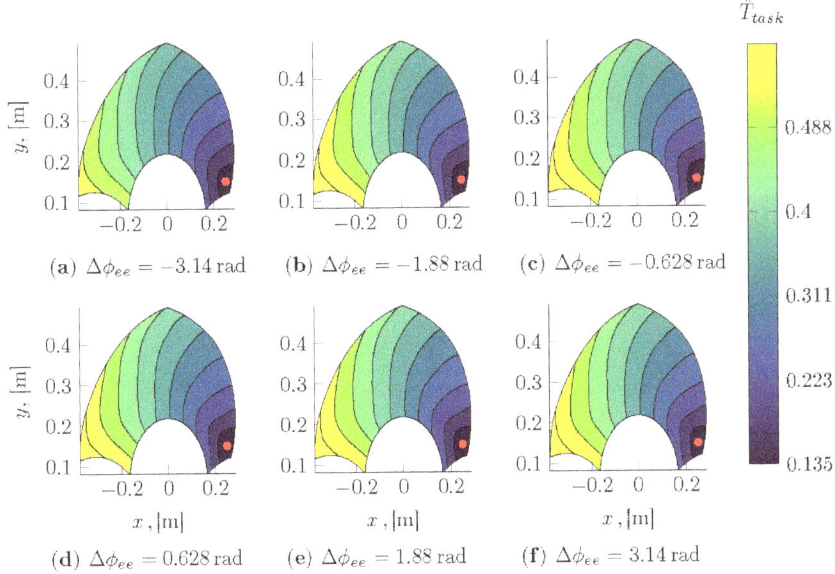

Figure 15. Stationary pick and place task time model evaluated at different end-effector rotational displacements $\Delta\phi_{ee}$; the initial configuration of the robot, shown as a red marker, has been fixed at $x_{ee} = 0.25$ m, $y_{ee} = 0.15$ m.

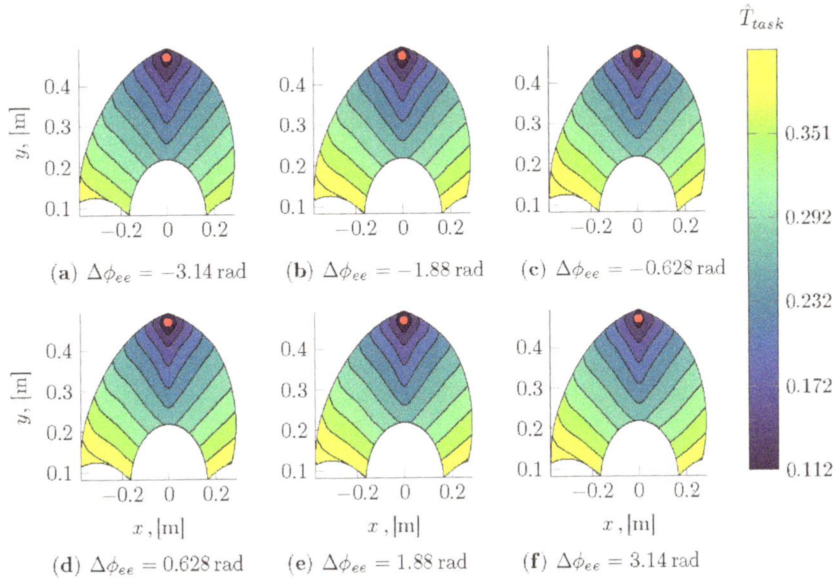

Figure 16. Stationary pick and place task time model evaluated at different end-effector rotational displacements $\Delta\phi_{ee}$; the initial configuration of the robot, shown as a red marker, has been fixed at $x_{ee} = 0.0$ m, $y_{ee} = 0.47$ m.

The resulting task time map covers the entire useful workspace of the robot; during the design phase, it enables therefore greater freedom in the definition of the layout of the elements surrounding the robot; it also provides insight into the more convenient locations in which to position, e.g., the transportation systems.

4.3. On-the-Fly Pick and Place Task Time Map

The task time map of on-the-fly pick and place operations naturally lends itself to the determination of the optimal intercept position for pick and place targets moving at a known speed profile. The need for this often arises when considering a system in which the items are picked and placed on continuously moving conveyors.

This type of planning has been illustrated qualitatively in Figure 17, which represents graphically Equation (10). In Figure 17a, the problem has been represented in the x-y plane in which the conveyor, depicted in green, is located; in the figure, the robot's workspace (represented as a grey area) can be seen, along with the initial positions of the manipulator's end-effector (red marker) and of the target (blue marker). As the target moves along the dotted blue line, the intercept will necessarily happen on a point belonging to it, with a final end-effector orientation and velocity (not depicted for simplicity) also dictated by the configuration of the target. The determination of the optimal intercept point is performed as shown in Figure 17b, where the time-displacement functions of the target and of the robot are displayed. The target function is a line, since its velocity is constant; on the other hand, the robot's function can be obtained from the evaluation of the task time map. The optimal intercept position is at the intersection of these two functions, as the operation can happen as fast as possible and without having to wait for the target. When an intersection is not present, it might mean that the robot cannot operate fast enough in order to reach the target (robot time function entirely above the target time function) or, conversely, that the robot necessarily has to wait for the target to enter the workspace (robot time function entirely below the target time function). In the first case, the operation is not feasible, while in the second, the intercept position that minimizes idle time is at the boundary of the workspace itself.

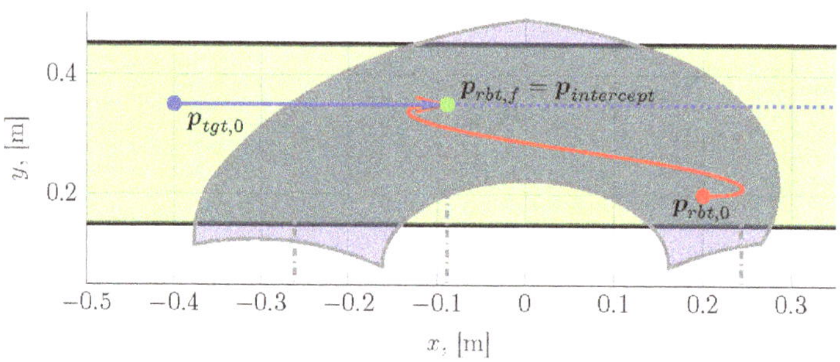

(**a**) Schematic representation of the optimal intercept problem seen in the x-y plane.

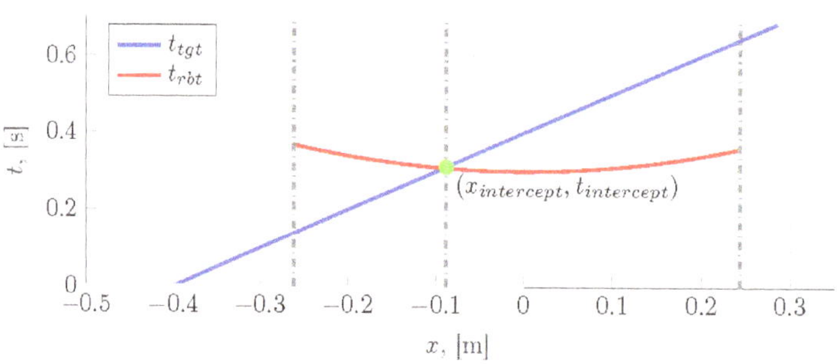

(**b**) Schematic representation of the optimal intercept problem seen in the x-t plane.

Figure 17. Graphical representation of an optimal intercept position computation.

A feedforward neural network having four hidden layers was selected; the main parameters of the network are summarized in Table 4.

Table 4. Main parameters of the neural network architecture used for the on-the-fly pick and place neural network model.

	Layer 1	Layer 2	Layer 3	Layer 4
Type	Fully connected	Fully connected	Fully connected	Fully connected
Activation	ReLU	ReLU	ReLU	Linear
Dropout probability	0.05	0.05	0.05	0.05
n_{in}	5	8192	8192	8129
n_{out}	8192	8192	8192	1

The minimum reciprocal condition number of the initial and final positions has been set to $\delta_\sigma = 0.2$ so that some additional space for the turning manoeuvers is always available.

The evaluation times of the on-the-fly pick and place model are:

- $\overline{T}_{eval,cpu} = 2.31$ ms on a single CPU core;
- $\overline{T}_{eval,gpu} = 73.42$ µs on the GPU;
- $\overline{T}_{plan} = 56.9$ ms on the CPU, using the C++ implementation of the GTP and TTO.

Figures 18–21 show the task time maps at selected initial positions and velocity boundary conditions. As expected, then, in these figures, it is clearly shown that motions with a monotonically increasing x_{ee} coordinate can be executed in a shorter amount of time. On the contrary, the turning manoeuvers required for the robot to move towards decreasing x_{ee} coordinates require a significant amount of time, thus making the task time function quite dissimilar from a simple overall displacement function.

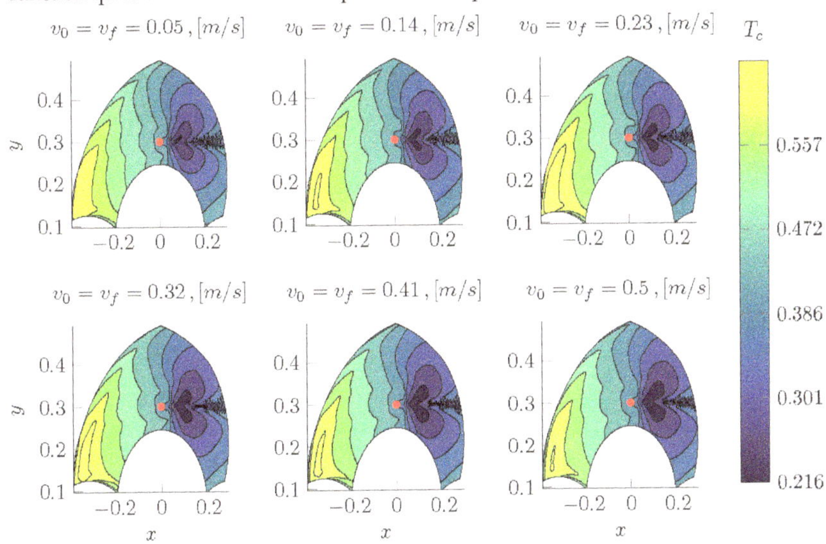

Figure 18. On-the-fly pick and place task time model evaluated for different end-effector initial and final velocities v_0 and v_f; the initial configuration of the robot, shown as a red marker, has been fixed at $x_{ee} = 0.0$ m, $y_{ee} = 0.25$ m.

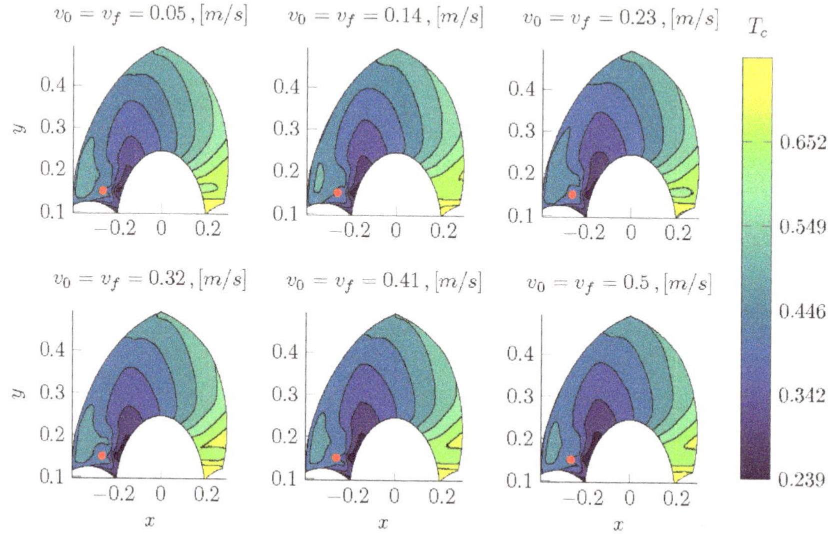

Figure 19. On-the-fly pick and place task time model evaluated for different end-effector initial and final velocities v_0 and v_f; the initial configuration of the robot, shown as a red marker, has been fixed at $x_{ee} = -0.25\,\mathrm{m}$, $y_{ee} = 0.15\,\mathrm{m}$.

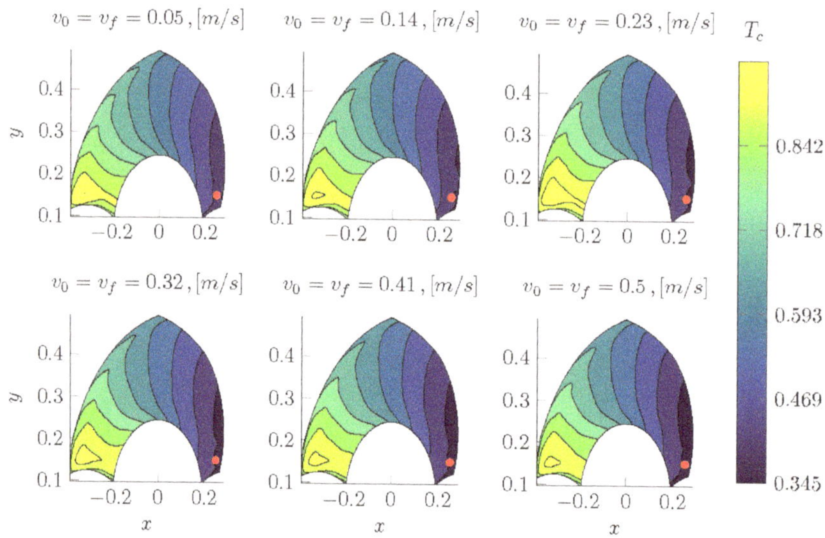

Figure 20. On-the-fly pick and place task time model evaluated for different end-effector initial and final velocities v_0 and v_f; the initial configuration of the robot, shown as a red marker, has been fixed at $x_{ee} = 0.25\,\mathrm{m}$, $y_{ee} = 0.15\,\mathrm{m}$.

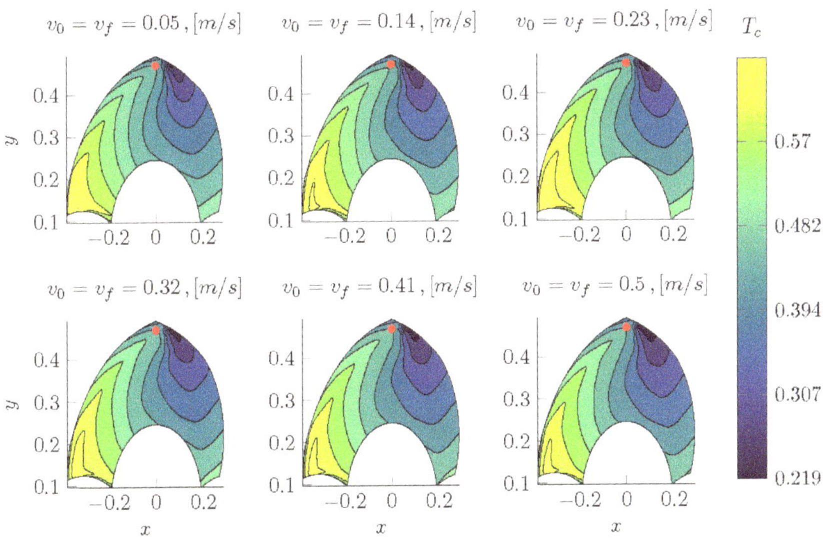

Figure 21. On-the-fly pick and place task time model evaluated for different end-effector initial and final velocities v_0 and v_f; the initial configuration of the robot, shown as a red marker, has been fixed at $x_{ee} = 0.0$ m, $y_{ee} = 0.47$ m.

The computational burden substantially increases together with the size of the network; still, it remains competitive with respect to the full planning time of the trajectory, especially if the GPU is used to evaluate the task time map. This increase in size of the network was dictated by the need to achieve sufficient accuracy. In Figure 22, the error distributions are shown; the lower accuracy obtained with the on-the-fly pick and place task time map (as compared to the stationary pick and place model) can be attributed to the higher dimensionality of the inputs and to the more complex shape of the trajectories; still, the error distribution remains reasonably concentrated around zero and is compatible to an adequate prediction of the actual task times.

In the case of on-the-fly pick and place tasks, the calculation of the intercept point depends on map task time $\hat{T}_{task}$. We can consider two possible application scenarios, as previously discussed; if the command to the robot uses the boundary conditions and the map task time $\hat{T}_{task}$, the robot executes the task in the given time and, therefore, the intercept error is null. If, on the contrary, the command to the robot uses only the boundary conditions, the actual and predicted task times will be different. Considering the worst case of a maximum prediction error of 20 ms and a maximum conveyor velocity of $0.5\,\mathrm{m\,s^{-1}}$, in this case, at the end of the intercept task, the target reaches a position different from the robot's, with a spatial error equal to 10 mm. This might be acceptable or not according to the picked item's geometry and size. Concerning the computational costs associated with the full planning pipeline, the following considerations can be made: Figures 13–16 and 18–21 show that even the shortest trajectories require more than 100 ms for their execution; on the other hand, in the histogram depicted in Figure 11, it can be seen that the execution of the complete motion planning pipeline, composed of the GTP and TTO, requires practically always less than 80 ms; this result, moreover, is tuneable by acting on the parameters of the optimization routines (chiefly the number of decision variables and constraints), sacrificing, if needed, accuracy for speed. Even in the most unfortunate cases (with comparatively short trajectory execution times and relatively long planning times), it is possible to plan the next trajectory while the current one is being executed. Since the conveyors move at quasi-constant velocities, it is indeed feasible to predict the motion of the targets with reasonable accuracy in order to plan in advance the next task assigned to the robot. This kind of plan-in-advance strategy is needed in any case for the on-the-fly pick and place tasks since the robot terminates its current trajectory with a non-null velocity and therefore immediately needs a new setpoint to safely prosecute its motion. As a further benefit, this strategy makes it unnecessary to consider, during the optimal intercept computation, the calculation times. The implementation of the overall planning system here briefly delineated should therefore allow the parallel execution of the control algorithms and of the planning operations, and additionally, it should perform short-term forecasts of the pick and place targets' states.

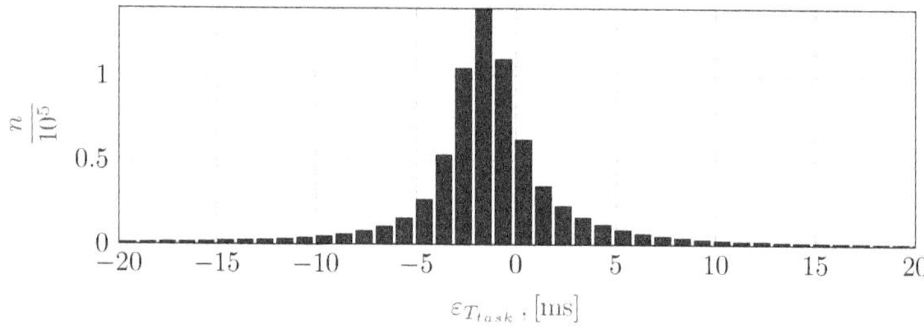

(a) Prediction error.

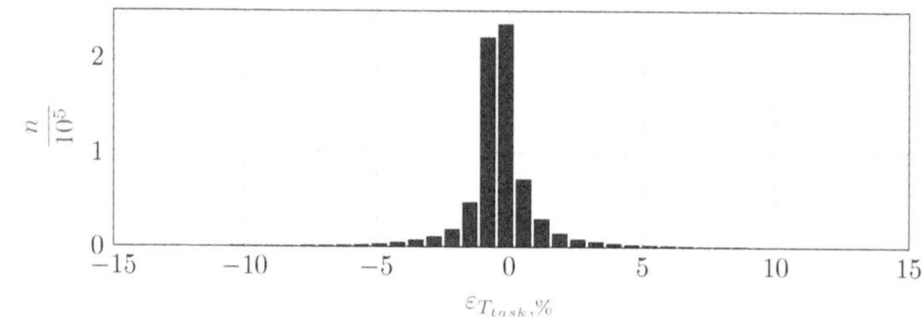

(b) Relative prediction error.

Figure 22. On-the-fly pick and place task time prediction errors calculated on the test dataset.

To highlight the usefulness of the proposed method, we show in Table 5 the computational costs associated with the solution of a few optimal intercept problems. The problem was solved thanks to an iterative root finding algorithm applied to Equation (10) using either the task time map or the direct evaluation of the GTP and TTO. In the table, the parameters concerning four different intercept tasks are shown. It can be seen that the solution of the intercept problem using either the task time map or the actual task time function requires a similar number of function evaluations. However, the difference lies in the cost of each evaluation. It can be seen that the total solution time is in the order of several hundreds of milliseconds using the direct computation approach. On the other hand, using the task time map, the problem can be solved within a a few tens of milliseconds (using the CPU) or in roughly one millisecond (using the GPU). Without the use of the task time map, therefore, the plan-in-advance strategy described above is not feasible since the computation times would be much longer than the trajectory execution times. On the contrary, the use of task time maps leads to calculation times compatible with a real application.

Table 5. Solution times for several optimal intercept problems using different methods.

Task Configuration Data	Task 1	Task 2	Task 3	Task 4
$x_{rbt,0}$, [mm]	−300	175	−300	175
$y_{rbt,0}$, [mm]	150	300	150	300
$\varphi_{rbt,0}$, [°]	0.0	0.0	0.0	0.0
$v_{rbt,0}$, [m s^{-1}]	0.2	0.2	0.2	0.2
$x_{tgt,0}$, [mm]	−150	−150	150	150
$y_{tgt,0}$, [mm]	400	400	175	175
$\varphi_{tgt,0}$, [°]	0.0	0.0	0.0	0.0
$v_{tgt,0}$, [m s^{-1}]	0.2	0.2	0.2	0.2
T_{task} evaluations	15	15	17	12
$\hat{T}_{task}$ evaluations	14	15	16	11
Solution time with GTP-TTO, [ms]	869	869	985	695
Solution time with task time map (CPU), [ms]	32	34	37	25
Solution time with task time map (GPU), [ms]	1.02	1.09	1.17	0.80

5. Conclusions

The paper presents a general methodology for mapping the minimal task times of industrial manipulators whose trajectories can have an arbitrary position in the workspace. The proposed methodology uses neural network models to approximate the task time function of a generic multi-DOF robot. Moreover, the work proposes a complete simulation-based workflow for determining the data needed to train the neural network that considers the manipulator's non-linear kinematics and dynamics and the actuation system's properties.

The neural network proposed for the task time mapping can be also trained using a different set of data with respect to those proposed by the authors. For instance, the data can be collected from simulations performed in the manufacturer's software environment or by the experimental measure of the task time obtained by a robot that executes a comprehensive set of trajectories from and to any points belonging to the workspace. The proposed workflow is however cheaper and does not require the setup of the experimental investigation, also introducing considerable time saving. The workflow presented for determining the training data considers trajectory optimization and advanced motion planning along it and is able to satisfy stationary and on-the-fly pick and place robot jobs. The paper presents these topics both theoretically and numerically.

The results show that the evaluation of the presented task time map is 600 to 8000 times faster than the direct use of the GTP-TTO algorithm. In more complex cases dealing with on-the-fly pick and place, the task time map evaluation is 20 to 780 times faster than the direct computational effort.

Even without specialized hardware, compared with the direct evaluation of the task times, the neural network model offers significant, orders-of-magnitude speedups, which can be further increased through the use of standard GPU computing or, if needed by more demanding applications, with dedicated AI hardware accelerators. Therefore, the neural network model has been found remarkably advantageous in terms of accuracy, computational efficiency, adaptability to diverse applications, and ease of deployment; the methodology proposed by the authors is therefore shown to be suitable for scheduling strategies requiring real-time evaluation of the minimal task time. Neural networks are demonstrated to constitute a convenient architecture for implementing the task time model, as they are inherently mesh-free and more readily scalable (unlike more traditional lookup tables). The case study here proposed can be immediately adapted to a wide array of practically relevant industrial processes, namely all those plants that feature straight conveyor belts and 4-DOF manipulators such as SCARA, Clavel Delta, linear Delta, palletizing, Adept Quattro, and Cartesian robots. In cases characterized by a higher number of degrees of freedom, we expect that the dimensions of the neural network would increase. For example, a similar application with pick and place operations also requiring general spatial rotations would involve a larger number of input parameters to be fed into the neural network; it can be anticipated that the training and use of the task time models would be more expensive; the overall approach would remain however unchanged. Indeed, the neural network models are easily applied to different robots and tasks while keeping a uniform software interface. In addition, composite or hierarchical tasks, such as the shown on-the-fly pick and place operations, can be treated using the proposed methodology through the mapping of each subtask time.

Application to an industrial robot would require the use of the motion planning strategies implemented by its manufacturer, both for the generation of the task time map and in general for the utilization of the manipulator. The core of this work, namely the method for task time mapping, would still be applicable. In particular, if the motion planning strategy is unknown, the method is applicable using a task time dataset generated through experimental execution of a sufficiently exhaustive set of trajectories.

As a future work, the authors are planning to experimentally validate the developed methodology on the actual manipulator, whose open architecture makes it straightforward to apply the techniques as here described.

Another prosecution of this research is related to the possibility of completely substituting the GTP and the TTO with a neural network that outputs the entire motion profile.

Finally, an additional future work will concern the application of the proposed method to a simulated multirobot plant, with a greater focus on the productivity advantages that can be obtained thanks to scheduling algorithms based on the task time map.

Author Contributions: Conceptualization, P.R. and F.C.; methodology, P.R. and F.C.; software, F.C.; validation, F.C.; investigation, P.R., R.S. and F.C.; writing—original draft preparation, P.R., R.S. and F.C.; supervision, P.R. All authors have read and agreed to the published version of the manuscript.

Funding: The research was funded by the University of Bergamo.

Data Availability Statement: Not applicable.

Conflicts of Interest: The authors declare no conflict of interest.

References

1. Daoud, S.; Chehade, H.; Yalaoui, F.; Amodeo, L. Solving a robotic assembly line balancing problem using efficient hybrid methods. *J. Heuristics* **2014**, *20*, 235–259. [CrossRef]
2. Humbert, G.; Pham, M.T.; Brun, X.; Guillemot, M.; Noterman, D. Comparative analysis of pick & place strategies for a multi-robot application. In Proceedings of the 2015 IEEE 20th Conference on Emerging Technologies & Factory Automation (ETFA), Luxembourg, 8–11 September 2015; pp. 1–8.
3. Humbert, G.; Brun, X.; Pham, M.T.; Guillemot, M.; Noterman, D. Development of a methodology to improve the performance of multi-robot pick & place applications: From simulation to experimentation. In Proceedings of the 2016 IEEE International Conference on Industrial Technology (ICIT), Taipei, Taiwan 14–17 March 2016; pp. 1960–1965.
4. Ferrari, G.; Ferrarini, L.; Petretti, A.; Pizzi, E. Modeling and design of an optimal line manager of a packaging system with MILP. In Proceedings of the IECON 2015—41st Annual Conference of the IEEE Industrial Electronics Society, Yokohama, Japan, 9–12 November 2015; pp. 005050–005056.
5. Pizzi, E.; Bouchrit, A.; Petretti, A.; Ferrarini, L. Performance improvement for online schedulers for packaging systems. In Proceedings of the 2016 IEEE International Conference on Automation Science and Engineering (CASE), Fort Worth, TX, USA, 21–24 August 2016; pp. 1243–1248.
6. Wang, P.; Ma, H.; Zhang, Y.; Cao, X.; Wu, X.; Wei, X.; Zhou, W. Trajectory Planning for Coal Gangue Sorting Robot Tracking Fast-Mass Target under Multiple Constraints. *Sensors* **2023**, *23*, 4412. [CrossRef] [PubMed]
7. Wilson, D.B.; Soto, M.A.T.; Goktogan, A.H.; Sukkarieh, S. Real-time rendezvous point selection for a nonholonomic vehicle. In Proceedings of 2013 IEEE International Conference on Robotics and Automation, Karlsruhe, Germany, 6–10 May 2013; pp. 3941–3946. [CrossRef]
8. Shin, I.S.; Nam, S.H.; Roberts, R.; Moon, S. A minimum-time algorithm for intercepting an object on a conveyor belt. *Ind. Robot.* **2009**, *36*, 127–137. [CrossRef]
9. Croft, E.A.; Fenton, R.G.; Benhabib, B. An on-line robot planning strategy for target interception. *J. Robot. Syst.* **1998**, *15*, 97–114. [CrossRef]

10. Jasour, A.M.; Farrokhi, M. Adaptive neuro-predictive control for redundant robot manipulators in presence of static and dynamic obstacles: A Lyapunov-based approach. *Int. J. Adapt. Control Signal Process.* **2014**, *28*, 386–411. [CrossRef]
11. Han, S.D.; Feng, S.W.; Yu, J. Toward fast and optimal robotic pick-and-place on a moving conveyor. *IEEE Robot. Autom. Lett.* **2019**, *5*, 446–453. [CrossRef]
12. Kröger, T. Opening the door to new sensor-based robot applications—The Reflexxes Motion Libraries. In Proceedings of the 2011 IEEE International Conference on Robotics and Automation, Shanghai, China, 9–13 May 2011; pp. 1–4.
13. Mattone, R.; Divona, M.; Wolf, A. Sorting of items on a moving conveyor belt. Part 2: Performance evaluation and optimization of pick-and-place operations. *Robot.-Comput.-Integr. Manuf.* **2000**, *16*, 81–90. [CrossRef]
14. Boschetti, G. A picking strategy for circular conveyor tracking. *J. Intell. Robot. Syst.* **2016**, *81*, 241–255. [CrossRef]
15. Zhou, M.; Jiang, R. Optimal Strategy for Pick-and-Place System with two Robots. *J. Phys.* **2022**, *2216*, 012022. [CrossRef]
16. Scalera, L.; Boscariol, P.; Carabin, G.; Vidoni, R.; Gasparetto, A. Enhancing energy efficiency of a 4-DOF parallel robot through task-related analysis. *Machines* **2020**, *8*, 10. [CrossRef]
17. Bobrow, J.E.; Dubowsky, S.; Gibson, J.S. Time-optimal control of robotic manipulators along specified paths. *Int. J. Robot. Res.* **1985**, *4*, 3–17. [CrossRef]
18. Pham, Q.C.; Stasse, O. Time-optimal path parameterization for redundantly actuated robots: A numerical integration approach. *IEEE/ASME Trans. Mechatronics* **2015**, *20*, 3257–3263. [CrossRef]
19. Pham, H.; Pham, Q.C. Time-optimal path tracking via reachability analysis. In Proceedings of the 2018 IEEE International Conference on Robotics and Automation (ICRA), Brisbane, Australia, 21–25 May 2018; pp. 3007–3012.
20. Verscheure, D.; Demeulenäre, B.; Swevers, J.; De Schutter, J.; Diehl, M. Practical time-optimal trajectory planning for robots: A convex optimization approach. *IEEE Trans. Autom. Control* **2008**, *53*, 28.
21. Pham, H.; Pham, Q.C. On the structure of the time-optimal path parameterization problem with third-order constraints. In Proceedings of the 2017 IEEE International Conference on Robotics and Automation (ICRA), Singapore, 29 May–3 June 2017; pp. 679–686.
22. Fan, W.; Gao, X.S.; Lee, C.H.; Zhang, K.; Zhang, Q. Time-optimal interpolation for five-axis CNC machining along parametric tool path based on linear programming. *Int. J. Adv. Manuf. Technol.* **2013**, *69*, 1373–1388. [CrossRef]
23. Zhang, Q.; Li, S.R.; Gao, X.S. Practical smooth minimum time trajectory planning for path following robotic manipulators. In Proceedings of the 2013 American Control Conference, Washington, DC, USA, 17–19 June 2013; pp. 2778–2783.
24. Gasparetto, A.; Boscariol, P.; Lanzutti, A.; Vidoni, R. Path planning and trajectory planning algorithms: A general overview. *Motion Oper. Plan. Robot. Syst. Backgr. Pract. Approaches* **2015**, *2015*, 3–27.
25. Braghin, F.; Cheli, F.; Melzi, S.; Sabbioni, E. Race driver model. *Comput. Struct.* **2008**, *86*, 1503–1516. [CrossRef]
26. Sabelhaus, D.; Röben, F.; zu Helligen, L.P.M.; Lammers, P.S. Using continuous-curvature paths to generate feasible headland turn manoeuvres. *Biosyst. Eng.* **2013**, *116*, 399–409. [CrossRef]
27. Zhao, H.; Zhu, L.; Ding, H. A real-time look-ahead interpolation methodology with curvature-continuous B-spline transition scheme for CNC machining of short line segments. *Int. J. Mach. Tools Manuf.* **2013**, *65*, 88–98. [CrossRef]
28. Tang, P.Y.; Lin, M.T.; Tsai, M.S.; Cheng, C.C. Toolpath interpolation with novel corner smoothing technique. *Robot. Comput. Integr. Manuf.* **2022**, *78*, 102388. [CrossRef]
29. Chen, G.; Luo, N.; Liu, D.; Zhao, Z.; Liang, C. Path planning for manipulators based on an improved probabilistic roadmap method. *Robot. Comput. Integr. Manuf.* **2021**, *72*, 102196. [CrossRef]
30. Righettini, P.; Strada, R.; Zappa, B.; Lorenzi, V. Experimental set-up for the investigation of transmissions effects on the dynamic performances of a linear PKM. In Proceedings of the Advances in Mechanism and Machine Science: Proceedings of the 15th IFToMM World Congress on Mechanism and Machine Science 15, Krakow, Poland, 30 June–4 July 2019; pp. 2511–2520.
31. Righettini, P.; Strada, R.; Cortinovis, F. Modal kinematic analysis of a parallel kinematic robot with low-stiffness transmissions. *Robotics* **2021**, *10*, 132. [CrossRef]
32. Righettini, P.; Strada, R.; Cortinovis, F. General Procedure for Servo-Axis Design in Multi-Degree-of-Freedom Machinery Subject to Mixed Loads. *Machines* **2022**, *10*, 454. [CrossRef]
33. Bourbonnais, F.; Bigras, P.; Bonev, I.A. Minimum-time trajectory planning and control of a pick-and-place five-bar parallel robot. *IEEE/ASME Trans. Mechatronics* **2014**, *20*, 740–749. [CrossRef]
34. Yu, H. Modeling and control of hybrid machine systems—A five-bar mechanism case. *Int. J. Autom. Comput.* **2006**, *3*, 235–243. [CrossRef]
35. Stellato, B.; Banjac, G.; Goulart, P.; Bemporad, A.; Boyd, S. OSQP: An operator splitting solver for quadratic programs. *Math. Program. Comput.* **2020**, *12*, 637–672. [CrossRef]
36. Takahasi, H.; Mori, M. Double exponential formulas for numerical integration. *Publ. Res. Inst. Math. Sci.* **1974**, *9*, 721–741. [CrossRef]

robotics

MDPI

Article

Mapping the Tilt and Torsion Angles for a 3-S$\underline{P}$S-U Parallel Mechanism

Swaminath Venkateswaran [1,*,†] and Damien Chablat [2,†]

[1] Research Center, Léonard de Vinci Pôle Universitaire, La Défense, 92916 Paris, France
[2] École Centrale Nantes, Nantes Université, CNRS, LS2N, UMR 6004, 44000 Nantes, France; damien.chablat@cnrs.fr
* Correspondence: swaminath.venkateswaran@devinci.fr
† These authors contributed equally to this work.

Abstract: This article presents the analysis of a parallel mechanism of type 3-S$\underline{P}$S-U. The usual singularity approach is carried out with respect to the Euler angles of the universal joint. However, this approach is computationally expensive especially when stacked structures are analyzed. Thus, the positioning of the mobile platform for the mechanism is analyzed using the theory of Tilt and Torsion (T&T). The singularity-free workspace and the tilt limits of the mechanism are disclosed through this method. These workspaces can then be mapped to the Euler angles of the universal joint and the relation between the T&T space and the Euler space is demonstrated and validated in this study. Initially, simulations are performed using the results of singularity analysis to have a comparison between the T&T space and the Euler space. Experimental validation is then carried out on the prototype of the mechanism to perform a circular trajectory, in line with the simulations. The outcome of this study will be helpful for the integration of the mechanism for a piping inspection robot and also for the analysis of stacked architectures.

Keywords: parallel mechanism; tilt and torsion; singularities; control

1. Introduction

Parallel kinematics machines (PKM) have interesting applications in industries over serial machines. PKMs offer better accuracy, lower mass/inertia properties and high structural stiffness [1]. The performance of parallel robots varies within their workspace, which is considerably smaller when compared to serial robots. At present, many researchers work on parallel robots as they offer simple and conventional designs with ease of calibration [2]. Moreover, the inverse kinematics, in general, is easier to solve for parallel robots, especially when prismatic joints are employed. With the presence of spherical joints in a parallel mechanism, the modified Euler angles, otherwise known as the "Tilt and Torsion" (T&T) angles, could be employed [3,4]. Bonev et al. [5] proposed the advantages of the T&T angles for the study of spatial parallel mechanisms in-depth. It was demonstrated that there exists a class of parallel mechanisms with zero torsion, which are referred to as zero-torsion parallel mechanisms [5,6]. Another interesting study which exploits the T&T angles is the 3-RRR spherical parallel mechanism, which was analyzed by Tao et al. [7]. This analysis helped in improving the workspace of the parallel mechanism and was applicable for architectures that have revolute joints. Generally, the applications of parallel or spherical mechanisms are found in industrial applications such as machining [8] and nano-alignment, [9]. However, there is another interesting application of a spherical-type parallel mechanism for a piping-inspection robot. A bio-inspired piping-inspection robot that moves like a caterpillar was designed and developed at LS2N, France [10]. This prototype was a rigid model and it was limited to straight pipe profiles. An articulation unit was proposed for this robot which addresses two main issues, namely, passive compliance and active compliance. Passive compliance is applicable for a robot working inside a pipeline having an elbow at 90° or 45°. In such scenarios, the articulation units usually work in passive modes. Some examples of such mechanisms that exist in the literature are the squirm-type robot of Zhang et al. [11] and the caterpillar-type robot of Kwon et al. [12]. On the other hand, for the active compliance issue, the articulation unit must be actuated in order to align the robot assembly with the pipeline axis or follow a certain direction inside the pipeline. This scenario could be imagined for a robot that passes through a junction or a T-branch inside a pipeline. Some examples of active compliance include the robot of Ryew et al. [13], which employs a dual-actuated universal joint for passing through junctions and a pantograph mechanism to ensure contact with pipeline walls during locomotion. The third criterion that must also be taken into account is the tilt limit of the articulation unit as it determines if the mechanism tilt is sufficient to overcome a bend or a junction. By taking these factors into account, a parallel mechanism that employs three tension springs and a universal joint was incorporated as an articulation unit for a bio-inspired piping inspection robot [10,14]. The positioning of the end-effector with respect to the fixed base plays an important role in identifying the tilt limits of this mechanism. The conventional approach involves analyzing the end-effector for the Euler angles of the universal joint [15]. As the universal joint has two degrees of freedom about the x and y axis, the workspace analysis involves studying the influence of both tilt angles for each position of the end-effector within the joint limits. However, there exists an alternate approach where the end-effector can be localized using the T&T angles. When compared to the Euler space, in the T&T space the singularity analysis appears much simpler as the tilt angle remains constant while the azimuth can vary from 0 to 2π radians or vice versa. This article demonstrates a relation that exists between the T&T space and the Euler space for the 3-S$\underline{P}$S-U parallel mechanism under study. This relation, also called the "mapping", has been demonstrated and validated for the mechanism through simulations and experiments.

The outline of this article is as follows. Section 2 presents the architecture and the geometrical equations of the mechanism using the T&T angles. In Section 3, the singularity analysis of the mechanism is presented in detail through algebraic methods. Following this, in Section 4, the mapping parameters are demonstrated for the T&T space to the Euler space. A numerical simulation is also presented to demonstrate this mapping. Section 5 presents the experimental validation of the mapping theory with the help of the prototype of the mechanism and control laws. Finally, in Section 6, closing discussions are made along with the future perspectives.

2. Architecture of the Parallel Mechanism

This section presents the architecture of the parallel mechanism that will be analyzed in this article. An overview of the final application of this mechanism is first presented. Followed by that, the geometrical equations that are necessary to analyze the singularities are presented.

2.1. A Flexible Mechanism for a Piping Inspection Robot

In collaboration with the enterprise AREVA in France, a rigid bio-inspired piping-inspection robot was designed and developed at LS2N, France. With the help of leg mechanisms and DC motors, the robot accomplishes the locomotion of a caterpillar to move inside a pipeline [10]. A 3D model of this robot and its 2D view are shown in Figure 1. The application of this robot is limited to straight pipelines, as the model is rigid. In order to work inside a pipeline that consists of bends and junctions, a flexible version of the robot was proposed in [14]. Two similar parallel mechanisms with two degrees of freedom (DOF) were introduced between each module of the first robot. These parallel mechanisms consist of a universal joint to constrain the mobility between the base and the mobile platform and three tension springs with or without cables for the actuation. This mechanism can potentially assist the piping-inspection robot to overcome a 90° bend or a junction. A digital model of the flexible piping-inspection robot with the parallel mechanism is shown in Figure 2.

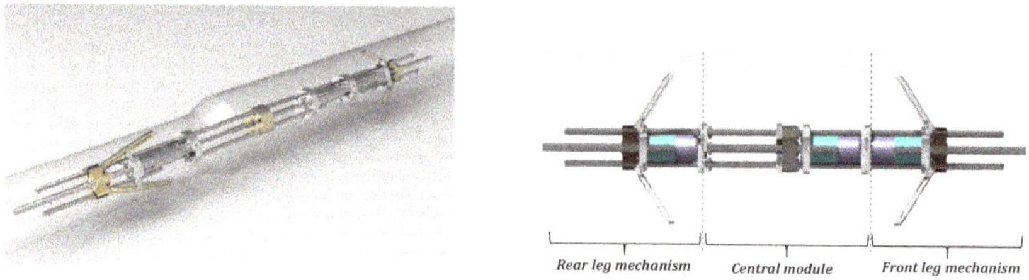

Figure 1. The 3D model (**left**) and the 2D view (**right**) of the rigid bio-inspired piping-inspection robot developed at LS2N, France [10].

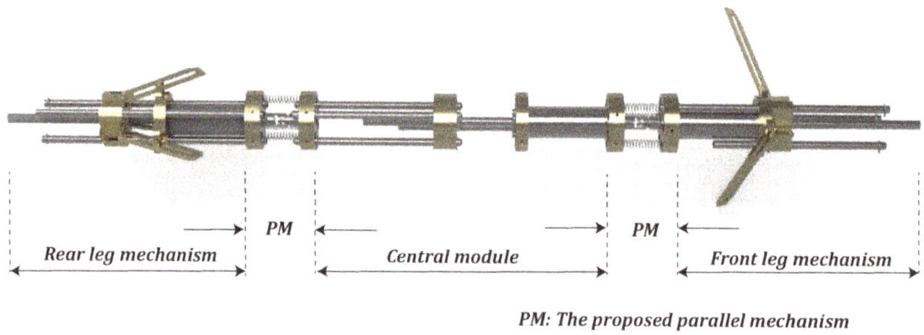

Figure 2. The 2D view of the flexible piping-inspection robot with the parallel mechanism [16].

The primary focus of this article is to analyze the singular configurations of this parallel mechanism which will provide the tilt limits.

2.2. Geometrical Equations of the Mechanism

The parallel mechanism under study was analyzed earlier through two approaches, namely, (i) using the Tilt and Torsion (T&T) angles and (ii) singularity analysis with respect to the Euler angles. The former was employed to study the joint limits where three types of architectures were analyzed to identify an inverse pendulum configuration [17]. On the other hand, the singularity analysis was carried out for the Euler angles, which led to the conclusion that the 3-S$\underline{P}$S-U architecture was not an ideal solution for addressing passive compliance owing to the limited tilt angles [15]. In this article, the singularity analysis is revisited for the 3-S$\underline{P}$S-U mechanism by using the T&T angles. The articulation unit that consists of the tension springs and universal joints is correlated to a parallel manipulator of type 3-S$\underline{P}$S-U [18,19]. The representation of the parallel mechanism under study is shown in Figure 3. In this mechanism, S indicates the spherical joint, U indicates the universal joint and $\underline{P}$ indicates the actuated prismatic joints or the springs. The mechanism also employs cables that pass through each spring from the base to the end-effector. The cable actuation is performed using external motors in the active modes and it does not resemble a classic prismatic link. Thus, under operation, the mobility of the end-effector is purely obtained through the two rotation angles of the universal joint. The home pose represented in Figure 3a is the position where the tilt angles, as well as the external forces on the mechanism, are equal to zero. The fixed coordinate frame of the base is represented by $\sum_0$. The spring mounting points on the fixed base are represented by B_1, B_2 and B_3 and they form an imaginary equilateral triangle of the manipulator base. The vector coordinates for the base mounting points can be written as:

$$\mathbf{b}_i^0 = \begin{bmatrix} r_f \cos\left(\frac{2\pi(i-1)}{3}\right) \\ r_f \sin\left(\frac{2\pi(i-1)}{3}\right) \\ -r_f h \end{bmatrix}, \quad \text{with} \quad i = 1,2,3 \tag{1}$$

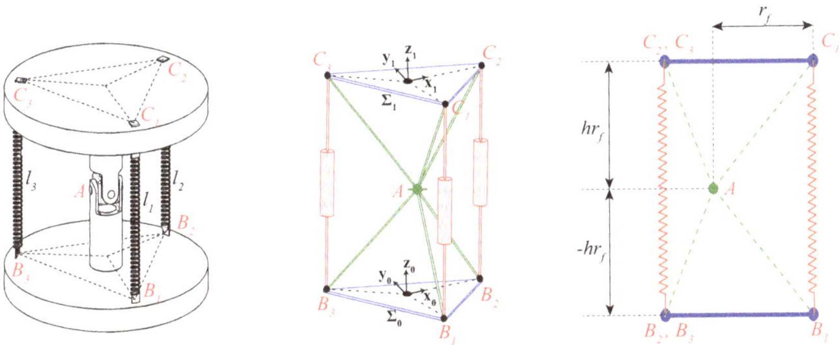

(a) **(b)** **(c)**

Figure 3. Representation of the (**a**) parallel mechanism at home pose, (**b**) 3D view of the correlation to a 3-SPS-U manipulator and (**c**) 2D view of the manipulator.

Equation (1) provides the coordinates of B_i in the fixed coordinate frame $\sum_0$. To determine the position of the end-effector, the Euler angles are usually employed [15], especially to study the singular configurations. In this case, the T&T angles are revisited from [17] for identifying singular configurations. It must be noted that in [17], the singularity analysis was not performed for the T&T space. As the torsion for the mechanism has been constrained, the tilt and azimuth angles are sufficient to analyze the architecture. The representation of the tilt and azimuth angles on the mechanism is revisited from [17] and it is shown below in Figure 4.

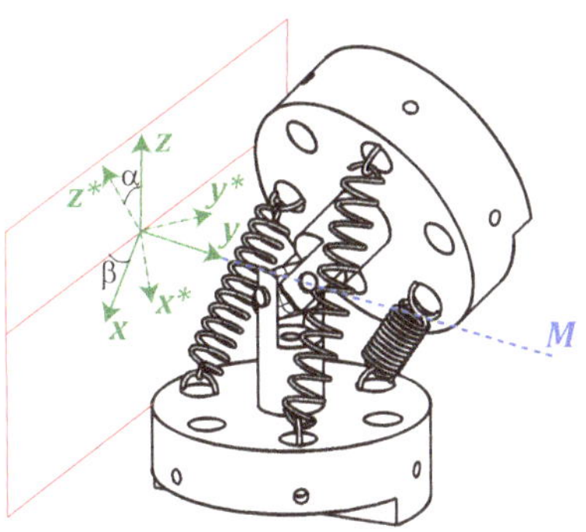

Figure 4. Representation of the tilt (α) and azimuth (β) angles on the parallel mechanism under study.

This representation is correlated with the study shown by Bonev et al. [3], where β represents the azimuth angle between the x-axis and the face of z-axis. A rotation about M causes a shift of the axis xyz to $x^*y^*z^*$. The angle between z and z^* represents the tilt angle, α. Here, the resultant rotation matrix **R** is a modified form of the equation presented in [3] by Bonev et al. but in this case without the torsion. The matrix is given by the equation:

$$\mathbf{R} = \mathbf{R}_z(\beta)\mathbf{R}_x(\alpha)\mathbf{R}_z(-\beta) \tag{2}$$

$$\mathbf{R} = \begin{bmatrix} \cos(\beta)^2 + \sin(\beta)^2\cos(\alpha) & \cos(\beta)\sin(\beta)(1-\cos(\alpha)) & \sin(\beta)\sin(\alpha) \\ \cos(\beta)\sin(\beta)(1-\cos(\alpha)) & \sin(\beta)^2 + \cos(\beta)^2\cos(\alpha) & -\cos(\beta)\sin(\alpha) \\ -\sin(\beta)\sin(\alpha) & \cos(\beta)\sin(\alpha) & \cos(\alpha) \end{bmatrix} \tag{3}$$

where $\mathbf{R}_x(\alpha) = \begin{bmatrix} 1 & 0 & 0 \\ 0 & \cos(\alpha) & -\sin(\alpha) \\ 0 & \sin(\alpha) & \cos(\alpha) \end{bmatrix}, \mathbf{R}_z(-\beta) = \begin{bmatrix} \cos(\beta) & \sin(\beta) & 0 \\ -\sin(\beta) & \cos(\beta) & 0 \\ 0 & 0 & 1 \end{bmatrix}$.

The moving coordinate frame of the end-effector is represented by $\sum_1$. The spring mounting points of the end-effector are represented by C_1, C_2 and C_3 for the mechanism. In order to obtain the position of the end-effector, Equation (2) is multiplied by the vector coordinates of C_i, which are given by:

$$\mathbf{c}_i^1 = \mathbf{R}\,\mathbf{o}_i^1, \text{ with } i = 1,2,3 \text{ \& } \mathbf{o}_i^1 = \begin{bmatrix} r_f \cos\left(\frac{2\pi(i-1)}{3}\right) \\ r_f \sin\left(\frac{2\pi(i-1)}{3}\right) \\ r_f h \end{bmatrix} \quad (4)$$

In Equation (4), $\mathbf{o}_i^1$ are the coordinates of C_i in the mobile reference frame $\sum_1$ of the end-effector. The length of the mechanism l_i, which provides the solution for the inverse kinematic problem (IKP), can be solved by computing the distance between the vector coordinates of the base and the end-effector. The constraint equations are given by:

$$\left\|\mathbf{b}_i^0 - \mathbf{c}_i^1\right\| = l_i \text{ where } \mathbf{b}_i^0 = [b_{ix}, b_{iy}, b_{iz}]^T, \mathbf{c}_i^1 = [c_{ix}, c_{iy}, c_{iz}]^T \text{ with } i = 1,2,3 \quad (5)$$

$$2\,rf^2\left((\cos(\beta))^2\cos(\alpha) - 2\sin(\beta)\sin(\alpha)h + \cos(\alpha)h^2 - (\cos(\beta))^2 + h^2 - \cos(\alpha) + 1\right) = l_1^2 \quad (6a)$$

$$-0.5\,rf^2(2\cos(\beta)\sin(\beta)\cos(\alpha)\sqrt{3} - 4\sqrt{3}\cos(\beta)\sin(\alpha)h + 2(\cos(\beta))^2\cos(\alpha)$$
$$-2\cos(\beta)\sin(\beta)\sqrt{3} - 4\sin(\beta)\sin(\alpha)h - 4h^2\cos(\alpha) - 2(\cos(\beta))^2 - 4h^2 + \cos(\alpha) - 1) = l_2^2 \quad (6b)$$

$$-0.5\,rf^2(-2\cos(\beta)\sin(\beta)\cos(\alpha)\sqrt{3} + 4\sqrt{3}\cos(\beta)\sin(\alpha)h + 2(\cos(\beta))^2\cos(\alpha)$$
$$+2\cos(\beta)\sin(\beta)\sqrt{3} - 4\sin(\beta)\sin(\alpha)h - 4h^2\cos(\alpha) - 2(\cos(\beta))^2 - 4h^2 + \cos(\alpha) - 1) = l_3^2 \quad (6c)$$

Equations (6a)–(6c) represents the solutions to the IKP. For known values of the tilt and azimuth angles as well as the design parameters r_f and h, these equations can be employed to calculate the distance between the base and end-effector of the mechanism.

3. Singularity Analysis and Workspace of the Mechanism in the T&T Space

This section focuses on the singularity analysis of the mechanism. With the presence of the universal joint, the end-effector coordinates are identified conventionally using the Euler XY rotation angles about the origin A shown in Figure 3. In this approach, for each position of the end-effector, the influence of both rotation angles must be taken into account. However, with the help of the tilt and azimuth angles, it is still possible to analyze the presence of singularities in the mechanism. Moreover, in the T&T space, the mechanism will be simpler to analyze wherein one of the angles remains fixed or constant while the other changes. In [17], only the analysis of joint limits was carried out by setting minimum and maximum values for each prismatic link. This analysis provided the tilt angles for the joint limits; however, the direct kinematic problem (DKP), which presents the singularity-free workspace in T&T space, was not solved in [17]. The singularity-free workspace is thus identified by analyzing the mechanism in the T&T space, which provides the solution to the DKP and the tilt limits.

3.1. Singularity Equations of the Mechanism

For a parallel mechanism, the singularity equation is given by the well-known equation [20]:

$$\mathbf{At} + \mathbf{B}\dot{\rho} = 0 \tag{7}$$

where $\mathbf{t}$ represents the angular velocity vector
and $\dot{\rho} = [\dot{l_1}, \dot{l_2}, \dot{l_3}]^T$ represents the joint velocity vector

The IKP Equations (6a)–(6c) are employed in Equation (7), where their time derivatives are taken to study the singularities. In Equation (7), $\mathbf{A}$ represents the direct-kinematics matrix or forward Jacobian matrix and $\mathbf{B}$ represents the inverse-kinematics matrix or inverse Jacobian matrix of the mechanism. The pose variables for the mechanism are the tilt and azimuth angles α and β. The articular variables are the lengths: l_1, l_2 and l_3. A parallel mechanism in general might encounter three types of singularities, namely, Type-1, Type-2 and Type-3 [21,22]. When the determinant of the direct kinematics matrix $\mathbf{A}$ loses its rank, the determinant of this matrix becomes zero and this scenario leads to parallel singularities or Type-2 singularities. From the number of pose variables and articular variables, it could be seen that matrix $\mathbf{A}$ does not correspond to an $n \times n$ square matrix. For identifying the singularity equations and to construct a square matrix, the 3-S$\underline{\text{P}}$S-U mechanism is split into three sets of 2-S$\underline{\text{P}}$S-U architecture, which comprise the length pairs $l_1 - l_2$, $l_2 - l_3$ and $l_1 - l_3$. The corresponding direct kinematic matrices $\mathbf{A}$ are created for the three sets by differentiating Equations (6a)–(6c) with respect to the pose variables. The matrices are given by:

$$\mathbf{A}_{1-2} = \begin{bmatrix} \frac{\partial L_1}{\partial \alpha} & \frac{\partial L_1}{\partial \beta} \\ \frac{\partial L_2}{\partial \alpha} & \frac{\partial L_2}{\partial \beta} \end{bmatrix}, \ \mathbf{A}_{2-3} = \begin{bmatrix} \frac{\partial L_2}{\partial \alpha} & \frac{\partial L_2}{\partial \beta} \\ \frac{\partial L_3}{\partial \alpha} & \frac{\partial L_3}{\partial \beta} \end{bmatrix}, \ \mathbf{A}_{1-3} = \begin{bmatrix} \frac{\partial L_1}{\partial \alpha} & \frac{\partial L_1}{\partial \beta} \\ \frac{\partial L_3}{\partial \alpha} & \frac{\partial L_3}{\partial \beta} \end{bmatrix} \tag{8}$$

In Equation (8), L_1, L_2 and L_3 represent the left-hand sides of Equations (6a)–(6c). The parallel singularities are analyzed by calculating the determinant of matrices in Equation (8) and equating them to zero. The determinant values are given by:

$$\begin{aligned} \det(\mathbf{A}_{1-2}),\ \mathcal{D}_{p1}:\ & -4\sin(\beta)(\cos(\alpha))^2(\cos(\beta))^2\sqrt{3}h + 2\sin(\beta)(\cos(\alpha))^2\sqrt{3}h^3 \\ & -2\sin(\alpha)\cos(\alpha)(\cos(\beta))^2\sqrt{3}h^2 + 6\sin(\beta)\sin(\alpha)\cos(\alpha)\cos(\beta)h^2 \\ & +8\sin(\beta)\cos(\alpha)(\cos(\beta))^2\sqrt{3}h + 2\sin(\alpha)(\cos(\beta))^2\sqrt{3}h^2 \\ & -6(\cos(\alpha))^2\cos(\beta)h^3 - 6\sin(\beta)\sin(\alpha)\cos(\beta)h^2 - 4\sin(\beta)(\cos(\beta))^2\sqrt{3}h \\ & -2\sin(\beta)\sqrt{3}h^3 + \sin(\alpha)\cos(\alpha)(\cos(\beta))^2\sqrt{3} - 3\sin(\alpha)\cos(\alpha)\sqrt{3}h^2 \\ & -3\sin(\beta)\sin(\alpha)\cos(\alpha)\cos(\beta) - 2\sin(\beta)\cos(\alpha)\sqrt{3}h - \sin(\alpha)(\cos(\beta))^2\sqrt{3} \\ & -\sin(\alpha)\sqrt{3}h^2 + 3(\cos(\alpha))^2\cos(\beta)h + 6\cos(\beta)h^3 + 3\sin(\beta)\sin(\alpha)\cos(\beta) \\ & +2\sin(\beta)\sqrt{3}h - \sin(\alpha)\cos(\alpha)\sqrt{3} + \sin(\alpha)\sqrt{3} - 3\cos(\beta)h = 0 \end{aligned} \tag{9}$$

$$\begin{aligned} \det(\mathbf{A}_{2-3}),\ \mathcal{D}_{p2}:\ & 8\sin(\beta)(\cos(\alpha))^2(\cos(\beta))^2h + 8\sin(\beta)(\cos(\alpha))^2h^3 \\ & -8(\cos(\beta))^2\sin(\alpha)h^2\cos(\alpha) - 16\sin(\beta)\cos(\alpha)(\cos(\beta))^2h \\ & +8\sin(\alpha)(\cos(\beta))^2h^2 - 6\sin(\beta)(\cos(\alpha))^2h + 8\sin(\beta)(\cos(\beta))^2h \\ & -8\sin(\beta)h^3 + 4(\cos(\beta))^2\sin(\alpha)\cos(\alpha) + 12\cos(\alpha)h^2\sin(\alpha) \\ & +4h\sin(\beta)\cos(\alpha) - 4(\cos(\beta))^2\sin(\alpha) - 4h^2\sin(\alpha) + 2\sin(\beta)h \\ & -\sin(\alpha)\cos(\alpha) + \sin(\alpha) = 0 \end{aligned} \tag{10}$$

$$\begin{aligned}\det(\mathbf{A}_{1-3})\ ,\ \mathcal{D}_{p3}:\ & 4\sin(\beta)(\cos(\alpha))^2(\cos(\beta))^2\sqrt{3}h - 2\sin(\beta)(\cos(\alpha))^2\sqrt{3}h^3 \\ & +2\sin(\alpha)\cos(\alpha)(\cos(\beta))^2\sqrt{3}h^2 + 6\sin(\beta)\sin(\alpha)\cos(\alpha)\cos(\beta)h^2 \\ & -8\sin(\beta)\cos(\alpha)(\cos(\beta))^2\sqrt{3}h - 2\sin(\alpha)(\cos(\beta))^2\sqrt{3}h^2 \\ & -6(\cos(\alpha))^2\cos(\beta)h^3 - 6\sin(\beta)\sin(\alpha)\cos(\beta)h^2 + 4\sin(\beta)(\cos(\beta))^2\sqrt{3}h \\ & +2\sin(\beta)\sqrt{3}h^3 - \sin(\alpha)\cos(\alpha)(\cos(\beta))^2\sqrt{3} + 3\sin(\alpha)\cos(\alpha)\sqrt{3}h^2 \\ & -3\sin(\beta)\sin(\alpha)\cos(\alpha)\cos(\beta) + 2\sin(\beta)\cos(\alpha)\sqrt{3}h + \sin(\alpha)(\cos(\beta))^2\sqrt{3} \\ & +\sin(\alpha)\sqrt{3}h^2 + 3(\cos(\alpha))^2\cos(\beta)h + 6\cos(\beta)h^3 + 3\sin(\beta)\sin(\alpha)\cos(\beta) \\ & -2\sin(\beta)\sqrt{3}h + \sin(\alpha)\cos(\alpha)\sqrt{3} - \sin(\alpha)\sqrt{3} - 3\cos(\beta)h = 0\end{aligned} \tag{11}$$

Using Equations (10) and (12), it is possible to identify the pose variables and understand the presence of parallel singularities in the mechanism. On the other hand, when the determinant of the inverse kinematics matrix **B** loses its rank, the determinant of this matrix becomes zero and this scenario leads to serial singularities or Type-1 singularities. Similar to the parallel singularity equations, the inverse kinematics matrix **B** is created by differentiating the right-hand side of Equations (6a)–(6c) with respect to the articular variables. Following this, the determinant of these matrices is computed and equated to zero for verifying the serial singularities. The matrices and the determinant values are given by:

$$\mathbf{B}_{1-2} = \begin{bmatrix} \frac{\partial R_1}{\partial l_1} & \frac{\partial R_1}{\partial l_2} \\ \frac{\partial R_2}{\partial l_1} & \frac{\partial R_2}{\partial l_2} \end{bmatrix}\ ,\ \mathbf{B}_{2-3} = \begin{bmatrix} \frac{\partial R_2}{\partial l_2} & \frac{\partial R_2}{\partial l_3} \\ \frac{\partial R_3}{\partial l_2} & \frac{\partial R_3}{\partial l_3} \end{bmatrix}\ ,\ \mathbf{B}_{1-3} = \begin{bmatrix} \frac{\partial R_1}{\partial l_1} & \frac{\partial R_1}{\partial l_3} \\ \frac{\partial R_3}{\partial l_1} & \frac{\partial R_3}{\partial l_3} \end{bmatrix} \tag{12}$$

$$\det(\mathbf{B}_{1-2})\ ,\ \mathcal{D}_{s1}:\ l_1 l_2 = 0\ ,\ \det(\mathbf{B}_{2-3})\ ,\ \mathcal{D}_{s2}:\ l_2 l_3 = 0\ ,\ \det(\mathbf{B}_{1-3})\ ,\ \mathcal{D}_{s3}:\ l_1 l_3 = 0 \tag{13}$$

In Equation (12), R_1, R_2 and R_3 represent the right-hand side of Equations (6a)–(6c). From Equation (13), when the determinant of the inverse kinematics matrices is equated to zero, serial singularities can occur only when the length of one or two prismatic springs becomes zero. This condition is not feasible for the mechanism under study as the prismatic springs cannot have a length of 0 mm. Thus, there exist no serial singularities in the mechanism. The Type-3 singularities exist when a parallel mechanism encounters both serial and parallel singularities. Since there exist no serial singularities in the mechanism, only the parallel singularities Equations (10)–(12) are analyzed to extract the feasible workspace of the mechanism.

3.2. Results of Singularity Analysis Using the CAD Algorithm

In order to extract the tilt limits for the mechanism from the workspace, the SIROPA library of Maple is employed [23]. The vector coordinates of base from Equation (1), the end-effector coordinates from Equation (4) and the inverse kinematics Equations (6a)–(6c) that map the base to the end-effector are used as inputs for the *CreateManipulator* function of the SIROPA library. This function virtually constructs the 3-SPS-U manipulator in Maple [23,24]. The tilt and azimuth angles are set as the pose variables and the lengths (l_1, l_2 and l_3) are set as the articular variables for computations. Then, using the *ParallelSingularities* function of SIROPA, the Type-2 singularities Equations (10)–(12) are generated. The actual prototype that will be incorporated with the piping-inspection robot has the design parameters r_f as 11 mm and h = 0.6 [16]. As an experimental validation is carried out in the upcoming section, a scaled value of r_f = 56.7 mm for h = 0.6 is chosen for the singularity analysis [25]. As the constraint equations require joint limits for analysis, the minimum and maximum lengths for each prismatic spring are set as l_{min} = 40 mm and l_{max} = 106 mm. The Cylindrical Algebraic Decomposition (CAD) algorithm is then employed to analyze the workspace of the mechanism. This approach was proposed for parallel robots by Chablat et al. [26] wherein the workspace, as well as the joint space, are decomposed into cells from $\mathcal{C}_1, ..., \mathcal{C}_k$. The steps involved in this computation are:

- $\mathcal{C}_k$ is an open connected subset of the workspace
- for all pose values in $\mathcal{C}_k$, the direct (resp. inverse) kinematics problem has a constant number of solutions
- $\mathcal{C}_k$ is maximal in the sense if $\mathcal{C}_k$ is contained in a set $\mathcal{E}$, then $\mathcal{E}$ does not satisfy the first or second condition.

In Maple, the SIROPA library is exploited to use the CAD algorithm for the parallel mechanism [26] and the analysis is carried out as follows:

- Computation of a subset of the workspace (resp. joint space) where the number of solutions changes: the *discriminant variety*
- Description of the complementary of the discriminant variety in connected cells: the *Generic Cylindrical Algebraic Decomposition*
- Connecting the cells that belong to the same connected component of the complementary of the discriminant variety: *interval comparisons*

Discriminant varieties can be computed using basic and well-known tools from computer algebra such as the Groebner bases [27]. A general framework for computing such objects is available through the *RootFinding*[*Parametric*] function of Maple. For extracting the workspace of the mechanism, the CAD algorithm combined with the parametric root-finding technique of Maple is employed to find solutions to the DKP for the tilt and azimuth angles [28]. This is carried out using the *CellDecompositionPlus* function of the SIROPA library in Maple [23,24]. The parallel singularity Equations (10)–(12) and the joint limit Equations (6a)–(6c) are transformed as inequalities for isolating the aspects around the home pose. The results of the workspace generated by the CAD algorithm for the tilt and azimuth angles are represented in Figure 5. In Figure 5a, the blue regions indicate the feasible workspaces for the mechanism where there exists one solution to the DKP. The white regions are the zones where there exists no solution. Around the home pose, a larger workspace could be observed. Beyond the singularity zones, it is also possible to identify solutions for the direct kinematics problem, especially between $-\pi$ to $-2\pi/3$ and $2\pi/3$ to π radians. However, in order to attain these positions, the mechanism has to cross singularity boundaries, which might lead to locking of the mechanism in singular modes. Thus, the workspace around the home pose is isolated and they are represented in Figure 5b. Based on observations from Figure 5b, the maximum tilt limits that could be attained or reached by the mechanism corresponds to $\pm\pi/6$ radians. These limits are represented by dotted orange lines in Figure 5b. Thus, for a maximum input tilt of $\alpha = \pm\pi/6$ radians, the mechanism is capable of following a circular trajectory around the home pose.

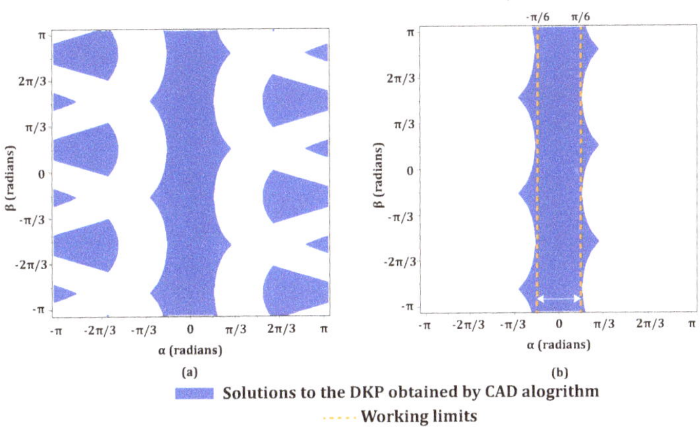

Figure 5. Representation of (**a**) aingularities and workspace zones for the mechanism at l_{min} = 40 mm and (**b**) extraction of feasible workspace around home pose at l_{min} = 40 mm.

4. Mapping the T&T Angles to the Euler Angles

The singularity analysis helped in understanding the tilt limits for the 3-S$\underline{\text{P}}$S-U parallel mechanism using the theory of T&T. However, these results must be mapped with respect to the Euler angles of the universal joint. The 3-S$\underline{\text{P}}$S-U architecture demonstrated in Figure 3a at the home pose has two degrees of freedom about the universal joint in the center. The conventional approach to identify the vector coordinates of C_i exploits the XY Euler angles about the origin A of the mechanism. In the Euler space, the two angles about the x and y axes are considered as η and ϕ. The transformation matrix $\mathbf{E}$ about the universal joint is given by:

$$\mathbf{E} = \mathbf{R}_x(\eta)\mathbf{R}_y(\phi) = \begin{bmatrix} \cos(\phi) & 0 & \sin(\phi) \\ \sin(\eta)\sin(\phi) & \cos(\eta) & -\sin(\eta)\cos(\phi) \\ -\cos(\eta)\sin(\phi) & \sin(\eta) & \cos(\eta)\cos(\phi) \end{bmatrix} \tag{14}$$

where $\mathbf{R}_x(\eta) = \begin{bmatrix} 1 & 0 & 0 \\ 0 & \cos(\eta) & -\sin(\eta) \\ 0 & \sin(\eta) & \cos(\eta) \end{bmatrix}$, $\mathbf{R}_y(\phi) = \begin{bmatrix} \cos(\phi) & 0 & \sin(\phi) \\ 0 & 1 & 0 \\ -\sin(\phi) & 0 & \cos(\phi) \end{bmatrix}$.

In Equation (14), $\mathbf{E} \in SE(3)$ represents the spatial transformation matrix obtained from the Euler angles of the universal joint. This matrix is used to identify the end-effector coordinates, which are given by:

$$\mathbf{c}_i^1 = \mathbf{E}\,\mathbf{o}_i^1, \quad \text{with} \quad i = 1,2,3 \tag{15}$$

In Equation (15), $\mathbf{o}_i^1$ are the coordinates of C_i in the mobile reference frame $\sum_1$ whose vector equation is given in Equation (4). The inverse kinematic model that determines the length of the prismatic springs is computed by calculating the distance between the base and end-effector at the home pose and working conditions. The constraint equations in the Euler space are given by:

$$2r_f^2(\cos(\phi)(h^2\cos(\eta) - 1) - h\sin(\phi)(\cos(\eta) + 1) + h^2 + 1) = l_1^2 \tag{16a}$$

$$\begin{aligned} 2r_f^2(\tfrac{\sqrt{3}}{2}\sin(\eta)\left(h\cos(\phi) + h + \tfrac{\sin(\phi)}{2}\right) + \cos(\phi)\left(h^2\cos(\eta) - \tfrac{1}{4}\right) + \tfrac{h\sin(\phi)}{2}(1 + \cos(\eta)) \\ +h^2 - \tfrac{3\cos(\eta)}{4} + 1) = l_2^2 \end{aligned} \tag{16b}$$

$$\begin{aligned} -2r_f^2(\tfrac{\sqrt{3}}{2}\sin(\eta)\left(h\cos(\phi) + h + \tfrac{\sin(\phi)}{2}\right) - \cos(\phi)\left(h^2\cos(\eta) - \tfrac{1}{4}\right) - \tfrac{h\sin(\phi)}{2}(1 + \cos(\eta)) \\ -h^2 + \tfrac{3\cos(\eta)}{4} - 1) = l_3^2 \end{aligned} \tag{16c}$$

In the T&T space, the mechanism is initially tilted to its maximum limit, i.e., $\alpha = \pi/6$ radians. Once this limit is reached, α remains a constant, whereas β changes from 0 to 2π radians, thereby achieving a complete revolution in the counter-clockwise direction. In the Euler space, the initial tilt is performed by keeping η as 0 radians and tilting $\phi = \pi/6$ radians. Unlike the T&T space, both the angles are modified as a function of the angle β in the Euler space to achieve a complete revolution. The actuation sequence of the mechanism in both spaces is provided in Table 1. This table demostrates the relation between the T&T space and the Euler space. For each position of β between 0 to 2π, each and every value of the angles η and ϕ is computed as a function of α and β as demonstrated in Table 1 to achieve the mapping in the Euler space. A simulation of this mapping was performed in MATLAB and the video link is provided at the bottom of this page (Video link for the simulation in MATLAB: Available online: https://drive.google.com/file/d/1xXPy4oU_Q04Sgcu1kcQn9qt_9WIT2N5E/view?usp=share_link (accessed on: 15 January 2023) 1× Speed, Available online: https://drive.google.com/file/d/1RrIgmipxXAAhKP71hspWaot4OM5XT4dL/view?usp=share_link (accessed on 15 January 2023) 0.4× Speed). The postures of the mechanism in both the spaces at the home pose and working conditions are represented in Figure 6.

Table 1. The actuation angles for the mechanism in the T&T space and the Euler space.

Movement Type	Tilt Space		Euler Space	
	α (Radians)	β (Radians)	η (Radians)	ϕ (Radians)
Initial tilt	$0 \rightarrow \pi/6$	0	0	$0 \rightarrow \pi/6$
Circular path	$\pi/6$	$0 \rightarrow 2\pi$	$\alpha \sin(\beta)$	$\alpha \cos(\beta)$
Return to home	$\pi/6 \rightarrow 0$	0	0	$\pi/6 \rightarrow 0$

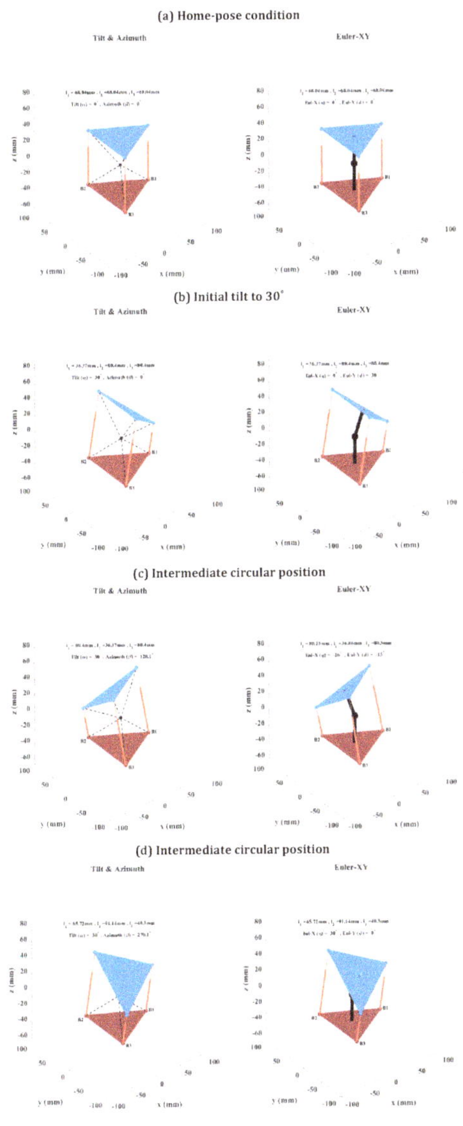

Figure 6. Postures of the 3-SPS-U mechanism in the T&T space and the Euler space.

5. Experimental Validation

In this section, the mapping theory is validated by performing a circular trajectory on the parallel mechanism. The prototype of the 3-SP̲S-U parallel mechanism is realized at LS2N, France with a universal joint of length 68 mm. As explained earlier, a value of r_f = 56.7 mm was considered for h = 0.6. The base and end-effector of the mechanism are realized by rapid prototyping. Using tension springs and supports, the base and end-effector are connected. The control of this mechanism is carried out using a BeagleBone (BB) black microcomputer. Three Maxon DC motors coupled with quadrature encoders are employed to tilt the mechanism. In order to have a static model and higher torques, a planetary gearhead with a reduction of 1621:1 is coupled to each DC motor. A 36/2 servo-controller is employed for each motor to control the position. The three DC motors are assembled on a platform, which is made by rapid prototyping. As the distance between the gearbox output shaft and the mechanism was higher, an extended shaft with a flange coupling is employed to connect to the gearbox shaft. A 40 mm diameter pulley is assembled on the extended shaft and a cable of 1.5 mm diameter is wound over this pulley. The other end of the cable connects to the end-effector mounting support by passing through each spring. For eliminating the effects of radial play in the pulley shaft, flange couplings are employed. The cables are assembled in such a way that at the home pose, the tension along the cables is high. The entire experimental setup and its digital model are represented below in Figure 7. The experiments are carried out for two orientations: vertical and horizontal under no-load conditions. The vertical orientation is in line with the orientation of the piping-inspection robot inside a vertical pipeline l and this posture is represented in Figure 7a. The horizontal orientation is in-line with the orientation of the robot inside a horizontal pipeline, as shown in Figure 1.

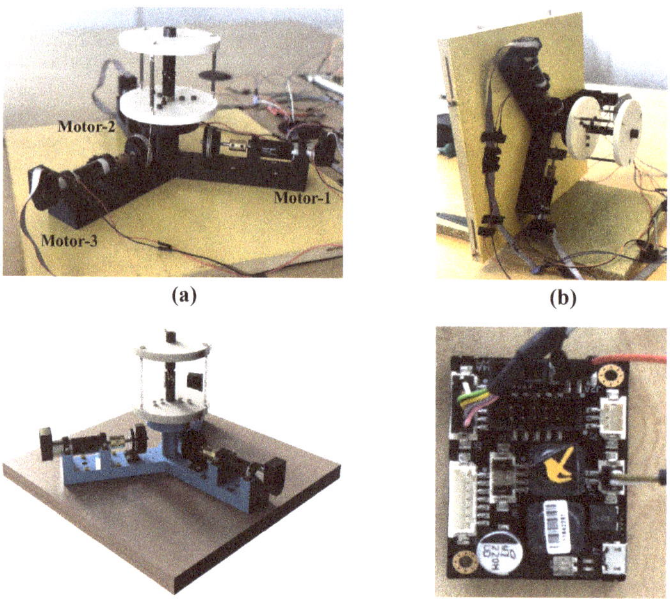

Figure 7. Representation of the (**a**) experimental setup of the mechanism in vertical orientation, (**b**) experimental setup of the mechanism in horizontal orientation, (**c**) digital model of the experimental setup and (**d**) one of the three ESCON 36/2 servo-controllers.

5.1. Trajectory Generation

For the experiments, at each instance of the control loop, the tilt angles of the Euler space η and ϕ were passed as inputs to solve the IKP. This actuation is carried out through the cables that pass through the springs of the mechanism. The pose variables and the articular variables for the mechanism are given by $\mathbf{q} = [\eta, \phi]$ and $\rho = [l_1, l_2, l_3]$ [25]. The length of the springs is calculated using Equations (16a)–(16c) in the Euler space. The IKP for each spring is converted into angular displacements of the pulley, which will be the target position for each DC motor. The equation for the desired angular position of the pulley with respect to the IKP is given by:

$$\theta_{di} = \frac{(l_i - l_{home})}{r}, \text{ with } i = 1,2,3 \tag{17}$$

In Equation (17), θ_{di} is the desired angular position calculated from IKP for given input tilt angles η and ϕ. The parameter r indicates the pulley radius, which is 20 mm. At the home position, the value of l_{home} is 68 mm. For every angular displacement of the DC motor, the encoder data are passed to the BB black ports. The angular displacement of the pulley from the encoder channel information can be calculated by:

$$\theta_{mi} = \frac{E_i \pi}{2CG}, \text{ with } i = 1,2,3 \tag{18}$$

In Equation (18), E_i indicates the output data of the encoder channels of Motor-i. The angle θ_{mi} is the measured angular displacement at the output shaft of the gearbox. C indicates the counts per revolution of the encoder, which is 500, and G is the reduction ratio of the gearbox, and these values are taken from the technical datasheet of the encoder and the gearbox unit. At a given tilt angle, the mechanism tries to attain the position θ_{di}. For the trajectory planning, the fifth-degree polynomial equation proposed by Khalil et al. [29] is employed. This equation is used to define the position $s(t)$, velocity $\dot{s}(t)$ and acceleration $\ddot{s}(t)$ for the mechanism and they are given by:

$$s(t) = 10\left(\frac{t}{t_f}\right)^3 - 15\left(\frac{t}{t_f}\right)^4 + 6\left(\frac{t}{t_f}\right)^5 \tag{19}$$

$$\dot{s}(t) = 30\left(\frac{t^2}{t_f^3}\right) - 60\left(\frac{t^3}{t_f^4}\right) + 30\left(\frac{t^4}{t_f^5}\right) \tag{20}$$

$$\ddot{s}(t) = 60\left(\frac{t}{t_f^3}\right) - 180\left(\frac{t^2}{t_f^4}\right) + 120\left(\frac{t^3}{t_f^5}\right) \tag{21}$$

An initial linear tilt is performed for the mechanism before performing the circular trajectory. For the trajectory generation, the simulation time t_f is estimated and it is given by:

$$t_f = \max(\gamma_V, \sqrt{\gamma_A}) \tag{22}$$

In Equation (22), γ_V is the ratio of the maximum joint velocity to the speed of the motor and γ_A is the ratio of the maximum joint acceleration to the acceleration of the motor [8]. In order to demonstrate the circular trajectory, the T&T angles must be mapped to the Euler space in the control law. Using Table 1, the mapping is incorporated in the control law and it is implemented using the equation:

$$\mathbf{D}(t) = \begin{bmatrix} \eta \\ \phi \end{bmatrix} = \begin{bmatrix} \alpha \sin(\beta s(t)) \\ \alpha \cos(\beta s(t)) \end{bmatrix} \tag{23}$$

Equation (23) represents the input tilt angles η and ϕ in the Euler space as a function of the tilt and azimuth angles of the T&T space. α represents the tilt angle and the azimuth value β goes from 0 to 2π radians for a constant tilt to perform a counter-clockwise circular trajectory in line with the simulation. The corresponding angular velocities and accelerations are given by:

$$\mathbf{V}(t) = \begin{bmatrix} \alpha\beta\dot{s}(t)\cos(\beta s(t)) \\ -\alpha\beta\dot{s}(t)\sin(\beta s(t)) \end{bmatrix} \quad (24)$$

$$\mathbf{A}(t) = \begin{bmatrix} \alpha\beta\ddot{s}(t)\cos(\beta s(t)) - \alpha\beta^2\dot{s}(t)^2\sin(\beta s(t)) \\ -\alpha\beta\ddot{s}(t)\sin(\beta s(t)) - \alpha\beta^2\dot{s}(t)^2\cos(\beta s(t)) \end{bmatrix} \quad (25)$$

In order to calculate the Cartesian velocities of the prismatic joints, the Jacobian matrices are employed. The equation is given by:

$$\begin{bmatrix} \dot{l_1} \\ \dot{l_2} \\ \dot{l_3} \end{bmatrix} = \mathbf{J}_c \begin{bmatrix} \dot{\eta} \\ \dot{\phi} \end{bmatrix}, \text{with } \mathbf{J}_c = \begin{bmatrix} \frac{\partial l_1}{\partial \eta} & \frac{\partial l_1}{\partial \phi} \\ \frac{\partial l_2}{\partial \eta} & \frac{\partial l_2}{\partial \phi} \\ \frac{\partial l_3}{\partial \eta} & \frac{\partial l_3}{\partial \phi} \end{bmatrix} \quad (26)$$

In Equation (26), $\mathbf{J}_c$ represents the direct kinematics matrix ($\mathbf{A}$). Using Equation (26), the joint velocities are estimated, following which the desired angular velocities of the output shaft can be estimated. The joint coordinate vector can be estimated by the equation:

$$\mathbf{q}(t) = \mathbf{f}\,(\mathbf{D}(t)) \quad (27)$$

In Equation (27), $\mathbf{f} = [l_1, l_2, l_3]^T$ is the vector that contains the solution of the IKP for the input tilt angles. The joint velocities can be computed with the help of the Jacobian matrix $\mathbf{J}_c$ using the equation:

$$\dot{\mathbf{q}}(t) = \mathbf{J}_c\mathbf{V}(t) \quad (28)$$

The accelerations in the joints can be computed using the Jacobian matrix and its time derivative, whose equation is given by:

$$\ddot{\mathbf{q}}(t) = \mathbf{J}_c\mathbf{A}(t) + \dot{\mathbf{J}}_c\mathbf{V}(t) \quad (29)$$

5.2. PID Control Law

In [10] a force control algorithm was implemented for the control of the piping-inspection robot represented in Figure 1. The same algorithm is employed here for the control of the parallel mechanism. By applying motor torques and current, the mechanism tilts to attain a certain position. Once the desired position is reached, the current and torque are cut off for attaining a static phase. For each angular position of the DC motor, the encoder position data are transmitted to the user by the BB black. However, the output data from the encoders are not directly used to calculate the solutions to the IKP as there exists a significant difference between the desired and measured positions. These differences are caused by factors such as motor inertia, frictional effects and inertial forces. In order to compensate for these factors, a closed-loop PID controller is employed, which tries to minimize the errors between the desired and measured positions. The PID control scheme as provided in [8,30] for calculating the motor torques is given by:

$$\Gamma = J\left(\ddot{\theta}_{di} + K_P(\theta_{di} - \theta_{mi}) + K_D(\dot{\theta}_{di} - \dot{\theta}_{mi}) + K_I\int_0^t(\theta_{di} - \theta_{mi})\right), \text{with } i = 1,2,3 \quad (30)$$

where θ_{di}, θ_{mi} are the desired and measured positions, $\dot{\theta}_{di}$, $\dot{\theta}_{mi}$ are the desired and measured velocities, $\ddot{\theta}_{di}$ is the desired acceleration.

In Equation (30), Γ represents the output torque at the gearbox after PID correction. From the technical datasheet of the gearbox unit, the inertia J is taken as 4.1×10^{-7} kg·m^2. The desired positions are computed using the mapping Equation (23) and the IKP relations. The desired velocity and acceleration of each joint is computed with the help of Equations (28) and (29). The measured positions are computed using Equation (18).

In Equation (30), K_P is the proportionality term, which examines the error term and responds proportionally. The integral term K_I tries to reduce the error at each time instant. The overshoot of value from the desired position is compensated by the difference term K_D. Thus, the PID in general takes into account the past step, the present step and the future step such that the error is reduced [31]. The values for the PID terms of Equation (30) are given by:

$$\omega = \frac{k_t k_e}{RJ} = 14 \text{ rad/s}$$
$$K_P = 3\omega^2 = 588\,,\ K_D = 3\omega = 42\,,\ K_I = \omega^3 = 2744 \tag{31}$$

In Equation (31), ω is a parameter that is a function of the torque constant (k_t), the speed constant (k_e), the motor resistance (R) and its inertia (J). It helps to determine the PID parameters K_P, K_D and K_I [32] as per the relation given in Equation (31). This approach was tested and validated on PID control of other robots at LS2N [8,33]. The closed-loop system that is incorporated for the control of the 3-S<u>P</u>S-U mechanism is represented below in Figure 8.

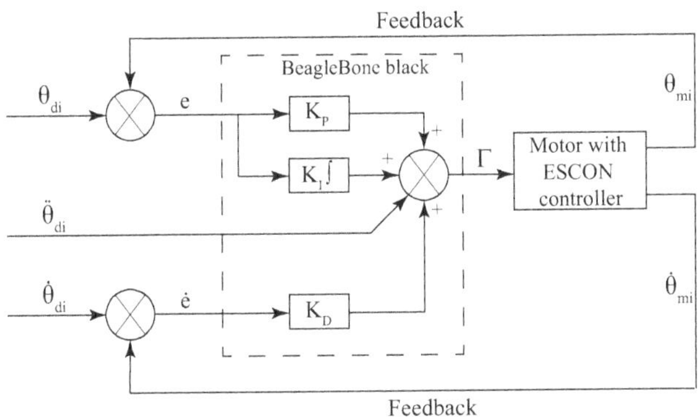

Figure 8. Closed-loop PID controller employed for the mechanism [8].

In Figure 8, e and ė indicate the error and its derivative, which are computed based on the difference between the desired and measured data. Before the start of experiments, the servo-controllers are calibrated in a computer. The saturation current is used as a reference for controlling the direction of rotation of the DC motors. For safer operations, 90% and 10% duty cycles are set as the limits for the rotation of motors in either direction. The PWM calibration of each servo-controller is carried out as per Table 2. The force control algorithm is executed in the C language using the BB black microcomputer. The link for the algorithm is provided at the bottom of this page (GitHub link for the algorithm: Available online: https://github.com/stvt1991/Control-of-3-SPS-U-Tensegrity-mechanism (accessed on 20 January 2023)).

Table 2. ESCON controller calibration for the force control algorithm.

PWM Duty	Nominal Current (A)	Motor Torque (mN·m)	Motor Speed (rpm)
10%	−0.177	−6.05	3720 (Counter-clockwise)
50% (idle)	0	0	0
90%	0.177	6.05	3720 (Clockwise)

5.3. Results of Experiments

For performing the circular trajectory, it is necessary to define the tilt angle α from the T&T space. An initial tilt along one of the motors (Motor-1) is performed. A circular path in a counter-clockwise direction is then traced by the mechanism. At the end of the circular trajectory, the mechanism is pulled back to its home pose. For the circular trajectory, a tilt of $\alpha = \pi/10$ radians within the workspace of the mechanism is chosen. The circular trajectory experiment is performed in the following sequence:

Home $\rightarrow$ Motor-1: linear tilt $\rightarrow$ Circular path (counter-clockwise) $\rightarrow$ Home: linear tilt

The linear tilt is performed before and after the circular trajectory for a period of 9 s. The circular path is performed for 35 s. Owing to the sleep routines in the control loop, the time taken to initiate the circular trajectory after the initial linear tilt is longer. The total time to perform the experiments is around 124 s. The plot of the joint positions at each instance of the experiments is represented in Figure 9 for both orientations of the mechanism. The initial linear tilt is performed from 0 to $\pi/10$ radians. The circular trajectory is created from this tilt position in the counter-clockwise direction. During the experimental cycle, the prismatic springs extended to a maximum length of 84.7 mm from the home position. The intermediate change in prismatic lengths can be observed in Figure 9a–f between 50 and 100 s for both orientations of the mechanism. The springs connected to each motor also reached a minimum length of 50 mm at $\pi/10$ radians. The zones between 75.9 and 84.6 mm describe the redundancy of the mechanism where one of the springs has limited effects in tilting the mechanism. The error between the measured and desired data can be calculated using the equation:

$$Error = \frac{(\theta_{mi} - \theta_{di})180}{\pi} \tag{32}$$

The plots of the error for both orientations of the mechanism are shown in Figure 10. Due to the frequency issues with the micro-computer, noise could be observed in the final plot. From Figure 10, the error data were found to be between -0.022° and 0.036° in the vertical orientation and between -0.025° and 0.025° in the horizontal orientation. These values were found to be closer to 0, thanks to the calibration of the PID parameters. Similarly, using the current flow data from the algorithm, the output torques were calculated using the motor torque constant. The plots of the motor torques in both orientations are shown in Figure 11. The torque values were between -0.0059 N·m and 0.0053 N·m in the vertical orientation and between -0.0057 N·m and 0.0058 in the horizontal orientation. It could be observed that these values were similar in both orientations. This is due to the fact that the self-weight of the parallel mechanism had a negligible influence on the motor torques, especially for the horizontal configuration. However, with the presence of external loading, the operating torques might vary significantly in both orientations. From Figure 11 it can also be observed that the operating torques were well within the nominal torque rating of the DC motor. A video link that demonstrates the generation of the circular trajectory in the Euler space of the mechanism is provided at the bottom of this page (Video link for the circular trajectory experiments: Available online: https://uncloud.univ-nantes.fr/index.php/s/b96nfCRTcBESzGo (accessed on 20 January 2023).

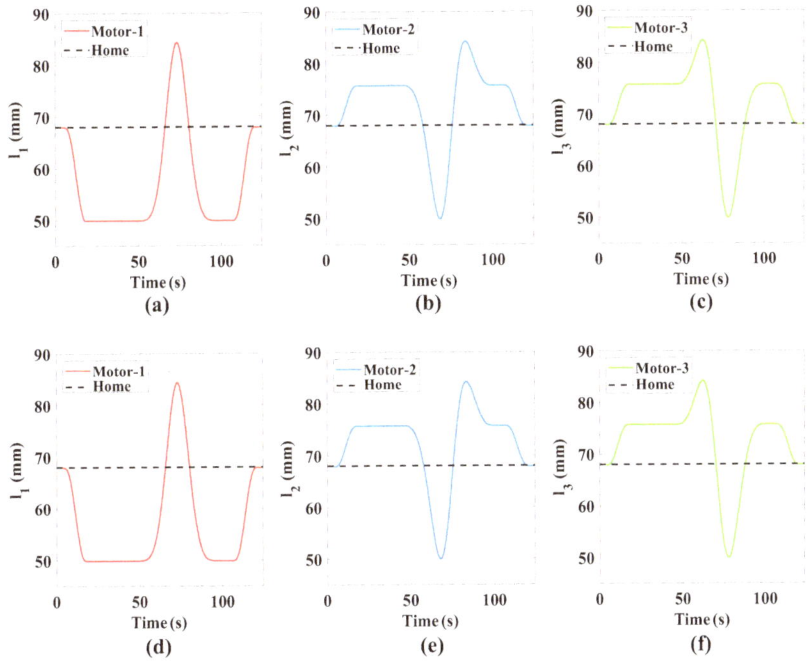

Figure 9. Position of prismatic springs along the circular trajectory in the (**a**–**c**): vertical and (**d**–**f**): horizontal orientations of the mechanism.

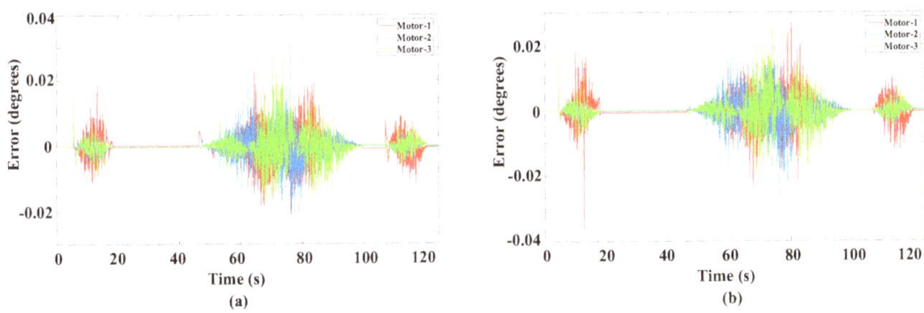

Figure 10. Joint position errors for the (**a**) vertical and (**b**) horizontal orientations of the mechanism.

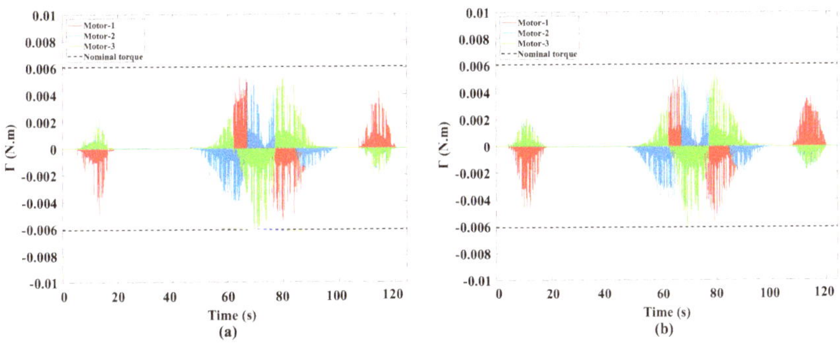

Figure 11. Motor torques generated on each motor during operation in (**a**) vertical and (**b**) horizontal orientations of the mechanism.

6. Discussion and Perspectives

This article presented the application of the theory of Tilt and Torsion (T&T) to a spherical parallel mechanism. The estimation of the vector coordinates as well as the inverse kinematics problem is solved conventionally using the Euler angles of the universal joint. However, the singularity approach using these angles will require the analysis of the workspace by considering the presence of both tilt angles, i.e., η and ϕ [15]. This computation was time-consuming and it will be even more complicated to study the workspaces of the mechanism when a stacked model is analyzed as there will be four tilt angles. In order to overcome this issue, the T&T space was considered in this study. The advantage of this method is that for a constant tilt angle, the azimuth can vary between 0 and 2π radians or vice versa. The computation was much simpler and consumes less time through this approach when compared to the Euler space. The mapping was successfully implemented from the T&T space to the Euler space. The numerical simulations were first validated for mapping using MATLAB. Following this, experiments were performed on the prototype of the mechanism. The T&T space was mapped to the Euler space by using the mapping equations and the mechanism was made to perform a counter-clockwise circular trajectory in line with the simulations. At each instance of the trajectory, the equivalent $X - Y$ Euler angles were mapped from the T&T space. The experiments also helped in calculating the error data as well as the torque generated on each motor. It was found that the operating torques were well within the nominal rating of the motor units employed. Since this parallel mechanism is proposed to be implemented in the piping-inspection robot that was presented in Figure 1, the circular trajectory can prove vital to align the axis of the robot assembly with the axis of the pipeline, especially when the robot encounters a pipe bend or a junction. This alignment can be performed using control systems that incorporate the mapping equations and thereby perform a circular path according to the profile of the pipe bend to ensure alignment.

The application of this mapping is proposed to be extended to the study of a two-stage (stacked) model of the 3-S$\underline{P}$S-U mechanism. The tilt limits generated by the single stage were around $\pm\pi/6$ radians. These limits were not sufficient to assist the piping-inspection robot to overcome a bend or a junction. With a stacked model, higher tilt limits could be obtained. The singularity analysis of the stacked structure in the Euler space will be computationally expensive due to the presence of four tilt angles. The T&T space will thus be employed and it will lead to the analysis of singularities with two tilt angles and a common azimuth. Moreover, the T&T equations facilitate an easier computational approach to calculate the value of the applied forces on the cables of the mechanism, as demonstrated in [17]. This will be very useful for the prototyping and control strategy of the stacked model in the future. Advanced industrial controllers such as the EtherCAT are proposed to be employed for the control strategy of the mechanism, as the existing BB black micro-computer has some frequency issues. This controller will also help provide an easy interfacing with the piping-inspection robot as well as the parallel mechanism and all these modules could be controlled from a central unit.

Author Contributions: Conceptualization, S.V. and D.C.; methodology, S.V.; numerical simulation, S.V.; experimental validation, S.V. and D.C.; investigation, S.V. and D.C.; writing—original draft presentation, S.V. and D.C.; writing—review and editing, S.V. and D.C. All authors have read and agreed to the published version of the manuscript.

Funding: This work has been funded by the CNRS through the "RobotVer" project within the framework of the project titled PEPS—"Ingénierie Inspirée par la Nature".

Data Availability Statement: Not applicable.

Acknowledgments: We would like to thank Stéphane Jolivet, Philippe Lemoine and Denis Creusot from LS2N for their extended support in the prototyping of the mechanism and the design of the circuit for the control part.

Conflicts of Interest: The authors declare no conflict of interest.

Abbreviations

The following abbreviations are used in this manuscript:

PKM	Parallel Kinematic Mechanism
T&T	Tilt and Torsion
IKP	Inverse Kinematics Problem
DKP	Direct Kinematics Problem
BB Black	BeagleBone Black

Nomenclature

$\mathbf{b}_i$	Vector coordinates for base B_i
$\mathbf{c}_i$	Vector coordinates of mobile platform C_i
r_f	Distance of spring mounting point in mm
h	Design constant of the mechanism (no unit)
$\mathbf{R}$	Tilt and Azimuth transformation matrix
l_i	Length of ith prismatic spring
α	Tilt angle in T&T space
β	Azimuth angle in T&T space
$\mathbf{A}$	Direct kinematics matrix
$\mathbf{B}$	Inverse kinematics matrix
$\mathcal{D}$	Determinant value of the direct kinematics matrix
$\mathbf{E}$	Euler angle transformation matrix
η, ϕ	XY tilt/rotation angles in the Euler space
θ_{di}, θ_{mi}	Desired and measured angular positions in rad
$\dot{\theta_{di}}, \dot{\theta_{mi}}$	Desired and measured angular velocities in rad/s
$\ddot{\theta_{di}}$	Desired angular acceleration in rad/s^2
ω	Pulsation in rad/s

References

1. Merlet, J.P. *Parallel Robots*; Springer Science & Business Media: Berlin/Heidelberg, Germany, 2006; Volume 128.
2. Pritschow, G. Parallel kinematic machines (PKM)–limitations and new solutions. *CIRP Ann.* **2000**, *49*, 275–280. [CrossRef]
3. Bonev, I.A.; Zlatanov, D.; Gosselin, C.M. Advantages of the modified Euler angles in the design and control of PKMs. In Proceedings of the 2002 Parallel Kinematic Machines International Conference, Chemnitz, Germany, 23–25 April 2002; pp. 171–188.
4. Bonev, I.A. *Geometric Analysis of Parallel Mechanisms*; Université Laval: Quebec City, QC, Canada, 2002.
5. Bonev, I.A.; Ryu, J. Orientation workspace analysis of 6-DOF parallel manipulators. In Proceedings of the International Design Engineering Technical Conferences and Computers and Information in Engineering Conference, Las Vegas, NV, USA, 12–16 September 1999; American Society of Mechanical Engineers: New York City, NY, USA, 1999; Volume 19715, pp. 281–288.
6. Briot, S.; Bonev, I.A. Singularity analysis of zero-torsion parallel mechanisms. In Proceedings of the 2008 IEEE/RSJ International Conference on Intelligent Robots and Systems, Nice, France, 22–26 September 2008; IEEE: Piscataway, NJ, USA, 2008; pp. 1952–1957.
7. Tao, Z.; An, Q. Interference analysis and workspace optimization of 3-RRR spherical parallel mechanism. *Mech. Mach. Theory* **2013**, *69*, 62–72. [CrossRef]
8. Caro, S.; Chablat, D.; Lemoine, P.; Wenger, P. Kinematic analysis and trajectory planning of the Orthoglide 5-axis. In Proceedings of the International Design Engineering Technical Conferences and Computers and Information in Engineering Conference, Boston, MA, USA, 2–5 August 2015; American Society of Mechanical Engineers: New York City, NY, USA, 2015; Volume 57144, p. V05CT08A004.
9. Yang, G.; Teo, T.J.; Chen, I.M.; Lin, W. Analysis and design of a 3-DOF flexure-based zero-torsion parallel manipulator for nano-alignment applications. In Proceedings of the 2011 IEEE International Conference on Robotics and Automation, Shanghai, China, 9–13 May 2011; IEEE: Piscataway, NJ, USA, 2011; pp. 2751–2756.
10. Venkateswaran, S.; Chablat, D.; Boyer, F. Numerical and experimental validation of the prototype of a bio-inspired piping inspection robot. *Robotics* **2019**, *8*, 32. [CrossRef]
11. Zhang, Y.; Zhang, M.; Sun, H.; Jia, Q. Design and motion analysis of a flexible squirm pipe robot. In Proceedings of the 2010 International Conference on Intelligent System Design and Engineering Application, Changsha, China, 13–14 October 2010; IEEE: Piscataway, NJ, USA, 2010; Volume 1, pp. 527–531.
12. Kwon, Y.S.; Lim, H.; Jung, E.J.; Yi, B.J. Design and motion planning of a two-moduled indoor pipeline inspection robot. In Proceedings of the 2008 IEEE International Conference on Robotics and Automation, Pasadena, CA, USA, 19–23 May 2008; IEEE: Piscataway, NJ, USA, 2008; pp. 3998–4004.

13. Ryew, S.; Baik, S.H.; Ryu, S.W.; Jung, K.M.; Roh, S.G.; Choi, H.R. In-pipe inspection robot system with active steering mechanism. In Proceedings of the 2000 IEEE/RSJ International Conference on Intelligent Robots and Systems (IROS 2000) (Cat. No. 00CH37113), Takamatsu, Japan, 31 October 2000–5 November 2000; IEEE: Piscataway, NJ, USA, 2000; Volume 3, pp. 1652–1657.
14. Venkateswaran, S.; Chablat, D. A new inspection robot for pipelines with bends and junctions. In *Advances in Mechanism and Machine Science, Proceedings of the 15th IFToMM World Congress on Mechanism and Machine Science, Krakow, Poland, 30 June–4 July 2019*; Springer International Publishing: Cham, Switzerland, 2019; pp. 33–42.
15. Venkateswaran, S.; Chablat, D. Singularity and Workspace Analysis of 3-SPSU and 4-SPSU Tensegrity Mechanisms. In *Advances in Robot Kinematics 2020*; Springer International Publishing: Cham, Switzerland, 2021; pp. 226–233.
16. Venkateswaran, S.; Chablat, D.; Hamon, P. An optimal design of a flexible piping inspection robot. *J. Mech. Robot.* **2021**, *13*, 035002. [CrossRef]
17. Venkateswaran, S.; Furet, M.; Chablat, D.; Wenger, P. Design and analysis of a tensegrity mechanism for a bio-inspired robot. In Proceedings of the International Design Engineering Technical Conferences and Computers and Information in Engineering Conference, Anaheim, CA, USA, 18–21 August 2019; American Society of Mechanical Engineers: New York City, NY, USA, 2019; Volume 59230, p. V05AT07A026.
18. Yigit, C.B.; Boyraz, P. Design and modelling of a cable-driven parallel-series hybrid variable stiffness joint mechanism for robotics. *Mech. Sci.* **2017**, *8*, 65–77. [CrossRef]
19. Alici, G.; Shirinzadeh, B. Topology optimisation and singularity analysis of a 3-SPS parallel manipulator with a passive constraining spherical joint. *Mech. Mach. Theory* **2004**, *39*, 215–235. [CrossRef]
20. Gosselin, C.; Angeles, J. Singularity analysis of closed-loop kinematic chains. *IEEE Trans. Robot. Autom.* **1990**, *6*, 281–290. [CrossRef]
21. Wenger, P.; Chablat, D. Definition sets for the direct kinematics of parallel manipulators. In Proceedings of the 1997 8th International Conference on Advanced Robotics, ICAR'97, Monterey, CA, USA, 7–9 July 1997; IEEE: Piscataway, NJ, USA, 1997; pp. 859–864.
22. Chablat, D.; Wenger, P. Working modes and aspects in fully parallel manipulators. In Proceedings of the 1998 IEEE International Conference on Robotics and Automation (Cat. No. 98CH36146), Leuven, Belgium, 20–20 May 1998; IEEE: Piscataway, NJ, USA, 1998; Volume 3, pp. 1964–1969.
23. Jha, R.; Chablat, D.; Baron, L.; Rouillier, F.; Moroz, G. Workspace, joint space and singularities of a family of delta-like robot. *Mech. Mach. Theory* **2018**, *127*, 73–95. [CrossRef]
24. Chablat, D.; Moroz, G.; Rouillier, F.; Wenger, P. Using Maple to analyse parallel robots. In *Maple in Mathematics Education and Research, Proceedings of the Third Maple Conference, MC 2019, Waterloo, ON, Canada, 15–17 October 2019*; Proceedings 3; Springer International Publishing: Cham, Switzerland, 2020; pp. 50–64.
25. Venkateswaran, S.; Chablat, D. Trajectory Planning for a 3-SPS-U Tensegrity Mechanism. In Proceedings of the International Design Engineering Technical Conferences and Computers and Information in Engineering Conference, Virtual, 17–20 August 2021; American Society of Mechanical Engineers: New York City, NY, USA, 2021; Volume 85451, p. V08BT08A004.
26. Chablat, D.; Moroz, G.; Wenger, P. Uniqueness domains and non singular assembly mode changing trajectories.In Proceedings of the 2011 IEEE International Conference on Robotics and Automation, Shanghai, China, 9–13 May 2011; IEEE: Piscataway, NJ, USA, 2011; pp. 3946–3951.
27. Cox, D.; Little, J.; OShea, D. *Ideals, Varieties, and Algorithms: An Introduction to Computational Algebraic Geometry and Commutative Algebra*; Springer Science & Business Media: Berlin/Heidelberg, Germany, 2013.
28. Chablat, D.; Ottaviano, E.; Moroz, G. A comparative study of 4-cable planar manipulators based on cylindrical algebraic decomposition. In Proceedings of the International Design Engineering Technical Conferences and Computers and Information in Engineering Conference, Washington, DC, USA, 28–31 August 2011; Volume 54839, pp. 1253–1262.
29. Khalil, W.; Dombre, E. *Modeling Identification and Control of Robots*; CRC Press: Boca Raton, FL, USA, 2002.
30. Jha, R.; Chablat, D.; Rouillier, F.; Moroz, G. Influence of the trajectory planning on the accuracy of the orthoglide 5-axis. In Proceedings of the International Design Engineering Technical Conferences and Computers and Information in Engineering Conference, Charlotte, NC, USA, 21–24 August 2016; American Society of Mechanical Engineers: New York City, NY, USA, 2016; Volume 50152, p. V05AT07A039.
31. Maung, M.M.; Latt, M.M.; Nwe, C.M. DC motor angular position control using PID controller with friction compensation. *Int. J. Sci. Res. Publ.* **2018**, *8*, 149. [CrossRef]
32. Lemoine, P. *Commande d'un Moteur éLectrique à Courant Continu Mettant en Mouvement une Charge Inertielle*; Technical Lab Report; Le Laboratoire des Sciences du Numérique de Nantes (LS2N): Nantes, France, 2019.
33. Fasquelle, B.; Furet, M.; Khanna, P.; Chablat, D.; Chevallereau, C.; Wenger, P. A bio-inspired 3-DOF light-weight manipulator with tensegrity X-joints. In Proceedings of the 2020 IEEE International Conference on Robotics and Automation (ICRA), Paris, France, 31 May 2020–31 August 2020; IEEE: Piscataway, NJ, USA, 2020; pp. 5054–5060.

robotics

MDPI

Article

A Reconfigurable Parallel Robot for On-Structure Machining of Large Structures

Abdur Rosyid [1], Cesare Stefanini [2] and Bashar El-Khasawneh [3,*]

[1] Center for Autonomous Robotic Systems (KUCARS), Khalifa University, Abu Dhabi P.O. Box 127788, United Arab Emirates
[2] The Biorobotics Institute and the Department of Excellence in Robotics and AI, Scuola Superiore Sant'Anna, 56127 Pisa, Italy
[3] Mechanical Engineering Department, Khalifa University, Abu Dhabi P.O. Box 127788, United Arab Emirates
* Correspondence: bashar.khasawneh@ku.ac.ae; Tel.: +971-23-124-157

Abstract: This paper presents a novel walking hybrid-kinematics robot that can be reconfigured to have three, five, and six degrees of freedom (DOFs) for adsorption machining of large structures. A symmetric 3PRPR or 3PRRR parallel mechanism with three translational (3T) DOFs is used to perform three-axis machining tasks. Three attachment pads connected to passive spherical joints are used to attach the robot to the surface of a large structure. Two or three rotational degrees of freedom can be added to the robot to adapt to a large structure's irregular surface geometry and perform five- or six-axis machining tasks. This is achieved through modular reassembly or joint locking that reconfigures the robot from a three-DOF robot to a five- or six-DOF robot. A serial module providing two rotational DOFs can be added to the 3T parallel mechanism to provide five DOFs. A parallel module, namely 3SPR or 3SU mechanism, can be added to the 3T parallel mechanism to provide six DOFs. The mobility, pose kinematics, differential kinematics, singularities, and workspace of the 3SPR and 3SU parallel mechanisms alone and combined with the 3T mechanism are discussed in this paper. It is shown that the singularities of the mechanism can be easily avoided by making the moving platform of the 3SPR or 3SU mechanism smaller than the base, limiting the range of some joints, and having an appropriate length of the links. Furthermore, a method to optimize the workspace of the mechanism was also discussed.

Keywords: walking robot; on-structure machining; machining of large structures; hybrid robot

1. Introduction

Large structures such as airplane and construction structures often need machining processes, including drilling. Robots have been used to perform those tasks to increase productivity and accuracy. Since the large structures are not practical and sometimes impossible to be moved, we need the robots to be mobile. To meet this need, there are already some existing mobile robots developed. However, many of them are based on serial ki-nematics which suffers from low compliance, particularly when the arms of such robots are significantly extended. In addition, these robots are often docked on ground when the machining tasks are performed. Hence, the rigidity of such robots when performing the machining tasks is compromised due to both the compliance of the serial robot arm and the compliance of the docking mechanism. To overcome this problem, we need a robot that is mobile and rigid at the same time. Added to these two criteria, we quite often need the robot to be able to move in confined spaces within the large structures. Hence, we also need the robot to have relatively small size. Although a serial robot can be mounted on a mobile base to be mobile, it cannot satisfy the rigidity and the small size at the same time. This is because the serial robot should be made bulky to have an acceptable rigidity. The three criteria can be satisfied by employing a parallel robot. To avoid the low compliance due to docking mechanism, on-structure parallel robots have been introduced. Such robots are attached directly to the large structures' surface while performing a machining task to the structures' surface. To change their positions, the robots also have locomotion capability. Various attachment methods such as magnetic and suction adhesion can be utilized to make the robot attach to large structures.

Among the developed on-structure robots are the walking hexapods [1–3] that require six actuators for manipulation and locomotion. A similar robot with a hybrid parallel mechanism that requires five actuators was also developed [4]. Meanwhile, typical legged robots to which direct attachment can also be potentially applied include quadrupeds and walking tree-like hexapod robots [5,6]. In addition, several legged robots in which the individual legs have parallel kinematics were also built [7–15].

A common drawback among the robots mentioned above is the use of many actuators, typically ranging from five to twelve. Although those robots can typically perform a five- or six-axis machining task with their many actuators, one often only needs a three-axis machining capability. In such a case, one should be able to use three actuators as a minimum to perform the task. For such a purpose, walking parallel robots with three degrees of freedom (DOFs), such as [16,17], were built. However, the use of such walking parallel robots is limited to only three-axis manipulation.

To provide flexibility in the type of machining tasks being able to be performed, including three-, five-, and six-axis machining tasks, this work proposes a novel on-structure robot that can be reconfigured to perform either a three-, five-, or six-axis machining task. The reconfigurability of the robot is achieved by modular reassembly or joint locking. With this reconfigurable design, one can optionally use a certain module based on the needs. For example, when only three-axis machining tasks, such as drilling perpendicular to a flat structural surface, are required, only the basic build of the robot with three actuators is needed. When a five- or six-axis machining is required, one can use the same robot with additional modules or DOFs. Hence, the advantage of the minimum number of actuators can be taken only when the basic build with three actuators is used. As soon as the additional modules or DOFs are used, the number of actuators turns to five or six, which is similar to some existing on-structure parallel robots.

Nevertheless, the proposed reconfigurable on-structure parallel robot has the following advantages in comparison with the existing on-structure parallel robots:

- Using reconfiguration by reassembly, we have a lighter robot when the basic build of the robot with three actuators is used. With other on-structure parallel robots having five or more actuators, the larger number of actuators results in larger weight. This is because such robots with five or more actuators have fixed topology, and hence, all their actuators should always be mounted on the robots although the robots are performing three-axis machining tasks.
- The modular design of the robot results in easier and less costly maintenance. When any additional module needs replacement, one can replace only that particular module, without a need to replace the whole robot. Replacement of such components is also easier due to the modular design.
- The modular robot can be sold as multiple packages using the same basic build. Each robot package is sold with its own topology and number of DOFs. The use of the same basic build for the various packages gives an advantage to the manufacturer in terms of the design, production, and sales. This is due to the modularity of the robot. On the user side, the purchase of the modular robot can be performed in multiple phases. One may start procuring the three-axis basic build with lower cost. This three-axis robot can already be used for three-axis machining tasks. At a later time, the additional modules can be procured when they are required.
- The joint motion planning in the three-axis motion, in both the reassembly and joint locking schemes, is simpler as the robot uses only three actuators. In contrast, other on-structure parallel robots with five or more actuators need to involve all of their actuators in the joint motion planning even when they are used for three-axis motion.

Hence, the contribution of this work is a novel reconfigurable design of an on-structure robot which enables the use of the robot for various types of machining tasks with weight adjustment capability through modularity, easier maintenance, lower design and manufacturing costs, modularity-based sales and procurement, and simpler three-axis motion planning.

Since the proposed design employs parallel kinematics with a typical drawback of limited workspace, this paper also discusses a method to optimize the workspace of the proposed robotic mechanism. As some performance measures of a parallel mechanism often contradict each other, a multi-objective optimization is often considered. There are three typical methods to handle the multi-objective optimization. The first is transforming the multiple objectives into a single objective with weights introduced to each objective; this is commonly called the weighted sum approach. The second is selecting the top-priority objective as the single objective while transforming the other objectives as the optimization constraints, such as in [18]. The third is implementing a Pareto optimality optimization such as in [19]; this provides multiple solutions ready for a designer to pick based on the designer's preference. References [20,21] compared some of these various methods. The optimization method used in this work combines the first and second me-thods mentioned above.

The remainder of this paper is organized as follows. Section 2 discusses the topology of the robot. Section 3 analyzes the mobility of the parallel mechanisms. Section 4 discusses the pose kinematics of the mechanisms. Section 5 derives the differential kinematics of the mechanisms and subsequently discusses the singularities of the mechanisms. Section 6 discusses the workspaces of the mechanisms. Section 7 describes how the mechanism is reconfigured. Section 8 discusses an optimization method to maximize the workspaces of the mechanisms. Section 9 presents some design considerations. Finally, Section 10 concludes the paper. In this paper, the terms "structure" and "workpiece" are used interchangeably. For convenience, we provide the list of abbreviations and symbols used in this paper in Appendices A and B.

2. Topology

The proposed three-axis robot consists of a pyramid frame and a symmetric parallel mechanism that can be either a 3PRPR topology, as shown in Figure 1, or a 3PRRR topology, as shown in Figure 2. The number three indicates that the mechanism has three symmetric limbs, whereas the letters P and R represent the prismatic and revolute joints, respectively. Both topologies provide 3T end-effector mobility without any dependent rotational motion. In both topologies, the proximal P joints and all the R joints are parallel with the three segments of the pyramid frame. The ends of the three segments of the pyramid frame are attached to spherical joints. The spherical joints are attached to attachment pads which can be implemented using vacuum suckers, magnetic pads, or other adhesion/attachment technologies. The attachment pads are to be placed on the workpiece surface. The robot is attached to the workpiece surface when the attachment pads are activated. While the robot is attached to the workpiece surface, the 3PRPR or 3PRRR mechanism makes a 3T motion to perform machining tasks such as drilling a hole perpendicular to the workpiece surface. The whole topology of the robot can be named 3A+3S+3PRPR or 3A+3S+3PRRR, where A and S, respectively, indicate the attachment and the spherical joint, whereas the + sign indicates a serial connection. For the purpose of walking motions, a support equipped with an attachment feature is attached to the moving platform. This support can be retractable. It is retracted when the robot is performing a manipulation task; it is extended and attached to the surface of a workpiece/structure when the robot is making a walking motion. The walking motions are performed by manipulating the 3PRPR or 3PRRR mechanism while either the robot is pivoting about one attachment pad as the other two attachment pads are released or the robot is translating as the retractable support and the attachment pads are, respectively, engaged and released in turn. The detailed description and the kinematics of the walking motions are beyond the scope of this paper. However, this paper has a Supplementary Video showing the robot's walking motions.

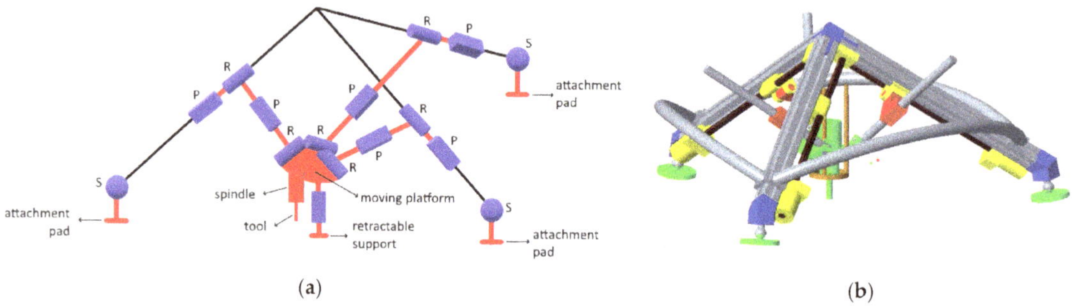

Figure 1. (**a**) Schematic and (**b**) implementation of the 3PRPR on-structure robot with 3T manipulation capability.

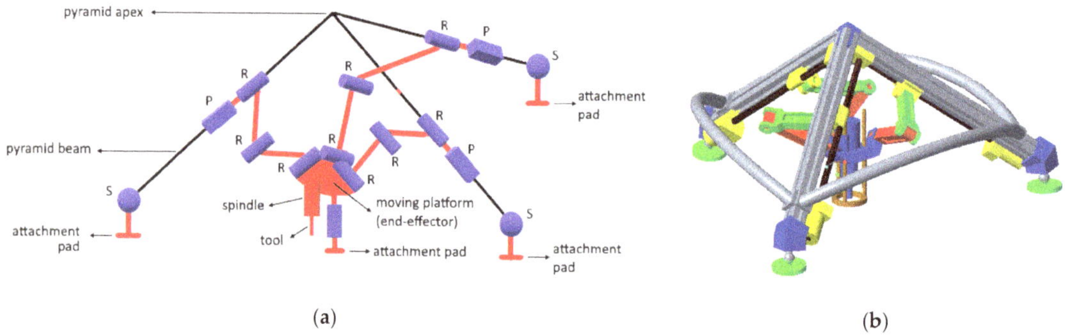

Figure 2. (**a**) Schematic and (**b**) implementation of the 3PRRR on-structure robot with 3T manipulation capability.

With the 3PRPR mechanism, the angle between two adjacent segments of the pyramid frame, in general, can be 90 degrees, less than 90 degrees, or more than 90 degrees. Making the angle 90 degrees gives several advantages, including a large workspace and moderate stability. With the 3PRRR mechanism, the angle between two adjacent segments of the pyramid frame must be 90 degrees to enable the mobility of the mechanism. Furthermore, in the 3PRPR mechanism, either the proximal or intermediate P joints can be actuated. When the 3PRRR mechanism is used, it is more practical to actuate the P joints. In the remainder of this paper, the illustration of both parallel mechanisms is represented by only one of them, the 3PRPR mechanism, for conciseness.

A typical application scenario is attaching a tool to the parallel mechanism's moving platform to perform specific tasks such as drilling or milling. Suppose the surface of the workpiece is flat or a section of a spherical surface. In that case, either the 3PRPR or 3PRRR mechanism will automatically orient the tool perpendicular to the tangent of the workpiece surface, as illustrated in Figure 3. In the case where the surface of the workpiece is not flat or not a section of a spherical surface, it is no longer guaranteed that the tool's orientation is perpendicular to the tangent of the workpiece surface. In such a case, the robot can be reconfigured by adding a module that adds two or three actuated rotational DOFs to compensate possible non-perpendicularity of the tool. In addition, the additional DOFs also provide five- or six-axis machining capability to the robot.

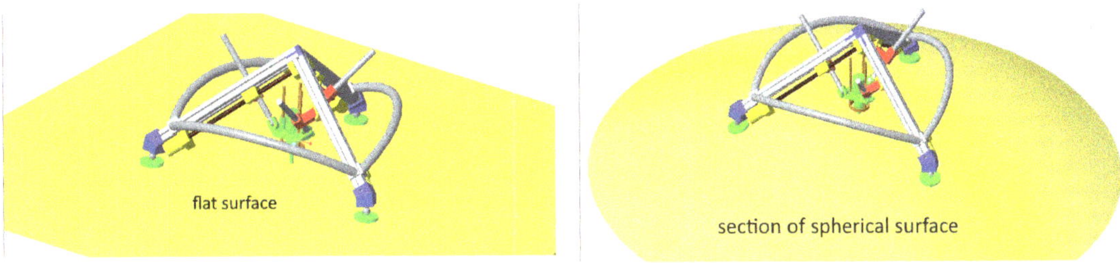

Figure 3. The three-axis robot is attached to (**a**) a flat surface and (**b**) a section of a spherical surface.

The first reconfiguration scheme is attaching a 2R module to the 3T moving platform, as illustrated in Figure 4. The 2R module can be a serial or parallel 2R module. The topo-logy of this scheme can be named 3A+3S+3PRPR+1RR or 3A+3S+3PRRR+1RR, depending on the 3T mechanism used. The transformation between the 3T and 3T2R robots is performed through reassembly or joint locking, which will be thoroughly discussed in Section 7.

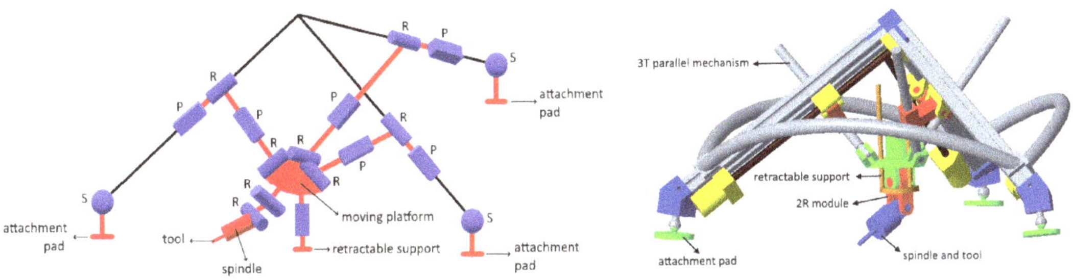

Figure 4. (**a**) The schematic and (**b**) implementation of the 3T2R robot achieved by adding an active 2R module to the moving platform.

The second reconfiguration scheme is adding a 3-DOF parallel mechanism that provides rotational mobility to the base of the 3T parallel mechanism. In other words, two parallel mechanisms, namely lower and upper parallel, are combined. In this scheme, the first joint in each limb of the lower parallel mechanism should be an S joint for modularity. The moving platform of the lower parallel mechanism serves as the base of the upper parallel mechanism. Hence, the whole mechanism is a 6-DOF hybrid-kinematics mechanism.

For the same reason, modularity, the lower parallel mechanism, should also be symmetric. Accordingly, two symmetric parallel mechanisms, namely 3SPR and 3SR⊥R (3SU), are proposed as the lower parallel mechanisms. The ⊥ sign in 3SR⊥R indicates that the two adjacent R joints are perpendicular. By adding any of the two lower parallel mechanisms to the 3T mechanism, the orientation of the tool can be adjusted within the tilting range of the lower parallel mechanisms. In the former topology, three PR kinematic chains, with the underlined letter indicating the actuated joint, are added between the lower ends of the pyramid frame and the corresponding S joints, as illustrated in Figure 5. Accordingly, the topology of the robot can be named 3A+3SPR+3PRPR or 3A+3SPR+3PRRR, depending on the 3T mechanism used. In the latter topology, three R⊥R kinematic chains (universal joints) are added between the lower ends of the pyramid frame and the corresponding S joints, as illustrated in Figure 6. Accordingly, the topology of the robot can be named 3A+3SR⊥R(3SU)+3PRPR or 3A+3SR⊥R(3SU)+3PRRR, depending on the 3T mechanism used.

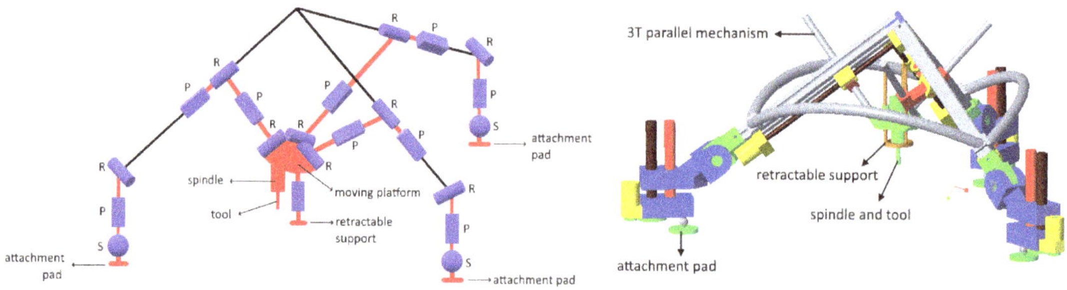

(**a**) (**b**)

Figure 5. (**a**) The schematic and (**b**) implementation of the 6-DOF robot achieved by adding SPR chains, with actuated P joints, to the 3T mechanism.

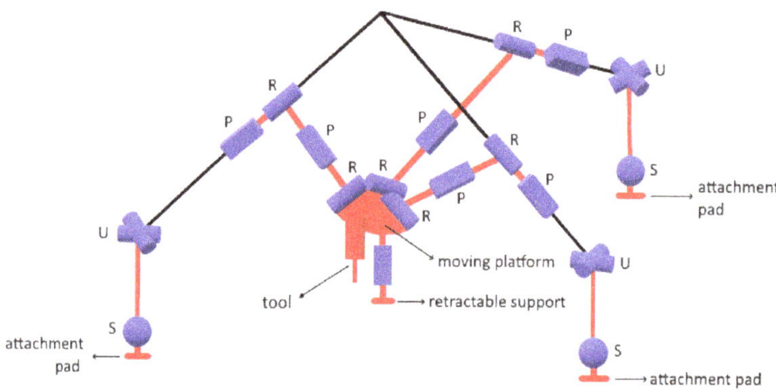

Figure 6. The schematic of the 6-DOF robot achieved by adding SR⊥R (SU) chains to the 3T mechanism.

The 3SR⊥R parallel mechanism can be implemented using two U joint configurations, as illustrated in Figures 7 and 8. In the first U joint configuration, shown in Figure 7a, the axes of the last R joints are always perpendicular to the segments of the pyramid frame. These axes create an equilateral (equidistant) triangle, as shown in Figure 7b. The actuation is applied to the last R joints of the SR⊥R kinematic chains. In its initial posture, the axes of the intermediate R joints intersect at a point, as shown in Figure 7a. However, as soon as the mechanism is displaced from its initial posture, the axes of the intermediate R joints are not necessarily intersecting at a point. In the second U joint configuration, shown in Figure 8a, the axes of the last R joints of the SR⊥R kinematic chains always intersect at a point. The actuation is applied to the intermediate R joints of the SR⊥R ki-nematic chains. In its initial posture, the axes of the intermediate R joints create an instantaneous equilateral triangle, as shown in Figure 8b. The kinematic chains using both U joint configurations can be written as 3SR⊥R and 3SR⊥R, respectively, with the underlined letters indicating the actuated joints.

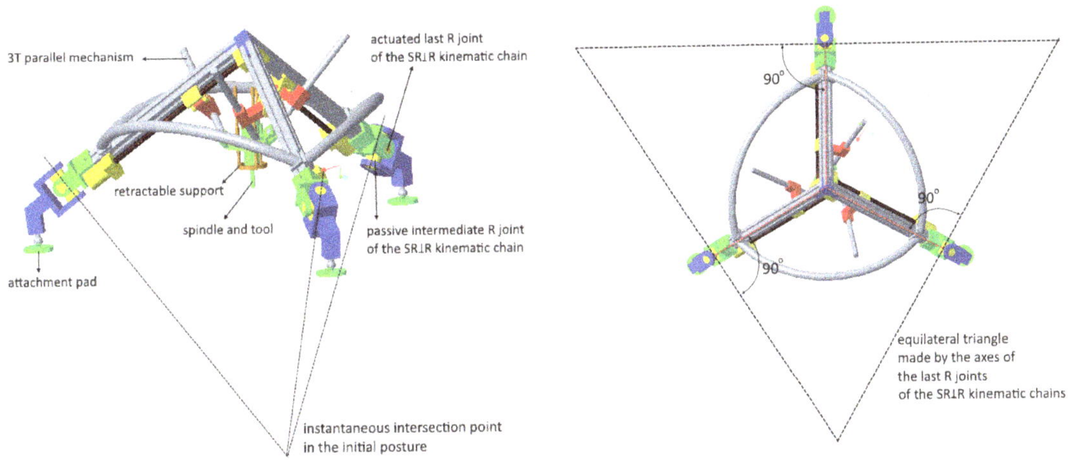

(a) (b)

Figure 7. The addition of SR⊥R (SU) chains to the 3T mechanism by using the first U joint configuration in which (**a**) the intermediate R joints of the SR⊥R kinematic chains intersect at a common point in their initial posture while the last R joints are actuated and (**b**) the axes of the last R joints of the SR⊥R kinematic chains are fixed in a perpendicular position with respect to the segments of the pyramid frame, making an equilateral triangle.

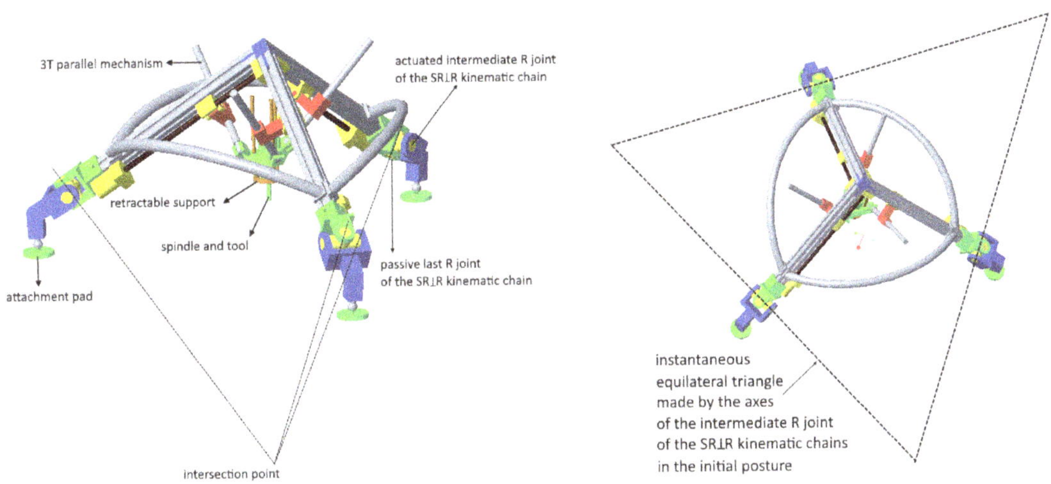

(a) (b)

Figure 8. The addition of SR⊥R (SU) chains to the 3T mechanism by using the second U joint configuration in which (**a**) the last R joints of the SR⊥R kinematic chains intersect at a common point while the intermediate R joints are actuated and (**b**) the axes of the intermediate R joints of the SR⊥R kinematic chains make an instantaneous equilateral triangle in the initial posture.

The robot can be transformed between 3T and 6-DOF configurations by either reassembly or joint locking, which will be thoroughly discussed in Section 7. The former method is performed by assembling the three PR or R⊥R kinematic chains to transform the robot to the 6-DOF configuration and disassembling them to transform it to the 3T configuration. The latter method is performed by unlocking the three PR or R⊥R kinematic chains to transform the robot to the 6-DOF configuration and locking/constraining them to transform the robot to the 3T configuration.

The walking scheme in all the configurations of the robot is the same, that is similar to that in the 3T configuration. When the robot is reconfigured to a 3T2R mechanism, the walking motion is performed by simply extending the retractable support to the structure's surface such that the serial RR mechanism is not engaged with the structure's surface. In other words, the serial RR mechanism is only used for manipulation, not for walking. When the robot is reconfigured to a 3SPR-3PRRR or 3SPR-3PRPR mechanism, the walking motion is performed as the active P joints are locked so that the robot becomes a 3T mechanism. When the robot is reconfigured to a 3SU-3PRPR or 3SU-3PRPR mechanism, the walking motion is performed as the active R joints in the U joints are locked so that the robot becomes a 3T mechanism. In all the 6-DOF configurations, the rotational DOFs are only used for manipulation, not for walking.

3. Mobility Analysis

It is well-known that the 3PRPR and 3PRRR mechanisms have 3T mobility. It is also evident that a serial RR mechanism has 2R mobility. Hence, only the mobility of the 3SPR and 3SU mechanisms are analyzed here. The mobility analysis is based on the screw theory. Since both mechanisms have symmetric topology, the formulation of the joint screws in all limbs can be simplified by using local frames attached to the limbs. In this case, the three limbs in each mechanism have an identical representation of the joint screws. After determining the reciprocal screw representing the constraint wrench in each limb, the constraint wrench applied to the moving platform can be inferred by looking at the geometrical condition of all limbs' constraint wrenches. The moving platform freedom is given by its unconstrained degrees of freedom.

3.1. *Mobility of 3SPR Mechanism*

Referring to Figure 9a, the joint screws in the i-th limb of the 3SPR mechanism can be written as follows:

$$\begin{array}{c} \$^i = \left[\$_1^i\ \$_2^i\ \$_3^i\ \$_4^i\ \$_5^i\right]^T \text{where :} \\ \$_1^i = (1,\ 0,\ 0,0,0,0), \\ \$_2^i = (0,\ 1,\ 0,0,0,0), \\ \$_3^i = (0,\ 0,\ 1,0,0,0), \\ \$_4^i = (0,\ 0,\ 0,0,0,1), \\ \$_5^i = (1,\ 0,\ 0,0,d_i,0). \end{array} \tag{1}$$

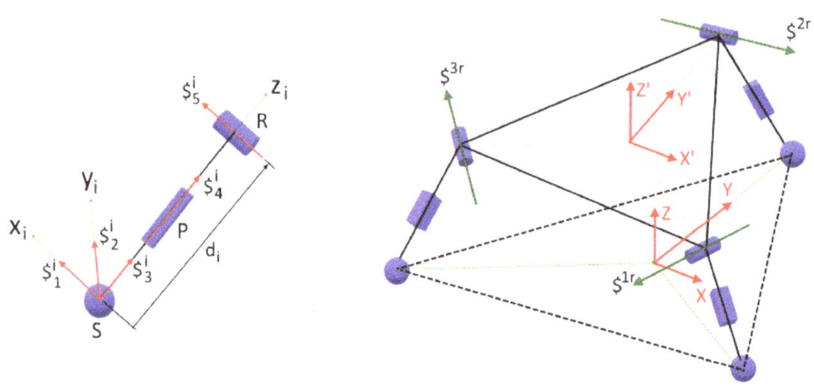

Figure 9. (**a**) The i-th limb joint screws and (**b**) constraint screws in the 3SPR mechanism.

These joint screws create a 1-system in each limb. Accordingly, the reciprocal screw in the *i*-th limb is given by:

$$\$^{ir} = (1,0,0,0,0,0) \quad (2)$$

which indicates a constraint force in the x_i direction. This constraint force implies that the tip of the *i*-th limb has the freedom to rotate about the x_i axis. Since the three limbs of the 3SPR mechanism have three constraint forces that are coplanar, non-parallel, and not intersecting at a common point, as depicted in Figure 9b, the mechanism has three linearly independent constraint wrenches, namely two translations in the plane, i.e., the XY plane, and one rotation about the normal to the plane, i.e., about the Z axis. This means that one of the three rotations is coupled with translation. Accordingly, the freedom of the moving platform is 1T2R, that is, one translation in the Z direction and two rotations about the X and Y axes.

3.2. Mobility of 3SU Mechanism

Referring to Figure 10a, the joint screws in the *i*-th limb of the 3SU mechanism can be written as follows:

$$\begin{aligned} \$^i &= \left[\$^i_1\ \$^i_2\ \$^i_3\ \$^i_5\ \$^i_4\right]^T \text{ for the first configuration of the U joints;} \\ \$^i &= \left[\$^i_1\ \$^i_2\ \$^i_3\ \$^i_4\ \$^i_5\right]^T \text{ for the sec ond configuration of the U joints,} \end{aligned} \quad (3)$$

where the joint screws in their initial posture are given by the following:

$$\begin{aligned} \$^i_1 &= (1,\ 0,\ 0,0,0,0), \\ \$^i_2 &= (0,\ 1,\ 0,0,0,0), \\ \$^i_3 &= (0,\ 0,\ 1,0,0,0), \\ \$^i_4 &= (1,\ 0,\ 0,0,L,0), \\ \$^i_5 &= (0,\ 1,\ 0,-L,0,\ 0). \end{aligned} \quad (4)$$

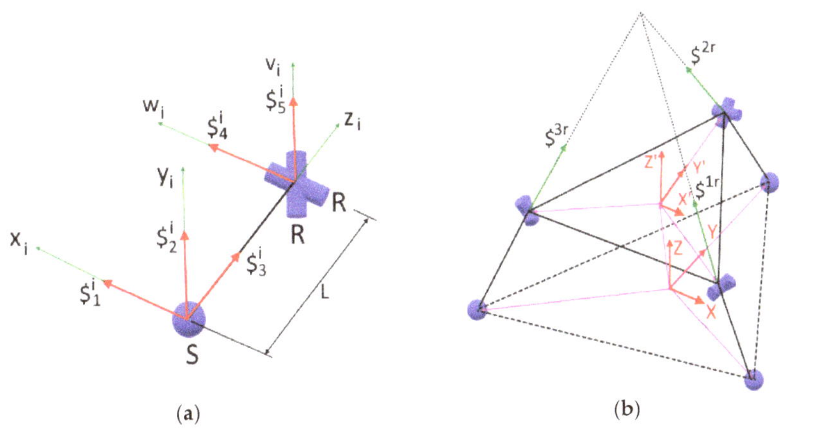

Figure 10. In its initial posture, (**a**) the *i*-th limb joint screws and (**b**) constraint screws in the 3SU mechanism.

These joint screws in both the U joint configurations create a 1-system in each limb. Accordingly, for both the U joint configurations, the reciprocal screw in the *i*-th limb is:

$$\$^{ir} = (0,0,1,0,0,0) \quad (5)$$

which indicates a constraint force in the z_i direction. In the initial posture, the three limbs of the 3SU mechanism have three constraint forces that are not coplanar and intersect at a common point, as depicted in Figure 10b. Therefore, the mechanism has three linearly independent constraint wrenches, namely three translations in the space. Accordingly, the freedom of the moving platform is 3R, that is, three rotations in the space. However, it turns out that this mobility is instantaneous and does not hold for the full-cycle motion of the mechanism. This is because the intersection at a point between the three constraint forces cannot be preserved in the full-cycle motion of the mechanism. Hence, after being displaced from its initial posture, the mobility of the 3SU mechanism turns from 3R to 1T2R, as presented in [22].

4. Pose Kinematics

4.1. Pose Kinematics of 3PRPR, 3PRRR, and Serial RR Mechanisms

The 3PRPR mechanism with the proximal P joints being actuated and the 3PRRR mechanism with the P joints being actuated have a unity mapping between the joint positions (x_1, x_2, and x_3) and end-effector position (x,y,z). In an actual implementation, there may be constant offsets between them. Since the kinematics of this mechanism is straightforward, it is not discussed in this paper.

The serial RR mechanism consists of a base, two R joints, and two links. The first link connects both the R joints, whereas the second link connects the last R joint with the tool center point (TCP). Since the pose kinematics of such a serial mechanism is well-known, it is not discussed here.

4.2. Pose Kinematics of 3SPR Mechanism

Figure 11 shows the 3SPR mechanism. Referring to the figure, the XYZ frame is fixed to the center of the mechanism base A, whereas the X′Y′Z′ frame is attached to the center of the moving platform. Both the base and the moving platform of the mechanism are equilateral triangles, namely $A_1A_2A_3$ and $B_1B_2B_3$, respectively. The circumradii of both triangles are, respectively, R and r. The positions of the points A_1, A_2, and A_3 in the XYZ frame are given by the vectors $\boldsymbol{a_1}$, $\boldsymbol{a_2}$, and $\boldsymbol{a_3}$, respectively, whereas the positions of the points B1, B2, and B3 in the X′Y′Z′ frame are given by the following vectors ${}^{P}\boldsymbol{b_1}$, ${}^{P}\boldsymbol{b_2}$, and ${}^{P}\boldsymbol{b_3}$, respectively.

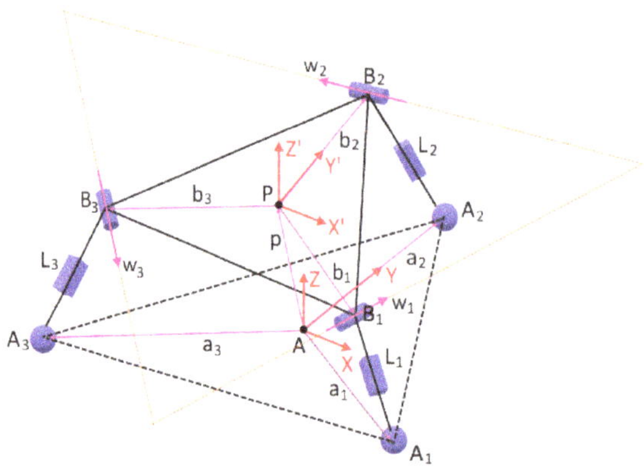

Figure 11. The schematic of the 3SPR mechanism.

The moving platform position and orientation are dictated by the lengths of the three active prismatic joints, namely L_1 , L_2 , and L_3 . The unit directional vectors of the active prismatic joints are $\boldsymbol{l_1}$, $\boldsymbol{l_2}$, and $\boldsymbol{l_3}$, respectively. These unit vectors represent the orientation of the legs with respect to the XYZ frame. Since the 3PRPR mechanism already gives the spatial position, the used DOFs of the 3SPR mechanism are the spatial rotation defined by three Euler angles, namely θ_x, θ_y, and θ_z, which represent the rotation of the moving platform about the X, Y, and Z axes. However, as discussed in the mobility analysis, the rotation about the Z axis is coupled with translation.

Let vector $\boldsymbol{p}$ denote the position of the center of the moving platform P with respect to point A. Let matrix $\boldsymbol{R}_A^P$ be a rotation matrix representing the rotation of the moving platform, i.e., the moving frame X′Y′Z′, with respect to the base, i.e., the XYZ frame. The closed chain constraints of the 3SPR mechanism can be written in the following relation:

$$\boldsymbol{p} - \boldsymbol{a}_i - L_i\,\boldsymbol{l}_i + \boldsymbol{R}_A^P\,{}^P\boldsymbol{b}_i = 0 \text{ where } i = 1,2,3, \tag{6}$$

where the superscript $\boldsymbol{P}$ preceding a vector indicates that the vector is expressed in the X′Y′Z′ frame. A vector not preceded by any superscript is expressed in the XYZ frame.

The axes of the passive revolute joints are fixed to the moving platform. The orientation of these axes are defined by the unit directional vectors $\boldsymbol{w}_1$, $\boldsymbol{w}_2$, and $\boldsymbol{w}_3$ which can be conveniently expressed in the X′Y′Z′ frame. The unit vectors $\boldsymbol{w}_1$, $\boldsymbol{w}_2$, and $\boldsymbol{w}_3$ are always perpendicular to the vectors $\boldsymbol{b}_1$, $\boldsymbol{b}_2$, and $\boldsymbol{b}_3$, respectively. Furthermore, each passive revolute joint constrains the motion of each leg I, i.e., the active prismatic joint, to be only a rotation about the axis of the passive revolute joint, while the leg changes its length L_i. Consequently, each leg is always perpendicular to the axis of the passive revolute joints. This perpendicularity constraint can be written as the following zero dot product:

$$\boldsymbol{l}_i{}^T\boldsymbol{R}_A^P\,{}^P\boldsymbol{w}_i = 0 \text{ where } i = 1,2,3. \tag{7}$$

As both the vector $\boldsymbol{b}_i$ and leg $\boldsymbol{l}_i$ are always perpendicular to the axis of the passive revolute joint i, each limb creates a plane, namely AA_iB_iP. Consequently, the sum of the vectors $\boldsymbol{p}$ and $\boldsymbol{b}_i$ is also always perpendicular to the axis of the passive revolute joint $\boldsymbol{w}_i$. This can be written as the following zero dot product:

$$\left(\boldsymbol{p} + \boldsymbol{R}_A^P\,{}^P\boldsymbol{b}_i\right)^T\boldsymbol{R}_A^P\,{}^P\boldsymbol{w}_i = 0 \text{ where } i = 1,2,3. \tag{8}$$

Equations (6)–(8) are the implicit kinematic equations of the 3SPR mechanism. The inverse kinematics problem can be stated as "given the orientation of the moving platform, find the corresponding active joint positions L_1, L_2, and L_3". Besides L_1, L_2, and L_3, there are twelve other unknowns in the inverse kinematics problem, namely the unit directional vectors of the three legs (each of the unit vectors contains three components in X, Y, and Z directions) and the position of the moving platform (x,y,z). On the other hand, the forward kinematics problem can be stated as "given the active joint positions L_1, L_2, and L_3, find the orientation of the moving platform". Besides the three Euler angles of the moving platform, there are twelve other unknowns similar to those in the inverse kinematics problem.

4.3. Pose Kinematics of the 3SU Mechanism

Figure 12 shows the 3SU mechanism. Referring to the figure, the XYZ and X′Y′Z′ frames, similar to those in Section 4.2, are used. Similar to the 3SPR mechanism, the base is defined by the vectors $\boldsymbol{a}_1$, $\boldsymbol{a}_2$, and $\boldsymbol{a}_3$ expressed in the XYZ frame, whereas the moving platform is defined by the vectors $\boldsymbol{b}_1$, $\boldsymbol{b}_2$, and $\boldsymbol{b}_3$ expressed in the X′Y′Z′ frame. The unit directional vectors of the three legs are $\boldsymbol{l}_1$, $\boldsymbol{l}_2$, and $\boldsymbol{l}_3$, respectively. These unit vectors represent the orientation of the legs with respect to the XYZ frame. All the legs have an identical, constant length L.

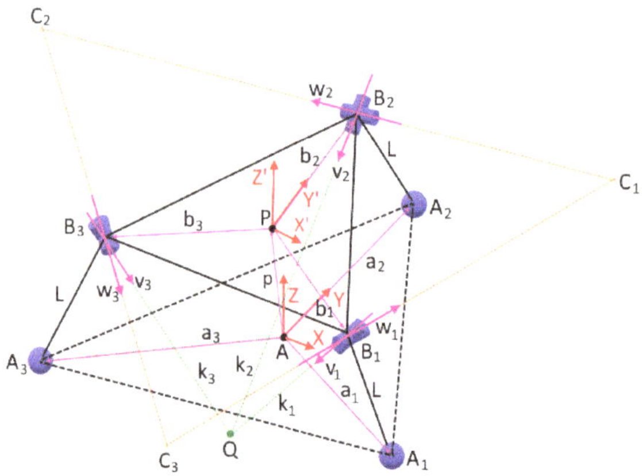

Figure 12. The schematic of the 3SU mechanism.

As discussed earlier, the 3SU mechanism can be implemented in two variations: the first variation in the 3SU mechanism, i.e., the 3SU mechanism with the first U joint configuration, and the second variation in the 3SU mechanism, i.e., the 3SU mechanism with the second U joint configuration. In the first variation, the R joints with the orientation given by the unit vectors v_i are connected to the legs, whereas the R joints with the orientation given by the unit vectors w_i are connected to the moving platform. In the second variation in the 3SU mechanism, the R joints with the orientation given by the unit vectors v_i are connected to the moving platform, whereas the R joints with the orientation given by the unit vectors w_i are connected to the legs. Hence, the unit vectors w_1, w_2, and w_3 in the second variation in the 3SU mechanism are always perpendicular to the unit directional vectors of the legs l_1, l_2, and l_3. This can be mathematically written as a zero dot product given in Equation (7), similar to the 3SPR mechanism. However, the lengths of the legs in the 3SU mechanism are fixed, whereas they are changed in the 3SPR mechanism. The first variation is actuated at the distal R joints, whereas the second is actuated at the intermediate R joints.

In each limb of both the variations, the axis v_i is always perpendicular to the axis w_i, as both axes represent two R joints composing a U joint. This perpendicularity can be written as the following dot product:

$$v_i{}^T R_A^{PP} w_i = 0 \text{ where } i = 1,2,3. \tag{9}$$

where the unit vector v_i is defined in the XYZ frame, whereas the unit vector ${}^P w_i$ is defined in the X′Y′Z′ frame.

The closed chain constraints of the first variation in the 3SU mechanism can be written in the following relation:

$$p + R_A^{PP} b_i - R_y(\theta_{ai}) R_x(\theta_{pi})^{B_i} l_i - a_i = 0;\ i = 1,2,3. \tag{10}$$

where the vector p denotes the position of the center of the moving platform P with respect to point A, whereas the matrix R_A^P is a rotation matrix representing the rotation of the moving platform, i.e., the moving frame X′Y′Z′, with respect to the base, i.e., the XYZ frame. Since the R joints in each U joint rotate about the X and Y axes of a local frame which has the origin B_i and the Z direction aligned with the leg direction, the rotation of both the R joints can be represented by the rotation matrices R_x and R_y that, respectively, represent the elementary rotation matrices about the X and Y axes. The angles denote the magnitude of the rotations θ_{ai} and θ_{pi} that, respectively, denote the angles of the active and passive

R joints of the U joint in the i-th limb. The vector ${}^{B_i}\boldsymbol{l}_i$ is the i-th leg vector expressed at its local frame that has the origin B_i, that is:

$$ {}^{B_i}\boldsymbol{l}_i = \begin{bmatrix} 0 & 0 & 1 \end{bmatrix}^T;\ i = 1,2,3. \tag{11} $$

On the other hand, the closed chain constraints of the second variation in the 3SU mechanism can be written in the following relation:

$$ \boldsymbol{p} + \boldsymbol{R}_A^{PP}\boldsymbol{b}_i - \boldsymbol{R}_x(\theta_{pi})\boldsymbol{R}_y(\theta_{ai})^{B_i}\boldsymbol{l}_i - \boldsymbol{a}_i = 0;\ i = 1,2,3. \tag{12} $$

The inverse kinematics problem can be stated as "given the orientation of the moving platform, find the corresponding active joint positions θ_{a1}, θ_{a2}, and θ_{a3}". Besides these active joint positions, there are six other unknowns in the inverse kinematics problem, namely the angles of the passive R joints of the i-th U joint (θ_{p1}, θ_{p2}, θ_{p3}) and the position of the moving platform (x,y,z). On the other hand, the forward kinematics problem can be stated as "given the active joint positions θ_{a1}, θ_{a2}, and θ_{a3}, find the orientation of the moving platform". Besides the three Euler angles of the moving platform, there are six other unknowns similar to those in the inverse kinematics problem.

4.4. Pose Kinematics of the Combined Mechanism

Figure 13 shows the combined 3SPR-3PRPR (or 3SPR-3PRRR) and 3SU-3PRPR (or 3SU-3PRRR) mechanisms consisting of the lower mechanisms serially connected with the pyramid structure of the upper 3PRPR or 3PRRR mechanism. The limbs of the 3PRPR or 3PRRR mechanism are not shown in the figure for more readability.

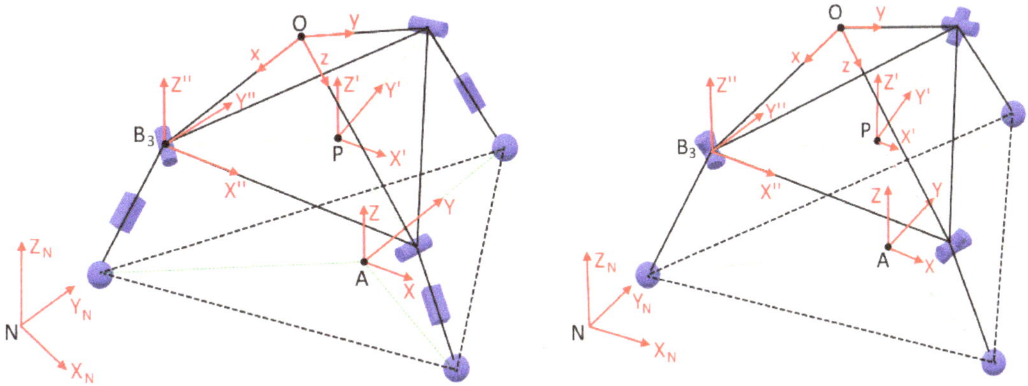

Figure 13. The schematic of (**a**) the 3SPR-3PRPR (or 3SPR-3PRRR) mechanism and (**b**) 3SU-3PRPR (or 3SU-3PRRR) mechanism.

The XYZ frame is attached to the center of the base of the lower mechanism, namely, point A located in the world frame $X_NY_NZ_N$. In other words, the position of the whole mechanism in the world frame is given by the vector $\overrightarrow{NA}$. This vector is constant when the whole mechanism is attached to the surface of the workpiece/structure, whereas the vector changes when the mechanism is walking. The X′Y′Z′ frame is attached to the moving platform of the 3SPR mechanism which serves as the base of the 3PRPR mechanism.

The pose of the end-effector of the 3PRPR or 3PRRR mechanism is conveniently expressed in the xyz frame. The intermediate X″Y″Z″ frame is defined to be used in conveniently formulating the kinematics of the walking motions of the mechanism and to conveniently transform from the xyz frame to the X′Y′Z′ frame. The Z′ and Z″ axes are always parallel with each other. There is no rotation between the X′Y′Z′ and X″Y″Z″ frames as both are attached to the same body, i.e., the moving platform of the 3SPR mechanism.

To express the position of the end-effector of the 3PRPR or 3PRRR mechanisms in the XYZ frame, first, the pose is defined in the xyz frame and, subsequently, transformed to the X″Y″Z″ frame, X′Y′Z′ frame, and, finally, to the XYZ frame. The orientation of the end-effector is identical to the orientation of the moving platform of the lower mechanism since the pyramid of the 3PRPR or 3PRRR mechanisms is axisymmetric. The pyramid has segments with an identical length L_p.

4.5. Numerical Example

A numerical simulation was conducted to implement the pose kinematics of the 6-DOF mechanisms utilizing the 3SPR and 3SU lower mechanisms. In this simulation, the circumradii of the lower mechanisms' base and moving platform are R = 0.5 m and r = 0.4 m, respectively. In the 3SPR mechanism, the legs' lengths are variable as they are the actuated P joints. In the 3SU mechanism, all the legs have an identical length of 0.3 m. The pyramid frame of the upper mechanism with 3PRPR topology has an angle of 90 degrees between its adjacent segments. Each segment has a length of 0.5 m. The upper mechanism is actuated at its proximal P joints.

In the inverse kinematics problem, the position of the end-effector E in both hybrid mechanisms was given to be coincident with the origin of X′Y′Z′ frame, namely point P. The orientation of the end-effector expressed as Euler angles with respect to the XYZ frame was given to be (0, 0, 0) in the 3SPR-3PRPR mechanism and (5, 0, 0) degrees in the 3SU-3PRPR mechanism. The forward kinematics of both the upper and lower mechanisms was solved by taking the solution of the inverse kinematics as the given actuator positions. The forward kinematics solution retrieved the end-effector pose given in the inverse kinematics problem. For each hybrid mechanism, the inverse and forward kinematics problems of the upper mechanism were solved analytically. In contrast, the lower mechanisms' inverse and forward kinematics problems were solved numerically using the Levenberg–Marquardt algorithm in MATLAB.

The actuator positions and the end-effector poses obtained in the inverse and forward kinematics problems are shown in Table 1. The actuator positions of the upper mechanism, namely x_1, x_2, and x_3, are expressed in the xyz frame. The actuator positions of the 3SPR mechanism, namely L_1, L_2, and L_3, are expressed as constants representing the lengths of the actuated P joints. The actuator positions of both the variations in the SRLR (SU) mechanism, namely θ_{a1}, θ_{a2}, and θ_{a3}, are expressed as constants representing the angles of the actuated R joints. The end-effector pose (x,y,z) and the end-effector orientation (θ_x, θ_y, θ_z) are expressed in the XYZ frame. Figures 14 and 15, respectively, show the plots of the 3SPR-3PRPR and 3SU-3PRPR mechanisms in the XYZ frame.

Table 1. Actuator Positions and End-Effector Poses.

Mechanism	Actuator Position	End-Effector Pose
3SPR-3PRPR	$x_1 = x_2 = x_3 = 0.1667$ m $L_1 = L_2 = L_3 = 0.3262$ m	$x = 0$, $y = 0$, $z = 0.3000$ m $\theta_x = \theta_y = \theta_z = 0$
3SRLR-3PRPR	$x_1 = x_2 = x_3 = 0.1667$ m $\theta_{a1} = -19.2743$ deg, $\theta_{a2} = 0$, $\theta_{a3} = 19.2743$ deg	$x = 0$, $y = 0.0948$ m, $z = 0.2651$ m $\theta_x = 5$ deg, $\theta_y = \theta_z = 0$
3SRLR-3PRPR	$x_1 = x_2 = x_3 = 0.1667$ m $\theta_{a1} = -16.7787$ deg, $\theta_{a2} = 0$, $\theta_{a3} = 16.7786$ deg	$x = 0$, $y = 0.0948$ m, $z = 0.2651$ m $\theta_x = 5$ deg, $\theta_y = \theta_z = 0$

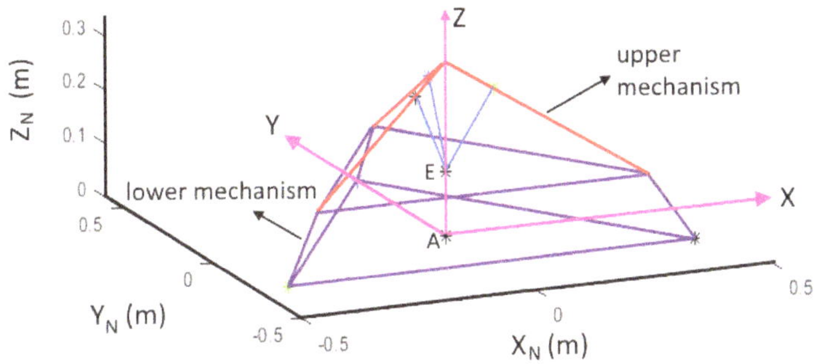

Figure 14. Three-dimensional view plot of the 3SPR-3PRPR mechanism with the end-effector E in an orientation of (0, 0, 0) and position coincident with the center of the moving platform of the lower mechanism.

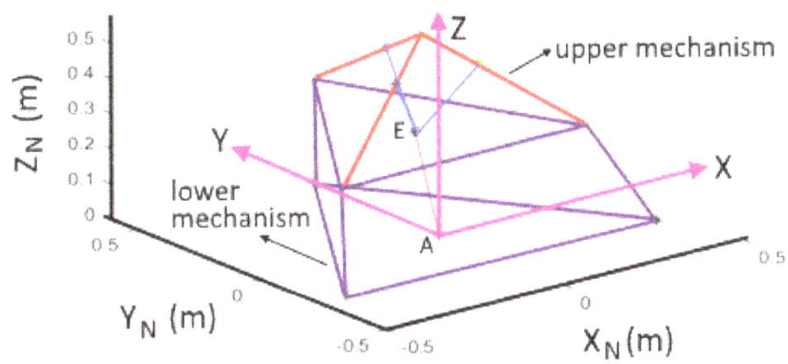

Figure 15. Three-dimensional view plot of the 3SU-3PRPR mechanism with the end-effector E in an orientation of (5, 0, 0) degrees and position coincident with the center of the pyramid base.

5. Differential Kinematics and Singularity Analysis

In the differential kinematics formulation presented in this section, the Plucker coordinates of the joint screws refer to Figures 9 and 10 used in Section 3, which discusses the mobility analysis.

5.1. Differential Kinematics and Singularity Analysis of 3PRPR and 3PRRR Mechanisms

With an orthogonal pyramid frame, the 3PRPR mechanism with the proximal P joints being actuated and the 3PRRR mechanism with the P joints being actuated have an identity 3×3 actuation Jacobian matrix. This implies a unity mapping between the twist of the moving platform and the twist of the actuators. When the intermediate P joints in the 3PRPR mechanism are actuated, it has a 3×3 actuation Jacobian matrix containing some trigonometric functions of the angles made by the intermediate P joint axes. The configuration singularities in both the 3PRPR and 3PRRR mechanisms only occur at their workspace boundary. Hence, they can be easily avoided. The 3PRPR and 3PRRR mechanisms have three constraint couples working on the moving platform that prevent the moving platform from rotating. Since these constraint couples are orthogonal, they never become dependent; hence, there is no constraint singularity in both mechanisms. Thus, there is no singularity inside the workspace of the 3PRPR and 3PRRR mechanisms.

5.2. Differential Kinematics and Singularity Analysis of 3SPR Mechanism

The instantaneous twist of the 3SPR moving platform is given by:

$$\$_P = \dot{\theta}_1^i \hat{\$}_1^i + \dot{\theta}_2^i \hat{\$}_2^i + \dot{\theta}_3^i \hat{\$}_3^i + \dot{d}_i \hat{\$}_4^i + \dot{\theta}_5^i \hat{\$}_5^i; \quad i = 1, 2, 3; \tag{13}$$

where $\dot{\theta}_1^i$, $\dot{\theta}_2^i$, $\dot{\theta}_3^i$, $\dot{d}_i$, and $\dot{\theta}_5^i$ are the joint screws' intensities, whereas the unit joint screws representing the five joints in the i-th limb are given by the following:

$$\hat{\$}_1^i = \begin{bmatrix} s_1^i \\ (b_i - d_i) \times s_1^i \end{bmatrix} \tag{14}$$

$$\hat{\$}_2^i = \begin{bmatrix} s_2^i \\ (b_i - d_i) \times s_2^i \end{bmatrix} \tag{15}$$

$$\hat{\$}_3^i = \begin{bmatrix} s_3^i \\ (b_i - d_i) \times s_3^i \end{bmatrix} \tag{16}$$

$$\hat{\$}_4^i = \begin{bmatrix} 0 \\ s_4^i \end{bmatrix} \tag{17}$$

$$\hat{\$}_5^i = \begin{bmatrix} s_5^i \\ b_i \times s_5^i \end{bmatrix} \tag{18}$$

Note that the vectors b_i is expressed in the base frame, i.e., the XYZ frame, and accordingly can be written as:

$$b_i = R_A^{P\,P} b_i \tag{19}$$

where R_A^P is the rotation matrix representing the rotation of the moving platform frame, i.e., the X′Y′Z′ frame, with respect to the base frame, whereas $^P b_i$ is the vector b_i expressed in the moving platform frame.

Since the axes of all the unactuated joints in each limb intersect the line passing A_i and B_i, a unique screw that is reciprocal to all the unactuated joint screws is:

$$\hat{\$}_4^{ir} = \begin{bmatrix} s_4^i \\ b_i \times s_4^i \end{bmatrix} \tag{20}$$

Taking the reciprocal product of both sides of Equation (13) with Equation (20) yields the following relationship between the instantaneous twist of the moving platform $\$_P = \begin{bmatrix} \omega_x \, \omega_y \, \omega_z \, \dot{x} \, \dot{y} \, \dot{z} \end{bmatrix}^T$ and the instantaneous twist of the actuators $\dot{q} = \begin{bmatrix} \dot{d}_1 & \dot{d}_2 & \dot{d}_3 \end{bmatrix}^T$:

$$J_x \$_P = J_q \dot{q} \tag{21}$$

where the forward Jacobian J_{x3x6} and the inverse Jacobian J_{q3x3} are given by:

$$J_x = \begin{bmatrix} (b_1 \times s_4^1)^T & (s_4^1)^T \\ (b_2 \times s_4^2)^T & (s_4^2)^T \\ (b_3 \times s_4^3)^T & (s_4^3)^T \end{bmatrix} \tag{22}$$

$$J_q = \begin{bmatrix} s_4^1 \cdot e_1 & 0 & 0 \\ 0 & s_4^2 \cdot e_2 & 0 \\ 0 & 0 & s_4^3 \cdot e_3 \end{bmatrix} \tag{23}$$

The vectors denote the unit directional vectors of the actuation. In this case, $e_1 = s_4^1$, $e_2 = s_4^2$, and $e_3 = s_4^3$. Hence, J_q is an identity matrix. The actuation Jacobian matrix J_{a3x6} is given by the following well-known equation:

$$J_a = J_q^{-1} J_x \tag{24}$$

On the other hand, taking the orthogonal product of both sides of Equation (13) with each of the reciprocal basis screws representing the constraint screws of the 3SPR mechanism yields the following relationship between the constraint Jacobian J_c and the instantaneous twist of the moving platform $\$_P$:

$$J_c \$_P = \mathbf{0} \tag{25}$$

which indicates that the constraint wrenches perform no work on the moving platform. In the 3SPR mechanism, the constraint Jacobian J_{c3x6} is given by:

$$J_c = \begin{bmatrix} (\boldsymbol{b_1} \times \boldsymbol{s_1^1})^T & (\boldsymbol{s_1^1})^T \\ (\boldsymbol{b_2} \times \boldsymbol{s_1^2})^T & (\boldsymbol{s_1^2})^T \\ (\boldsymbol{b_3} \times \boldsymbol{s_1^3})^T & (\boldsymbol{s_1^3})^T \end{bmatrix} \tag{26}$$

As it is well-known, the overall Jacobian $J_{6\times 6}$ is obtained by stacking the actuation Jacobian and the constraint Jacobian, that is:

$$J = \begin{bmatrix} J_a \\ J_c \end{bmatrix} \tag{27}$$

The relationship between the overall Jacobian and the instantaneous twist of the moving platform is given by:

$$J\$_P = \dot{q}_0 \tag{28}$$

For the SPR mechanism, the vector $\dot{q}_0$ is given by:

$$J\$_P = \dot{q}_0 \text{ where } \dot{q}_0 = \begin{bmatrix} \dot{d}_1 & \dot{d}_2 & \dot{d}_3 & 0 & 0 & 0 \end{bmatrix}^T \text{ for the 3SPR mechanism.} \tag{29}$$

The architectural singularities occur in the 3SPR mechanism in three geometric conditions. First, an architectural singularity occurs if the three actuation forces lie on a common plane and intersect at one point. This happens when the moving platform collapses to the base. This singularity can be easily avoided by limiting the range of the S joints and adjusting the minimum limits of the P joints, namely L_1, L_2, and L_3, such that the moving platform cannot collapse to the base. Second, an architectural singularity occurs if the three actuation forces are parallel, as depicted in Figure 16a. This only occurs if the moving platform and the base have identical geometry and dimensions. Thus, this singularity can be easily avoided by making the moving platform smaller than the base. Third, an architectural singularity occurs if any of the P joint axes are collinear with the vector $\boldsymbol{b}_i$ of the moving platform, as illustrated in Figure 16b.

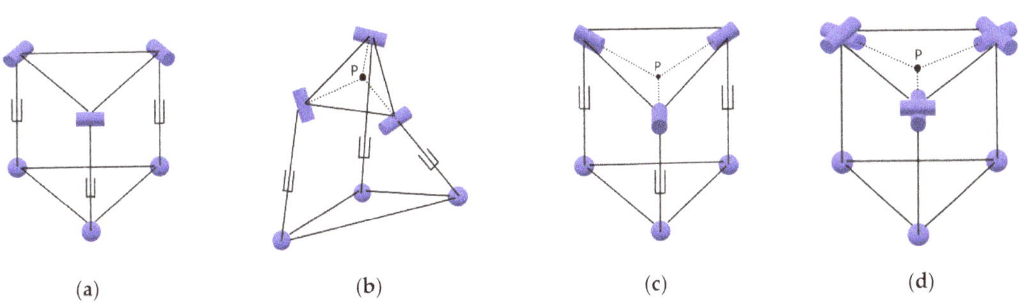

Figure 16. (**a**,**b**) Two configurations having architectural singularities in the 3SPR mechanism, (**c**) the constraint singularity configuration in the 3SPR mechanism, and (**d**) an architectural singularity in the 3SU mechanism.

The constraint singularity occurs when the three constraint forces, that are in the direction of the R joint axes, lie on a common plane and intersect at one point, as illustrated in Figure 16c. However, this geometrical condition is not the geometrical arrangement of the 3SPR mechanism at hand since the R joint axes are always perpendicular to the vector $\boldsymbol{b}_i$. Hence, this constraint singularity never occurs.

5.3. Differential Kinematics and Singularity Analysis of 3SU Mechanism

The instantaneous twist of the 3SU moving platform is given by:

$$\$_P = \dot{\theta}_1^i \hat{\$}_1^i + \dot{\theta}_2^i \hat{\$}_2^i + \dot{\theta}_3^i \hat{\$}_3^i + \dot{\theta}_4^i \hat{\$}_4^i + \dot{\theta}_5^i \hat{\$}_5^i; i = 1, 2, 3; \tag{30}$$

where $\dot{\theta}_1^i$, $\dot{\theta}_2^i$, $\dot{\theta}_3^i$, $\dot{\theta}_4^i$, and $\dot{\theta}_5^i$ are the joint screws' intensities, whereas the unit joint screws of the i-th spherical joint, namely $\hat{\$}_1^i$, $\hat{\$}_2^i$, and $\hat{\$}_3^i$, are identical to those in the 3SPR mechanism, i.e., Equations (14)–(16). The other unit joint screws corresponding to the i-th U joint are given by:

$$\hat{\$}_4^i = \begin{bmatrix} \boldsymbol{s}_4^i \\ \boldsymbol{b}_i \times \boldsymbol{s}_4^i \end{bmatrix} \tag{31}$$

$$\hat{\$}_5^i = \begin{bmatrix} \boldsymbol{s}_5^i \\ \boldsymbol{b}_i \times \boldsymbol{s}_5^i \end{bmatrix} \tag{32}$$

Note that the vectors $\boldsymbol{b}_i$ are given by Equation (19).

When the R joints parallel with $\boldsymbol{w}_1$, $\boldsymbol{w}_2$, and $\boldsymbol{w}_3$ axes are actuated, the instantaneous twist of the actuators is given by $\dot{q} = \begin{bmatrix} \dot{\theta}_4^1 & \dot{\theta}_4^2 & \dot{\theta}_4^3 \end{bmatrix}^T$. On the other hand, when the R joints with $\boldsymbol{v}_1$, $\boldsymbol{v}_2$, and $\boldsymbol{v}_3$ axes are actuated, the instantaneous twist of the actuators is given by $\dot{q} = \begin{bmatrix} \dot{\theta}_5^1 & \dot{\theta}_5^2 & \dot{\theta}_5^3 \end{bmatrix}^T$. The forward Jacobian $J_{x3\times6}$ and the inverse Jacobian $J_{q3\times3}$ are given by:

$$J_x = \begin{bmatrix} (\boldsymbol{b}_1 \times \boldsymbol{s}_3^1)^T & (\boldsymbol{s}_3^1)^T \\ (\boldsymbol{b}_2 \times \boldsymbol{s}_3^2)^T & (\boldsymbol{s}_3^2)^T \\ (\boldsymbol{b}_3 \times \boldsymbol{s}_3^3)^T & (\boldsymbol{s}_3^3)^T \end{bmatrix} \tag{33}$$

$$J_q = \begin{bmatrix} \boldsymbol{s}_3^1 \cdot \boldsymbol{e}_1 & 0 & 0 \\ 0 & \boldsymbol{s}_3^2 \cdot \boldsymbol{e}_2 & 0 \\ 0 & 0 & \boldsymbol{s}_3^3 \cdot \boldsymbol{e}_3 \end{bmatrix} \tag{34}$$

The vectors denote the unit directional vectors of the actuation. If the R joints parallel with the unit vectors $\boldsymbol{w}_1$, $\boldsymbol{w}_2$, and $\boldsymbol{w}_3$ are actuated, $\boldsymbol{e}_1 = \boldsymbol{s}_4^1$, $\boldsymbol{e}_2 = \boldsymbol{s}_4^2$, and $\boldsymbol{e}_3 = \boldsymbol{s}_4^3$. On the other hand, if the R joints parallel with the unit vectors $\boldsymbol{v}_1$, $\boldsymbol{v}_2$, and $\boldsymbol{v}_3$ are actuated, $\boldsymbol{e}_1 = \boldsymbol{s}_5^1$, $\boldsymbol{e}_2 = \boldsymbol{s}_5^2$, and $\boldsymbol{e}_3 = \boldsymbol{s}_5^3$. The actuation Jacobian matrix J_{a3x6} is given by Equation (24).

The constraint Jacobian J_{c3x6} is given by:

$$J_c = \begin{bmatrix} (\boldsymbol{b}_1 \times \boldsymbol{s}_3^1)^T & (\boldsymbol{s}_3^1)^T \\ (\boldsymbol{b}_2 \times \boldsymbol{s}_3^2)^T & (\boldsymbol{s}_3^2)^T \\ (\boldsymbol{b}_3 \times \boldsymbol{s}_3^3)^T & (\boldsymbol{s}_3^3)^T \end{bmatrix} \tag{35}$$

The overall Jacobian $J_{6\times6}$ is given by Equation (27), whereas the relationship between the overall Jacobian and the instantaneous twist of the moving platform is given by Equation (28).

The 3SU mechanism undergoes an architectural singularity in two geometric conditions. First, the architectural singularity occurs if the unit vectors $\boldsymbol{v}_1$, $\boldsymbol{v}_2$, and $\boldsymbol{v}_3$ lie on a common plane and intersect at one point. The latter condition occurs all the time, but the former only occurs in the architectural singularity. This singularity happens when the moving platform collapses to the base or when the three links are parallel, as illustrated in Figure 16d. As in the 3SPR mechanism, the collapse of the moving platform to the base can be easily avoided by limiting the range of the S joints and making the link length L long enough. The parallelism of the three links only occurs when the moving platform and the base have identical geometry and dimensions. This can be easily avoided by making the moving platform smaller than the base. The second geometric condition leading to an architectural singularity is a condition in which the unit vectors w_1, w_2, and w_3 lie on a common plane and intersect at one point. The former condition occurs at all times, but the latter condition never occurs since the unit vectors w_1, w_2, and w_3 are always perpendicular to the vectors $\boldsymbol{b}_i$.

The constraint singularity occurs in the 3SU mechanism if the three constraint forces lie on a common plane or are parallel. The former condition occurs when the moving platform collapses to the base. The latter condition occurs when the three links of the mechanism are parallel to each other, i.e., when the moving platform and the base have identical geometry and dimensions. Thus, the constraint singularity can be avoided easily.

6. Workspace

6.1. Workspace of the 3T Mechanism

With a pyramid frame having an angle of 90 degrees between its adjacent segments, the upper mechanism, i.e., the 3PRPR or 3PRRR mechanism, has a rectangular cuboid workspace with a size mainly dictated by the motion limits of the actuators, as illustrated in Figure 17a.

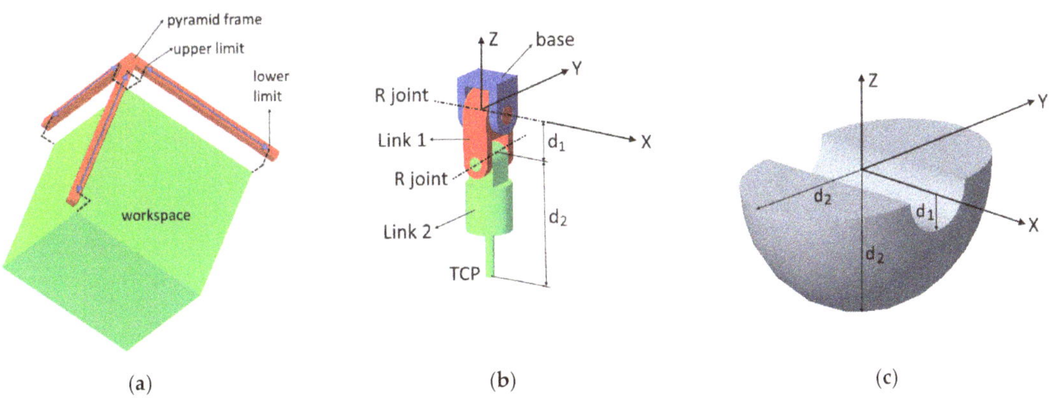

Figure 17. (**a**) The workspace of the 3T mechanism, (**b**) serial RR mechanism with two non-zero links, and (**c**) total orientation workspace of the serial RR mechanism shown in (**b**) given the tilting range of both the R joints between −90 and 90 degrees.

6.2. Workspace of the Serial RR Mechanism

The serial RR mechanism in the 3T2R mechanism has a large orientation workspace, mainly limited by the collision among the mechanical components, cabling, and motors' rotation limits. In the serial RR mechanism, as depicted in Figure 17b, its orientation is coupled with dependent translation due to the non-zero distance between both the R joints' axes and the non-zero distance between the TCP and the axis of the last R joint. In this case, the dependent translation should be compensated by translating the 3T mechanism. The serial RR mechanism's total orientation workspace is shown in Figure 17c, given the R joints' tilting range between −90 and 90 degrees. The orientation workspace in the 3T robot combined with the serial RR module is identical across the position workspace of the 3T robot unless there is a collision around the boundary of the position workspace that may reduce the actual tilting capability of the serial RR module.

6.3. Workspace of the 3SPR and 3SU Mechanisms

The 3SPR and 3SU mechanisms have a more limited orientation workspace. However, their parallel kinematics are expected to provide higher stiffness than the serial RR mechanism. Considering this, a five-axis machining task that typically requires a large orientation workspace is more convenient to perform using the 3T2R mechanism while the overall stiffness is compromised. On the other hand, the 6-DOF mechanisms are more appropriate for machining workpiece/structure with an irregular surface while higher stiffness is prioritized.

The workspace of the 3SPR mechanism can be determined graphically by first finding the intersection of three hollow spheres created by the three SP chains of the mechanism. To determine a constant orientation workspace, the reachable points previously obtained need to be shifted to the center of the moving platform at a specific orientation. To determine the total orientation workspace, the moving platform is rotated about the three R joints to create the outer layer of the workspace. The limits of the S, P, and R joints also dictate the reachable workspace. Figure 18a,b shows the total orientation workspaces of 3SPR mechanisms with two different sizes of the fixed base.

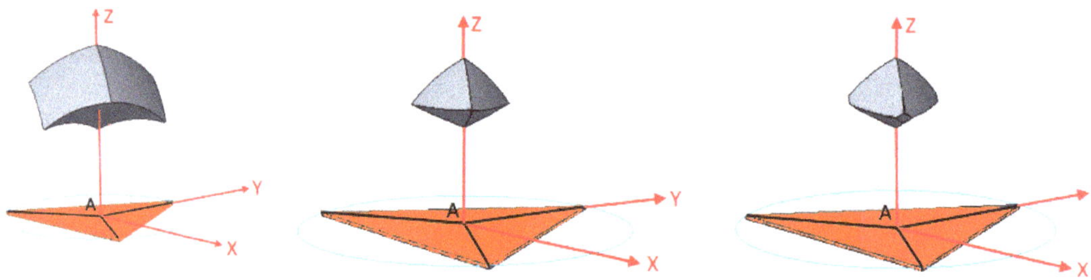

(**a**) (**b**) (**c**)

Figure 18. The total orientation workspaces of the 3SPR mechanism with (**a**) $R = 0.3$ m, $L_i, min = 0.2$ m, $L_i, max = 0.6$ m, and (**b**) $R = 0.5$ m, $L_i, min = 0.2$ m, $L_i, max = 0.6$ m, and (**c**) the total orientation workspace of the 3SU mechanism with $R = 0.5$ m, $r = 0.3$ m. All the mechanisms use the S joints with a rotational motion range between −30 and 30 degrees.

The workspace of the 3SU mechanism can be determined graphically by first finding the centers of rotation of the U joints, that is, lying on three spherical surfaces created by the tips of the three links of the mechanism. Subsequently, assuming that both the R joints in each U joint have axes intersecting at a point, the constant orientation workspace can be created by shifting the previously obtained reachable points to the center of the moving platform at a specific orientation. In contrast, the total orientation workspace is obtained by rotating both the R joints of each U joint about the known centers of rotation lying on the previously created surface. The limits of the S and U joints in all three limbs also affect the workspace. Figure 18c shows the total orientation workspace of a 3SU mechanism.

When a 3SPR or 3SU mechanism is combined with the 3T mechanism, the workspace of the combined robot is a superposition of the workspace of the 3SPR or 3SU mechanism and the workspace of the 3T mechanism. The workspace of the 3SPR or 3SU mechanism combined with the 3T mechanism is the rectangular cuboid of the 3T mechanism being tilted by the 3SPR or 3SU mechanism. However, it is not a purely tilting motion due to the dependent translational motion in the 3SPR or 3SU mechanism. Since the 3T mechanism does not have any rotational DOF, it only contributes to the translational components. Since the 3SPR or 3SU mechanism has dependent translational motion, its translation should be added to the translation of the 3T mechanism to obtain the combined translation.

7. Reconfiguration Schemes

Modularity and joint locking can be used to achieve reconfigurability. In the modularity scheme, two possible scenarios can be implemented, as illustrated in Figure 18. With the 3T mechanism as the basic build, modular components can be added to make a 3-DOF, 5-DOF, or 6-DOF robot. In Figure 19a, the attachment pad and spherical joint are included in each module. This requires less time for assembly but requires more attachment pads and spherical joints. In Figure 19b, the attachment pad and spherical joint are made as separate modules, namely AS modules. Accordingly, the same set of AS modules can be used with various modules such as serial RR, PR, and SU modules. This requires a lower number of AS modules but requires more time for assembly. For both the modularity scenarios, the design of interfaces between mating parts should be made to enable the self-alignment of the mating parts. If required, a calibration routine can be run after an assembly to increase the robot's accuracy.

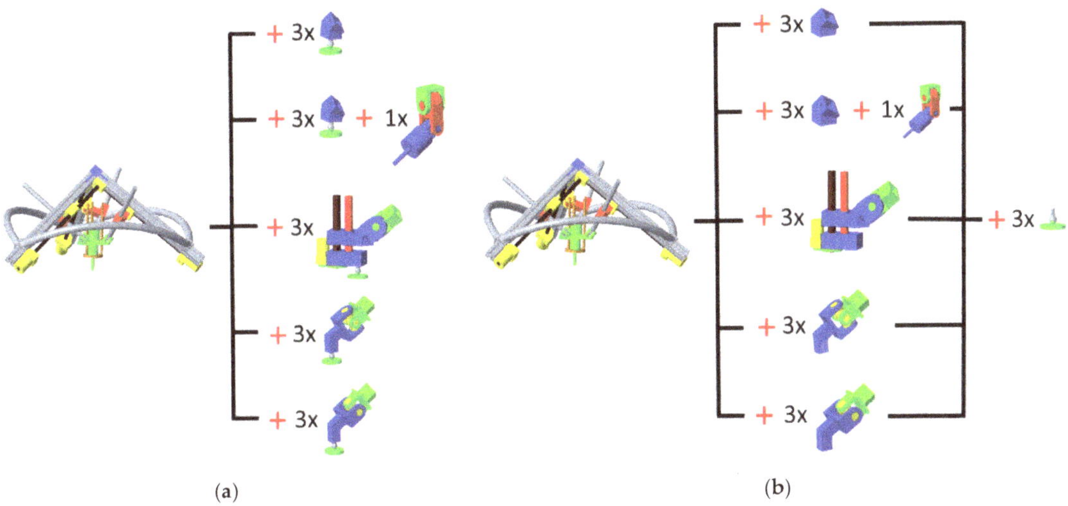

Figure 19. Two possible reconfiguration scenarios using modularity: (**a**) the attachment pad and spherical joint are included in each module, and (**b**) the attachment pad and spherical joint are made as separate modules.

It is not always necessary to have all the modules. One only needs certain modules based on the need. For three-axis machining, only the modules for the 3-DOF robot are needed. For five-axis machining, only the modules for the 5-DOF robot are needed. For six-axis machining, either the modules for the 3SPR or SU mechanism are needed. If a capability to switch between three-axis and five-axis machining or between three-axis and six-axis machining is required, one only needs the required modules.

In the joint locking scheme, reassembly is not required since a module is always assembled. One can simply lock some joints for specific mobility to change the robot's DOF. A switch between three-axis and five-axis machining can be performed by using lockable serial RR mechanism, as illustrated in Figure 20a. Since both the R joints in the serial RR mechanism are actuated, locking both the R joints to change the robot to a 3T mobility can be simply achieved by locking the R joints' actuators. A switch from a six-axis robot with a 3SPR mechanism to a 3T robot can be performed by locking the active P joints and the passive R joint. As depicted in Figure 20b, locking the active P joints can be performed by simply locking their actuators, whereas locking the passive R joints can be achieved by using a certain design of a lockable passive R joint. A switch from a six-axis robot with a 3SU mechanism to a 3T robot can be performed by locking the actuated R joints and the passive R joints. As illustrated in Figure 20c,d, locking the actuated R joints can be simply achieved by locking their actuator, whereas locking the passive R joints can be achieved by using a certain design of a lockable passive R joint. The design of lockable passive R joints is beyond the scope of this paper.

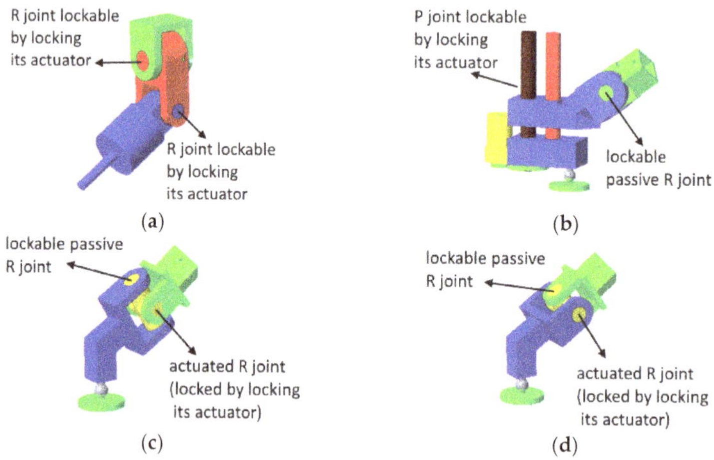

Figure 20. Reconfiguration by joint locking in the (**a**) serial RR chain, (**b**) SPR chain, (**c**) first variant of SU chain, and (**d**) second variant of SU chain.

8. Dimensional Optimization

It is widely known that the main drawback of parallel mechanisms is their limited workspace. In this section, we discuss a method to optimize some dimensions of the proposed mechanisms to obtain the optimal workspace.

8.1. Optimization of the 3T Mechanism

Since the motion limits of the actuators mainly dictate the workspace of the 3T mechanism, its volume can be easily maximized by increasing the length of the pyramid segments such that its linear actuators have a longer sliding span. Since this implies increasing the size of the mechanism, one may compromise it with its resulting footprint and weight. Since this optimization is straightforward, it is not discussed in detail here.

8.2. Optimization of the Serial RR Mechanism

The serial RR mechanism theoretically can make a full rotation about its two R joints' axes. However, due to cabling and collision among components, the tilting limits of the mechanism in the real practice are typically between −90 and 90 degrees. In general, this range is sufficient to perform any five-axis machining. To minimize the translational compensation that the 3T mechanism should provide due to the dependent translation of the serial RR mechanism, it is preferred to minimize the length of the two links in the serial RR mechanism as far as possible.

8.3. Optimization of the 3SPR and 3SU Mechanisms

8.3.1. Design Variables

Since the translational workspace of the 3SPR and 3SU mechanisms are provided by the intersection between the workspaces of each limb, the volume of the intersection is dictated by the position of the three S joints of each mechanism, that is defined by the circumradius of the base, $|\boldsymbol{a}_i|$. The total orientation workspace is affected by the size of the moving platform, that is defined by its circumradius, $|\boldsymbol{b}_i|$. Therefore, both $|\boldsymbol{a}_i|$ and $|\boldsymbol{b}_i|$ are considered the design variables in the optimization of the 3SPR ad 3SU mechanisms. In the 3SPR mechanism, the volume of the intersection is also dictated by the range of the active P joints, that is the span between their lower and upper limits, namely L_i, min and L_i, max. It is quite intuitive to say that the larger the range of the active P joints, the higher the tilting capability of the 3SPR mechanism. Accordingly, maximizing the tilting capability of the 3SPR mechanism can be performed by simply enlarging the range of their active P joints. Since this is trivial, L_i, min and L_i, max are not included as the design va-riables in the optimization of the 3SPR mechanism. Instead, they are determined a priori based on the design consideration. On the other hand, in the 3SU mechanism, the volume intersection is also dictated by the link length, L_i. This parameter is considered a design variable in the optimization of the 3SU mechanism. All the mentioned design parameters are to be optimized to obtain an optimal workspace.

8.3.2. Optimization Constraints

The optimization constraints consist of design, technological, and interference constraints. The design constraints are the constraints that involve the design variables. Assuming that the base and the moving platform have equidistant triangles, $|\boldsymbol{a}_1| = |\boldsymbol{a}_2| = |\boldsymbol{a}_3| = R$ and $|\boldsymbol{b}_1| = |\boldsymbol{b}_2| = |\boldsymbol{b}_3| = r$. For design consideration, we can determine the lower and upper bounds of R and r, which can be written as:

$$\begin{aligned} R_{min} \leq R \leq R_{max} \\ r_{min} \leq r \leq r_{max} \end{aligned} \tag{36}$$

For the 3SU mechanism, the lower and upper bounds of the link length L_i can be written as:

$$L_{i,min} \leq L_i \leq L_{i,max} \tag{37}$$

The base circumradius should not be equal to the moving platform circumradius to avoid singularity, as discussed earlier. For better stability of the mechanism, let us constraint the base circumradius to be larger than the moving platform circumradius. This can be written as the following inequality:

$$-R + r \leq 0 \tag{38}$$

The technological constraints do not involve the design variables. They are determined based on the technological limitations of some components of the mechanism. In this case, the limits of the S, R, and U joints are the technological constraints of the optimization. In fact, these joints' limits also affect the workspace.

The interference constraints are the constraints determined to avoid collision bet-ween the mechanism components. In this case, we constraint the mechanism to avoid collision between the moving platform and the base. Considering that the moving platform with the circumradius r can tilt at angles θ_x and θ_y about the X and Y axes, respectively, this can be written as the following inequality constraints:

$$|r \sin \theta_x| - z < 0 \tag{39}$$

$$|r \sin \theta_y| - z < 0 \tag{40}$$

The constraints in Equations (39) and (40) also avoid the mechanism from its singular configuration associated with the collapse of the moving platform to the base.

8.3.3. Objective Function

Since the 3SPR and 3SU mechanisms' main purpose is to provide tilting capability, the objective of the workspace optimization should be to maximize the orientation workspace. Since the orientation workspace is defined by the moving platform angles θ_x and θ_y, maximizing the range of each angle can be written as maximizing the difference between the maximum and minimum limits of each angle, i.e., $\theta_{x,max} - \theta_{x,min}$ and $\theta_{y,max} - \theta_{y,min}$. Since there are two ranges of angles to be maximized, it is more practical in the optimization algorithm to combine both as a single objective given by the sum of them. Hence, the single objective function can be written as the following weighted sum:

$$f_{maximization} = g_1(\theta_{x,max} - \theta_{x,min}) + g_2(\theta_{y,max} - \theta_{y,min}) \quad (41)$$

where g_1 and g_2 denote the weights of both terms in the right hand side of Equation (41). Since the ranges of both the angles θ_x and θ_y are equally important, we use $g_1 = g_2 = 1$.

If the optimization is written as a minimization problem, the constrained optimization formulation is to minimize the negative of the aforementioned objective function, i.e.:

$$f_{minimization} = -f_{maximization} \quad (42)$$

8.3.4. Optimization Formulation

The workspace optimization problem can be formulated as a constrained optimization problem. Accordingly, the workspace optimization of the 3SPR mechanism can be stated as follows:

Find the optimum design parameters R and r
that maximize the objective function $f_{maximization}$ given in Equation (41)
subject to:

- the lower and upper bounds of R as written in Equation (36);
- the lower and upper bounds of r as written in Equation (36);
- the inequality constraint written in Equation (38);
- the limits of the S joints, i.e., $\theta^{is}_{pj,min} \leq \theta^{is}_{pj} \leq \theta^{is}_{pj,max}$ for i = 1,2,3 and j = 1,2,3;
- the limits of the R joints, i.e., $\theta_{pi,min} \leq \theta_{pi} \leq \theta_{pi,max}$ for i = 1,2,3;
- the interference constraints written in Equations (39) and (40).

On the other hand, the workspace optimization of the 3SU mechanism can be stated as follows:

Find the optimum design parameters R, r, and L_i
that maximize the objective function $f_{maximization}$ given in Equation (48)
subject to:

- the lower and upper bounds of R as written in Equation (36);
- the lower and upper bounds of r as written in Equation (36);
- the lower and upper bounds of L_i as written in Equation (37);
- the inequality constraint written in Equation (38);
- the limits of the S joints, i.e., $\theta^{i}_{pj,min} \leq \theta^{i}_{pj} \leq \theta^{i}_{pj,max}$ for i = 1,2,3 and j = 1,2,3;
- the limits of the U joints, i.e., $\theta_{ai,min} \leq \theta_{ai} \leq \theta_{ai,max}$ and $\theta_{pi,min} \leq \theta_{pi} \leq \theta_{pi,max}$ for i = 1,2,3;
- the interference constraints written in Equations (39) and (40).

8.3.5. Optimization Algorithm

The objective function is computed using the following routine:

- Step 1: Run the forward kinematics of the mechanism by inputting the active joints' positions at a specific discretization interval within their ranges. The tilting angles of the moving platform θ_x and θ_y corresponding to the input active joints' positions are obtained from the forward kinematics solution. It is worth mentioning that the lower and upper limits of the active P joints, $L_{i,min}$ and $L_{i,max}$, determined a priori in the 3SPR mechanism, can be considered pre-determined constraints that are not part of the optimization. Similarly, the limits of the active joints in the 3SU mechanism that are determined a priori can also be considered pre-determined constraints that are not part of the optimization.
- Step 2: Extract the minimum and maximum values of the obtained tilting angles of the moving platform θ_x and θ_y. This gives us $\theta_{x,min}$, $\theta_{x,max}$, $\theta_{y,min}$, and $\theta_{y,max}$.

- Step 3: Compute the objective function written in Equations (41) or (42), depending on whether it is maximization or minimization.

The constrained optimization problem formulated above was solved by using genetic algorithm optimizer, namely *ga*, in MATLAB. It is a global optimizer that can avoid getting trapped in a local optimum.

8.3.6. Numerical Example

Here, we present an example of the optimization of the 3SPR and 3SU mechanisms. As described earlier, the design variables in the optimization of the 3SPR mechanism are R and r, whereas the design variables in the optimization of the 3SU mechanism are R, r, and L_i The lower and upper bounds of the design variables are $0.3\ \text{m} \leq R \leq 0.6\ \text{m}$, $0.2\ \text{m} \leq r \leq 0.5\ \text{m}$, and $0.2\ \text{m} \leq L_i \leq 0.6\ \text{m}$. The inequality constraints written in Equations (38)–(40) are imposed. The lower and upper rotational limits of the S, R, and U joints are -70 deg and 70 deg, unless specified otherwise.

The forward kinematics of the 3SPR and 3SU mechanisms were solved by iterating the active joints positions from their minimum values to their maximum values at a certain discretization interval. The smaller the interval, the finer the visualization of the mechanism workspace, but the longer the time required for the optimization, and vice versa. In the optimization, the parameters of the genetic algorithm optimizer are shown in Table 2.

Table 2. Parameters of the MATLAB genetic algorithm optimizer.

Parameter	Value
Population size	50
Maximum generations	100 the number of design variables
Creation function	Random initial population satisfying bounds and linear constraints
Mutation function	Mutation Adapt Feasible
Selection function	Stochastic Uniform
Crossover function	CrossoverIntermediate (weighted average of the parents)
Crossover fraction	0.8
Elite count	3
Function tolerance	1×10^{-6}
Constraint tolerance	1×10^{-3}
Maximum stall generations	50
Maximum stall time	∞
Maximum time	∞

Using a computer having 3 GHz Intel® Xeon® processor and 64 GB RAM, the optimization of the 3SPR with forward kinematics discretization interval of 0.01 m took 340 min. On the other hand, the optimization of the 3SU mechanisms with the forward kinematics discretization interval of 0.01 deg took 1125 min. The optimization stopped after the average change in the fitness value is less than the function tolerance. The optimization of the 3SPR mechanism gives the following optimized design variables: R = 0.549 m and r = 0.205 m. On the other hand, the optimization of the 3SU mechanism gives the following optimized design variables: R = 0.533 m, r = 0.234 m, and $L_1 = L_2 = L_3 = 0.493$ m.

Figure 21 shows the orientation workspace of the optimized 3SPR mechanism in multiple views. The discretization interval of 0.01 m, which is considered quite large, does not provide a smooth boundary of the workspace visualization. It is shown that the optimized 3SPR mechanism can be tilted from approximately -25 deg to 12.5 deg about the X axis and from approximately -20 deg to 20 deg about the Y axis. The dots in the figure, regardless of their color, represent the reachable tilting angles in the 3SPR mechanism. It can be seen in Figure 21b that the tilting range about the Y axis is symmetric. This is because the mechanism is symmetric with respect to the Y axis. In contrast, the tilting range about the X axis is larger in the negative direction than in the positive direction as the mechanism is not symmetric with respect to the X axis. The moving platform can be tilted more to a negative angle about the X axis since this tilting requires only one link to be lowered. On the other hand, the moving platform can be tilted less to a positive angle about the X axis since this tilting requires two links to be lowered. Figure 22 shows some postures of the tilted 3SPR mechanism.

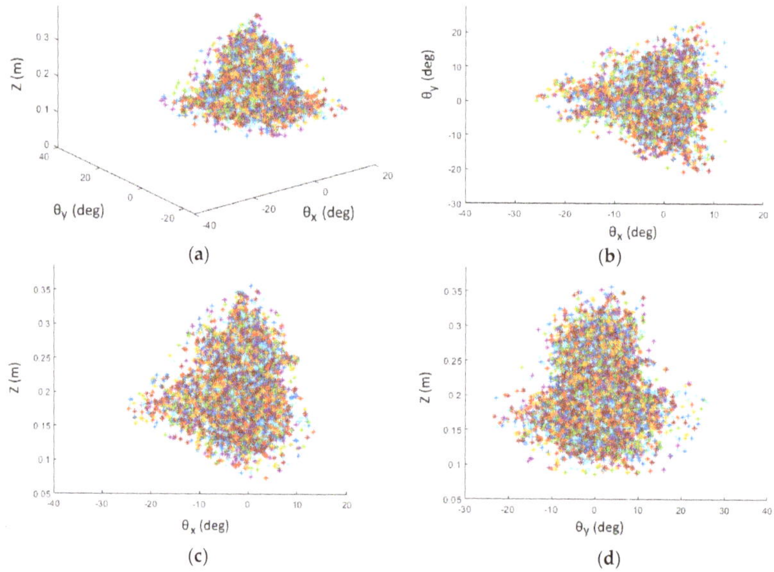

Figure 21. (**a**) Three-dimensional and (**b–d**) two-dimensional plots of the orientation workspace of the optimized 3SPR mechanism. The dots, regardless of their color, represent the reachable tilting angles.

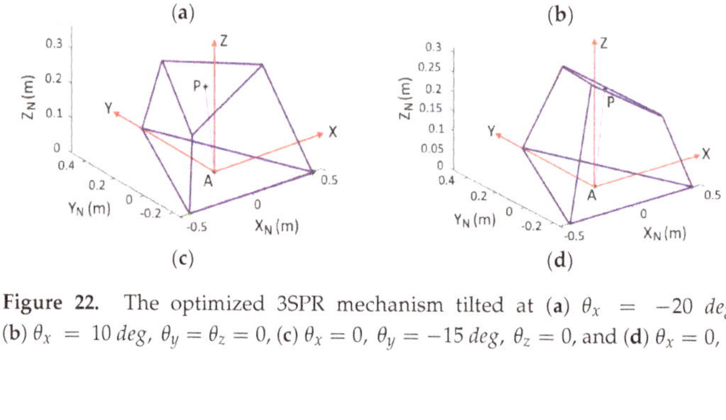

Figure 22. The optimized 3SPR mechanism tilted at (**a**) $\theta_x = -20\ deg$, $\theta_y = \theta_z = 0$, (**b**) $\theta_x = 10\ deg$, $\theta_y = \theta_z = 0$, (**c**) $\theta_x = 0$, $\theta_y = -15\ deg$, $\theta_z = 0$, and (**d**) $\theta_x = 0$, $\theta_y = 15\ deg$, $\theta_z = 0$.

In a similar fashion, Figure 23 shows in multiple views the orientation workspace of the optimized 3SU mechanism given the lower and upper limits of the S and U joints of −80 deg and 80 deg. The discretization interval of 0.01 deg, which is considered quite small, provides a smooth boundary of the workspace visualization. It is shown that the optimized 3SPR mechanism can be tilted from approximately −80 deg to 60 deg about the X axis and from approximately −40 deg to 40 deg about the Y axis. The dots in the figure, regardless of their color, represent the reachable tilting angles in the 3SU mechanism. The symmetry of the tilting range about the Y axis and the non-symmetry of the tilting range about the X axis have a similar reason to the case in the 3SPR mechanism. Furthermore, it is quite intuitive to say that longer links in the SU mechanism lead to larger tilting angles since the moving platform can be tilted at a larger angle without colliding with the base. Figure 24 shows some postures of the tilted 3SU mechanism.

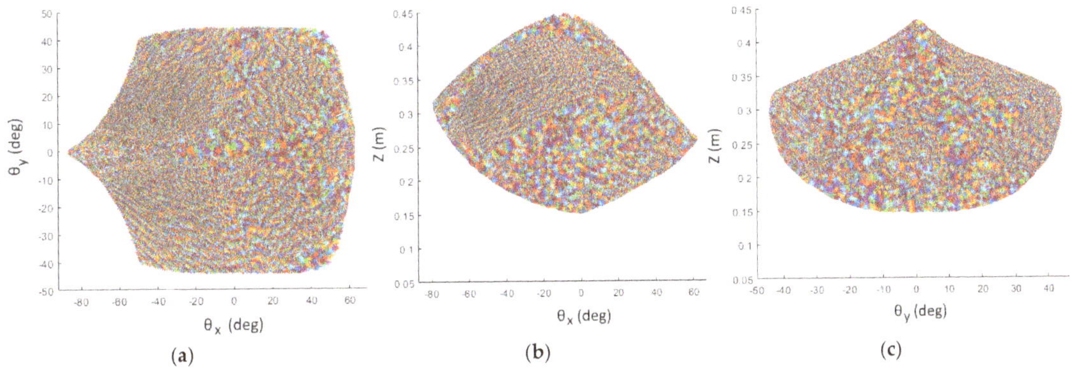

Figure 23. (**a**–**c**) Two-dimensional plots of the orientation workspace of the optimized 3SU mechanism with the rotational range of the S and U joints between −80 deg and 80 deg. The dots, regardless of their color, represent the reachable tilting angles.

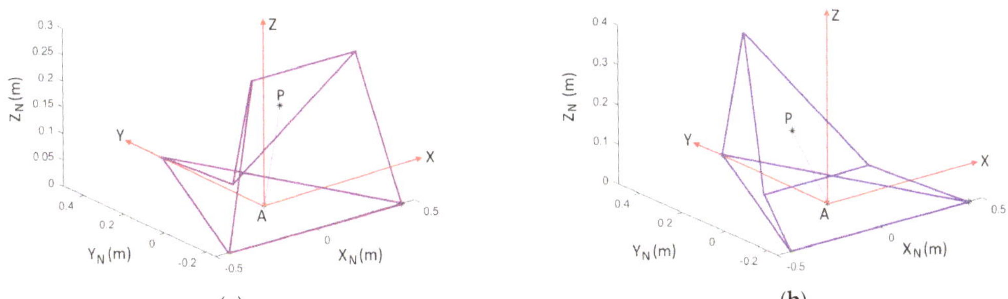

Figure 24. The optimized 3SU mechanism with the rotational range of the S and U joints between −80 deg and 80 deg tilted at (**a**) $\theta_x = -70\ deg,\ \theta_y = \theta_z = 0$ and (**b**) $\theta_x = 60\ deg,\ \theta_y = \theta_z = 0$.

However, when the rotational ranges of the S and U joints are tightened, the orientational workspace of the 3SU mechanism becomes smaller. The tighter the ranges of the S and U joints, the smaller the orientation workspace. For example, when the lower and upper limits of the S and U joints are changed to −30 deg and 30 deg, the orientation workspace of the 3SU mechanism is shrunk to that shown in Figure 25. It is clear that this smaller orientation workspace is a subset of the larger one shown in Figure 23. It is shown in the figure that the optimized 3SPR mechanism can be tilted from approximately –30 deg to 17.5 deg about the X axis and from approximately −15 deg to 15 deg about the Y axis. Figure 26 shows some postures of the tilted 3SU mechanism.

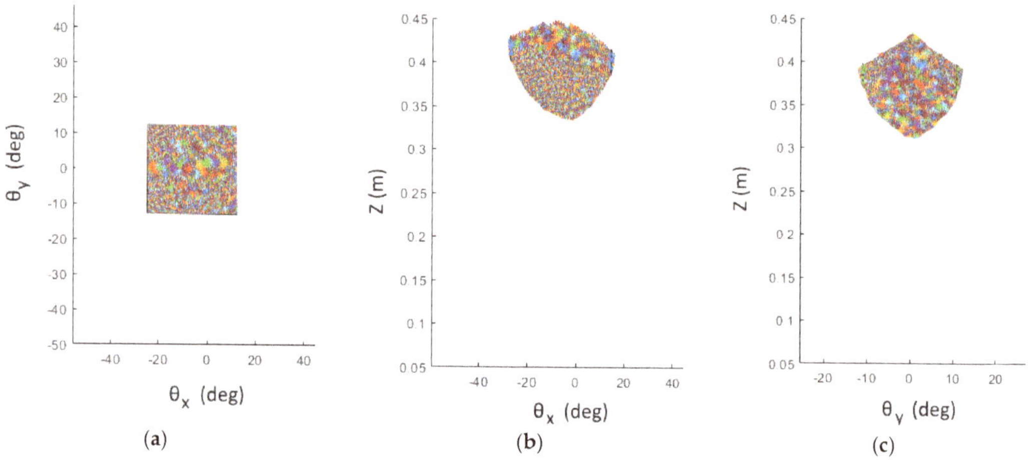

Figure 25. (**a–c**) Two-dimensional plots of the orientation workspace of the optimized 3SU mechanism with the rotational range of the S and U joints between −30 deg and 30 deg. The dots, regardless of their color, represent the reachable tilting angles.

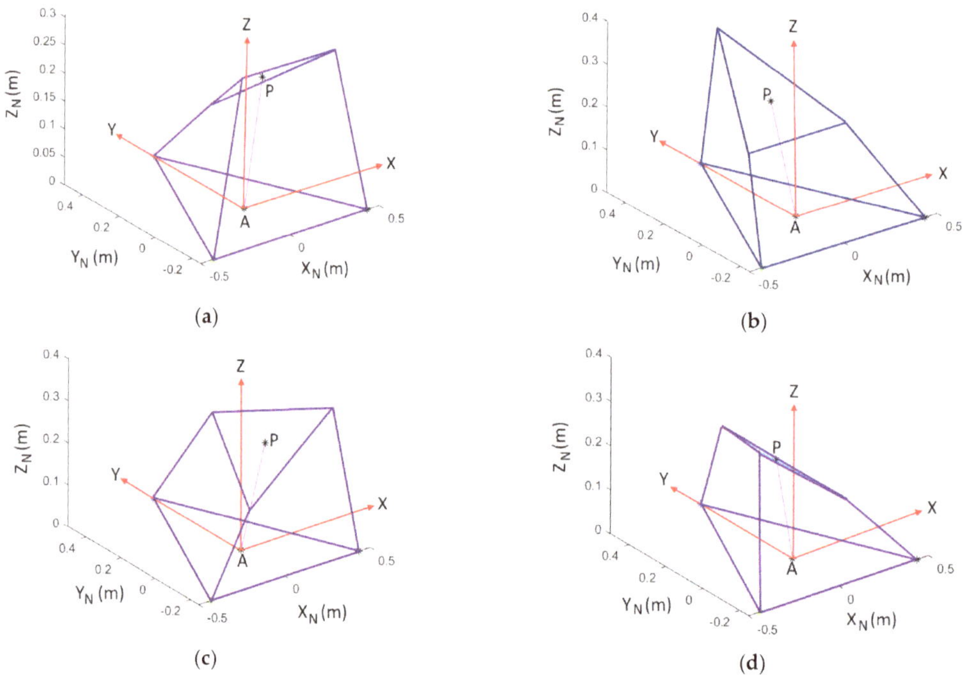

Figure 26. The optimized 3SU mechanism with the rotational range of the S and U joints between −30 deg and 30 deg tilted at (**a**) $\theta_x = -10\ deg,\ \theta_y = \theta_z = 0$, (**b**) $\theta_x = 10\ deg,\ \theta_y = \theta_z = 0$, (**c**) $\theta_x = 0,\ \theta_y = -10\ deg, \theta_z = 0$, and (**d**) $\theta_x = 0,\ \theta_y = 10\ deg, \theta_z = 0$.

Comparing the orientation workspace of the 3SPR and 3SU joints, we observed that tilting capability of the 3SPR mechanism is much dictated by the range of the active P joints' motion besides the ranges of the S and R joints. The 3SPR mechanism provides a larger tilting angle when one of links is shortened to its minimum while the other links are lengthened to their maximum, or when two links are shortened to their minimum while the other link is lengthened to its maximum. On the other hand, the orientation workspace of the 3SU mechanism is mainly limited by the rotational range of the S and U joints. Although the 3SU mechanism theoretically has a quite large orientation workspace, the technological limitation of the S and U joints typically provides a quite tight rotational ranges.

9. Design Considerations

Although parallel mechanisms in general are theoretically more rigid than serial mechanisms due to their closed chains, the number of joints in the parallel mechanisms is typically larger than those in serial mechanisms. This is often believed to be among the reasons why the parallel mechanisms in general often cannot outperform well-designed and well-manufactured serial mechanisms, particularly the Cartesian/gantry serial machines. Nevertheless, compared to articulated serial robots, well-designed and well-manufactured parallel robots are typically more rigid, despite their more limited workspace. Hence, the proposed parallel robot is to compete with the articulated serial robots, not the Cartesian/gantry machines. For the machining of large structures, the use of Cartesian/gantry machines are often not feasible as the use of such machines requires the structures/workpieces to be inserted within the machines' volume. Hence, in such scenarios, the articulated serial robots are currently often used. Such robots suffer a lot from low rigidity when it is significantly extended, unless the robots are built with bulky and heavy arms. The size and weight of on-structure (adsorption) parallel robots, such as the one proposed in this paper, are much lower with comparable and even better stiffness. The high stiffness is due to two reasons: (1) the closed chain of the parallel mechanisms, and (2) the direct attachment of the mechanism to the structure/workpiece.

To overcome possible low accuracy and stiffness due to the presence of many joints in parallel mechanisms, one should have a good detail design and engineering practice to physically implement the joints, for example, by using preloaded bearings and precision bearings with good assembly fit and tolerance. When the joint locking scheme is used for the reconfiguration, the lockable joints should be designed carefully to avoid significant clearance or backlash that may deteriorate the robot performance. Finally, the manufacturing and assembly of the robot links should be performed with appropriate fit and to-lerance to avoid the resulting inaccuracy due to manufacturing and assembly errors.

10. Conclusions

A novel modular/reconfigurable walking robot topology with 3T2R manipulation capability, namely, 3A+3S+3PRPR+1RR (or 3A+3S+3PRRR+1RR), and two novel modular/reconfigurable walking robots with six-DOF manipulation capability, namely, 3A+3SPR+3PRPR (or 3A+3SPR+3PRRR) and 3A+3SR⊥R(3SU)+3PRPR (or 3A+3SR⊥R(3SU)+3PRRR), are proposed. Furthermore, the 3A+3SR⊥R(3SU)+3PRPR (or 3A+3SR⊥R(3SU)+3PRRR) topology is presented in two variations. The 3T module of all the robot topologies is a 3PRPR or 3PRRR parallel mechanism. Attachment pads are serially connected with S joints to the base frames of the hybrid-kinematics mechanisms, while a support is attached to the moving platform of the mechanisms. It was shown that the 3T robot can be reconfigured to a 3T2R or six-DOF robot capable of adjusting the orientation of its tool to be perpendicular to the irregular surface of a workpiece/structure and capable of performing five- or six-axis machining tasks. This reconfiguration can be performed by either reassembly or joint locking. It was shown that the 3T2R robot utilizing a serial RR module to provide the rotational DOFs has larger orientation workspace compared to the six-DOF robots utilizing the 3SPR and 3SU modules. Therefore, the 3T2R robot is more appropriate to be used in a five-axis machining task due to its larger orientation workspace, whereas the six-DOF robots are appropriate to be used for applications that require an angular adjustment due to an irregular or uneven surface of large workpieces/structures.

11. Patent

The content of this paper is a part of the authors' pending patent application US 63/184.505 entitled "Reconfigurable Walking Robot".

Supplementary Materials: The supplementary video can be downloaded at: https://kudrive.ku.ac.ae/oc-shib/index.php/s/DM6bvzFewLRnYNb (accessed date 15 August 2022).

Author Contributions: Conceptualization, A.R., B.E.-K. and C.S.; investigation, formal analysis, and visualization, A.R.; writing—original draft preparation, A.R.; writing—review and editing, B.E.-K. and A.R.; supervision, B.E.-K. and C.S.; project administration, B.E.-K. and C.S. All authors have read and agreed to the published version of the manuscript.

Funding: This work was funded by Khalifa University's internal fund RC1-2018-KUCARS awarded to Khalifa University Center for Autonomous Robotic Systems (KUCARS).

Data Availability Statement: Not applicable.

Conflicts of Interest: The authors declare no conflict of interest.

Appendix A

Table A1. List of Abbreviations.

Abbreviation	Meaning
DOF	Degree of freedom
P, R, U, S	Prismatic, revolute, universal, and spherical joints
A	Attachment
3T	Three-translation
2R	Two-rotation
3R	Three-rotation
3T2R	Three-translation and two-rotation
1T2R	One-translation and two-rotation
TCP	Tool Center Point

Appendix B

Table A2. List of Symbols.

Symbol	Meaning
$\perp$	Perpendicularity between two adjacent joint axes
Underline	Actuated joint
$\$^i$	Joint screws in the *i*-th limb
$\$^i_j$	Screw of the *j*-th joint in the *i*-th limb
$\$^{ir}$	Reciprocal screw in the *i*-th limb
x_1, x_2, x_3	Positions of the three active P joints in the 3T mechanism
x,y,z	End-effector position
$\mathrm{x_i}y_\mathrm{i}\mathrm{z_i}$	Coordinate frame of the *i*-th limb
xyz	Coordinate frame attached to the apex of the pyramid of 3T mechanism
XYZ	Coordinate frame attached to the center of the base of the lower mechanism

Table A2. *Cont.*

Symbol	Meaning
X′Y′Z′	Coordinate frame attached to the moving platform of the lower mechanism
X″Y″Z″	Coordinate frame attached to a tip of the pyramid segments of the 3T mechanism
$X_NY_NZ_N$	Inertial (world) coordinate frame
p	Position vector of the moving platform with respect to the center of the base
$\boldsymbol{R}_A^P$	Rotation matrix of the moving platform with respect to the base
$\boldsymbol{R}_x$, $\boldsymbol{R}_y$	Elementary rotation matrix about X and Y axes
θ_x, θ_y, θ_z	Tilting angles of the moving platform with respect to X, Y, and Z axes
$\theta_{x,min}$, $\theta_{x,max}$	Minimum and maximum values of θ_x
$\theta_{y,min}$, $\theta_{y,max}$	Minimum and maximum values of θ_y
θ_{ai}, θ_{pi}	Angles of the active and passive R joints of the R or U joint in the i-th limb
$\theta_{ai,min}$, $\theta_{ai,max}$	Minimum and maximum values of θ_{ai}
$\theta_{pi,min}$, $\theta_{pi,max}$	Minimum and maximum values of θ_{pi}
θ_{pj}^{is}	Angle of the *i*-th rotational DOF of the S joint in the *j*-th limb
$\theta_{pj,min}^{is}$, $\theta_{pj,min}^{is}$	Minimum and maximum values of θ_{pj}^{is}
R	Circumradius of the base
R_{min}, R_{max}	Minimum and maximum values of R
r	Circumradius of the moving platform
r_{min}, r_{max}	Minimum and maximum values of r
${}^A\boldsymbol{r}$	Position vector **r** expressed with respect to point A
$\boldsymbol{a}_i$	Position vector of the *i*-th S joint with respect to the base center
$\boldsymbol{b}_i$	Position vector of the *i*-th distal joint with respect to the moving platform center
$\boldsymbol{d}_i$	Position vector of the active P joint in the 3SPR mechanism
L_p	Length of each of the pyramid segments
L_i	Stroke length of the *i*-th actuator in the 3SPR mechanism
$L_{i,min}$, $L_{i,max}$	Minimum and maximum values of L_i
L	Constant length of the link connecting the S and U joints in the SU mechanism
$\boldsymbol{l}_i$	Unit directional vector of the *i*-th active P joint in the 3SPR mechanism
$\boldsymbol{w}_i$	Unit directional vector of the R joint in the *i*-th U joint perpendicular to $\boldsymbol{b}_i$
$\boldsymbol{v}_i$	Unit directional vector of the R joint in the *i*-th U joint perpendicular to $\boldsymbol{w}_i$
$\$_P$	Instantaneous twist of the moving platform
ω_x, ω_y, ω_z	Angular velocities of the moving platform with respect to X, Y, and Z axes
$\dot{x}$, $\dot{y}$, $\dot{z}$	Translational velocities of the moving platform in the X, Y, and Z directions
$\dot{q}$	Twist of the actuators
$\dot{\theta}_j^1$, $\dot{d}_j$	Intensity of the *j*-th rotational and translational joints in the *i*-th limb

Table A2. *Cont.*

Symbol	Meaning
$\hat{\$}_j^i$	j-th unit joint screw in the i-th limb
s_j^i	**s** vector of the j-th unit joint screw in the i-th limb
e_i	i-th actuation vector
J_x	Forward Jacobian matrix
J_q	Inverse Jacobian matrix
J_a	Actuation Jacobian matrix
J_c	Constraint Jacobian matrix
J	Overall Jacobian matrix
$f_{maximization}$	Maximization objective function
$f_{minimization}$	Minimization objective function
g_i	Weight of the i-th objective in the weighted sum

References

1. Axinte, D.A.; Allen, J.M.; Anderson, R.; Dane, I.; Uriarte, L.; Olara, A. Free-leg Hexapod: A novel approach of using parallel kinematic platforms for developing miniature machine tools for special purpose operations. *CIRP Ann.* **2011**, *60*, 395–398. [CrossRef]
2. Rushworth, A.; Cobos-Guzman, S.; Axinte, D.A.; Raffles, M. Pre-gait analysis using optimal parameters for a walking machine tool based on a free-leg hexapod structure. *Robot. Auton. Syst.* **2015**, *70*, 36–51. [CrossRef]
3. Olarra, A.; Axinte, D.; Uriarte, L.; Bueno, R. Machining with the WalkingHex: A walking parallel kinematic machine tool for in situ operations. *CIRP Ann.* **2017**, *66*, 361–364. [CrossRef]
4. Chen, J.; Xie, F.; Liu, X.; Bi, W. Stiffness Evaluation of an Adsorption Robot for Large-Scale Structural Parts Processing. *ASME J. Mech. Robot.* **2021**, *13*, 040902. [CrossRef]
5. Tedeschi, F.; Carbone, G. Design Issues for Hexapod Walking Robots. *Robotics* **2014**, *3*, 181–206. [CrossRef]
6. Tedeschi, F.; Carbone, G. Design of a Novel Leg-Wheel Hexapod Walking Robot. *Robotics* **2017**, *6*, 40. [CrossRef]
7. Wang, M.F.; Ceccarelli, M.; Carbone, G. A feasibility study on the design and walking operation of a biped locomotor via dynamic simulation. *Front. Mech. Eng.* **2016**, *11*, 144–158. [CrossRef]
8. Russo, M.; Ceccarelli, M. Kinematic design of a tripod parallel mechanism for robotic legs. In *International Workshop on Computational Kinematics*; Springer International Publishing: Cham, Switzerland, 2018; pp. 121–130.
9. Russo, M.; Herrero, S.; Altuzarra, O.; Ceccarelli, M. Kinematic analysis and multi-objective optimization of a 3-UPR parallel mechanism for a robotic leg. *Mech. Mach. Theory* **2018**, *120*, 192–202. [CrossRef]
10. Giewont, S.; Sahin, F. Delta-Quad: An omnidirectional quadruped implementation using parallel jointed leg architecture. In Proceedings of the 12th System of Systems Engineering Conference (SOSE), Waikoloa, HI, USA, 18–21 June 2017; pp. 1–6.
11. Li, L.; Fang, Y.; Guo, S.; Qu, H.; Wang, L. Type synthesis of a class of novel 3-DOF single-loop parallel leg mechanisms for walking robots. *Mech. Mach. Theory* **2020**, *145*, 103695. [CrossRef]
12. Lin, R.; Guo, W.; Li, M. Novel Design of Legged Mobile Landers With Decoupled Landing and Walking Functions Containing a Rhombus Joint. *ASME J. Mech. Robot.* **2018**, *10*, 061017. [CrossRef]
13. Lin, R.; Guo, W. Creative Design of Legged Mobile Landers With Multi-Loop Chains Based on Truss-Mechanism Transformation Method. *ASME J. Mech. Robot.* **2021**, *13*, 011013. [CrossRef]
14. Yang, H.; Krut, S.; Pierrot, F.; Baradat, C. Locomotion Approach of REMORA: A Reonfigurable Mobile Robot for Manufacturing Applications. IROS: Intelligent Robots and Systems. In Proceedings of the 2011 IEEE/RSJ International Conference on Intelligent Robots and Systems, San Francisco, CA, USA, 25–30 September 2011; pp. 5067–5072. [CrossRef]
15. Figliolini, G.; Rea, P.; Conte, M. Mechanical Design of a Novel Biped Climbing and Walking Robot. In *ROMANSY 18 Robot Design, Dynamics and Control*; Parenti Castelli, V., Schiehlen, W., Eds.; CISM International Centre for Mechanical Sciences; Springer: Vienna, Austria, 2010; Volume 524. [CrossRef]
16. Dunlop, G.R. Foot Design for a Large Walking Delta Robot. In *Experimental Robotics VIII*; Springer Tracts in Advanced, Robotics; Siciliano, B., Dario, P., Eds.; Springer: Berlin/Heidelberg, Germany, 2003; Volume 5. [CrossRef]
17. Li, R.; Meng, H.; Bai, S.; Yao, Y.; Zhang, J. Stability and Gait Planning of 3-UPU Hexapod Walking Robot. *Robotics* **2018**, *7*, 48. [CrossRef]
18. Brahmia, A.; Kelaiaia, R.; Chemori, A.; Company, O. On Robust Mechanical Design of a PAR2 Delta-Like Parallel Kinematic Manipulator. *ASME J. Mech. Robot.* **2022**, *14*, 011001. [CrossRef]

19. Kelaiaia, R.; Zaatri, A.; Company, O. Multiobjective Optimization of 6-dof UPS Parallel Manipulators. *Adv. Robot.* **2012**, *26*, 1885–1913. [CrossRef]
20. Kelaiaia, R.; Zaatri, A.; Company, O.; Chikh, L. Some investigations into the optimal dimensional synthesis of parallel robots. *Int. J. Adv. Manuf. Technol.* **2016**, *83*, 1525–1538. [CrossRef]
21. Rosyid, A.; El-Khasawneh, B.; Alazzam, A. Genetic and hybrid algorithms for optimization of non-singular 3PRR planar parallel kinematics mechanism for machining application. *Robotica* **2018**, *36*, 839–864. [CrossRef]
22. Rubbert, L.; Charpentier, I.; Henein, S.; Renaud, P. Higher-order continuation method for the rigid-body kinematic design of compliant mechanisms. *Precis. Eng.* **2017**, *50*, 455–466. [CrossRef]

robotics

MDPI

Review

A Review of Parallel Robots: Rehabilitation, Assistance, and Humanoid Applications for Neck, Shoulder, Wrist, Hip, and Ankle Joints

Victoria E. Abarca *,† and Dante A. Elias †

Biomechanics and Applied Robotics Research Laboratory, Pontificia Universidad Católica del Perú, Lima 15088, Peru; delias@pucp.pe

* Correspondence: victoria.abarca@pucp.edu.pe

† These authors contributed equally to this work.

Abstract: This review article presents an in-depth examination of research and development in the fields of rehabilitation, assistive technologies, and humanoid robots. It focuses on parallel robots designed for human body joints with three degrees of freedom, specifically the neck, shoulder, wrist, hip, and ankle. A systematic search was conducted across multiple databases, including Scopus, Web of Science, PubMed, IEEE Xplore, ScienceDirect, the Directory of Open Access Journals, and the ASME Journal. This systematic review offers an updated overview of advancements in the field from 2012 to 2023. After applying exclusion criteria, 93 papers were selected for in-depth review. This cohort included 13 articles focusing on the neck joint, 19 on the shoulder joint, 22 on the wrist joint, 9 on the hip joint, and 30 on the ankle joint. The article discusses the timeline and advancements of parallel robots, covering technology readiness levels (TRLs), design, the number of degrees of freedom, kinematics structure, workspace assessment, functional capabilities, performance evaluation methods, and material selection for the development of parallel robotics. It also examines critical technological challenges and future prospects in rehabilitation, assistance, and humanoid robots.

Keywords: assistance; exoskeletons; parallel robots; prosthetics; rehabilitation

1. Introduction

Rehabilitation and assistance for human body joints play crucial roles in people's health, well-being, and quality of life. These activities are essential for helping individuals recover or improve their functionality, mobility, and autonomy following injuries, surgeries, or diseases affecting the joints. In recent years, technological advances have opened up new possibilities in the use of parallel robots in the fields of rehabilitation, assistive technologies, and humanoid systems. These robotic systems are specifically designed to facilitate the recovery of motor and functional skills. Featuring a parallel mechanical structure, these robots offer greater precision, stability, and adaptability to meet the individual rehabilitation or assistance needs of patients.

Rehabilitation technologies aid in the recovery or improvement of motor function after an injury or illness. Utilizing parallel robots in rehabilitation enables specialists to help patients regain strength and mobility in affected joints more rapidly and efficiently than traditional therapeutic methods. The scope of these technologies includes rehabilitation devices for patients with head and neck injuries [1–4]; pediatric rehabilitation devices for the arm [5]; wearable rehabilitation devices for the arm [6]; exoskeletons for the arms of patients with stroke and spinal cord injuries [7]; and devices specifically designed for wrist [8–10], ankle [11–16], and foot rehabilitation [17].

Parallel robots in assistive devices help individuals carry out activities of daily life, thereby providing greater autonomy and independence. Examples include prostheses designed to offer functional mobility by replacing missing limbs in amputees [18], as well as shoulder disarticulation arm prostheses [19,20], wrist prostheses [21], and disarticulated hip prostheses [22].

The integration of humanoid robots into medical rehabilitation and assistance offers exciting opportunities. These include personalized therapy, precise motivation and tracking, assisted mobility, and the objective assessment of patients' progress.

Parallel robots have advanced significantly in the fields of architectural [23] and mathematical modeling, particularly in kinematic [24–26] and dynamic analyses [27]. These robots can move at high speeds due to their lightweight and simple structural design [28]. They are also highly rigid, making them ideal for tasks requiring substantial force or pressure [29]. Their high-precision design minimizes unwanted movements and vibrations [30]. Furthermore, they can support heavier loads [31,32] and offer better positioning accuracy due to their high rigidity and low weight, which ensure minimal deformation [33]. Utilizing sensor technology and control algorithms, these parallel robots can automatically adjust to meet each patient's specific needs, whether for rehabilitation or assistance, thus delivering a highly personalized experience.

However, there is a gap in the current literature concerning the use of parallel robots in the fields of rehabilitation, assistance, and humanoid systems. This review aims to address this gap by summarizing the available evidence on the utilization of parallel robots.

Therefore, this article aims to review the state of parallel robot technology as applied to rehabilitation, assistance, and humanoid systems, focusing on joints with three degrees of freedom in the human body: the neck, shoulder, wrist, hip, and ankle. Initially, the search strategy, inclusion and exclusion criteria, quality assessment, and data extraction methods are defined. Subsequently, the search results are presented, along with a detailed description of the biomechanics of the joints, parallel robots, and applications in the medical field. The article then discusses the timeline and advancements of parallel robots between 2012 and 2023 and examines the technology readiness levels (TRLs), design, number of degrees of freedom, kinematics structure, workspace assessment, functional capabilities, performance methods, and material selection in the development of parallel robotics, as well as the critical technological challenges and future prospects in rehabilitation, assistance, and humanoids. Finally, conclusions are presented.

2. Methodology

2.1. Search Strategy

Before initiating the search, a research question was formulated: "Are there parallel robots designed for rehabilitation, assistance, and humanoid applications that target joints such as the neck, shoulder, wrist, hip, and ankle?" This question helped focus the search for pertinent information. Subsequently, several databases were identified for the search, including Scopus, Web of Science, PubMed, IEEE Xplore, ScienceDirect, the Directory of Open Access Journals (DOAJ), and the ASME Journal. Following this, search terms were determined, incorporating keywords related to the topic such as "parallel robot and neck", "parallel robot and shoulder", "parallel robot and hip", "parallel robot and wrist", and "parallel robot and ankle".

2.2. Inclusion Criteria

The search was conducted from 2012 to 2023, and articles were selected based on the following inclusion criteria:

- Articles in English that discuss parallel robots for the rehabilitation of the neck, shoulder, wrist, hip, and ankle.
- Articles in English that discuss parallel robots for assistance related to the neck, shoulder, wrist, hip, and ankle.
- Articles in English that discuss parallel robots for humanoid applications focusing on the neck, shoulder, wrist, hip, and ankle.
- Parallel robots with three or more degrees of freedom.

- Articles discussing parallel robots at either the conceptual or prototype level of technological development.

2.3. Exclusion Criteria

The article search, conducted between 2012 and 2023, was performed based on the following exclusion criteria:

- Articles that do not discuss parallel robots.
- Articles unrelated to rehabilitation involving parallel robots.
- Articles unrelated to assistance involving parallel robots.
- Articles unrelated to humanoid applications involving parallel robots.

2.4. Quality Assessment

Articles were identified and selected based on their relevance to the research question. Initially, the titles and abstracts were reviewed to determine if they met the inclusion criteria. Subsequently, the content of each article was thoroughly read. The quality of the selected articles was then evaluated based on the reputation and trustworthiness of the scientific journals in which they were published.

2.5. Data Extraction

Data from the selected articles were synthesized to answer the research question. This involved conducting an analysis of various factors concerning the development of parallel robots, including the year of publication, country of development, type of rehabilitation targeted at specific joints, level of technological maturity, types of mechanisms, degrees of freedom, types of movement, types of actuators, kinematic models, and simulation tools used.

2.6. Search Performance

The search yielded a total of 846 articles: Scopus contributed 128, Web of Science 103, PubMed 21, IEEE Xplore 63, ScienceDirect 271, DOAJ 101, and ASME Journal 159. After eliminating 23 duplicate articles, 823 articles remained. Following the application of the exclusion criteria, 93 papers were selected for review. These included 13 articles related to the neck joint, 19 related to the shoulder joint, 22 related to the wrist joint, 9 related to the hip joint, and 30 related to the ankle joint, as illustrated in Figure 1.

2.7. Systematic Review

The 93 selected articles were analyzed and categorized by joint type: neck, shoulder, wrist, hip, and ankle. Key aspects such as the year of publication, country of origin, type of rehabilitation targeted by joint, mechanism type, degrees of freedom, types of movement, type of actuator, mathematical models, and simulation tools were considered.

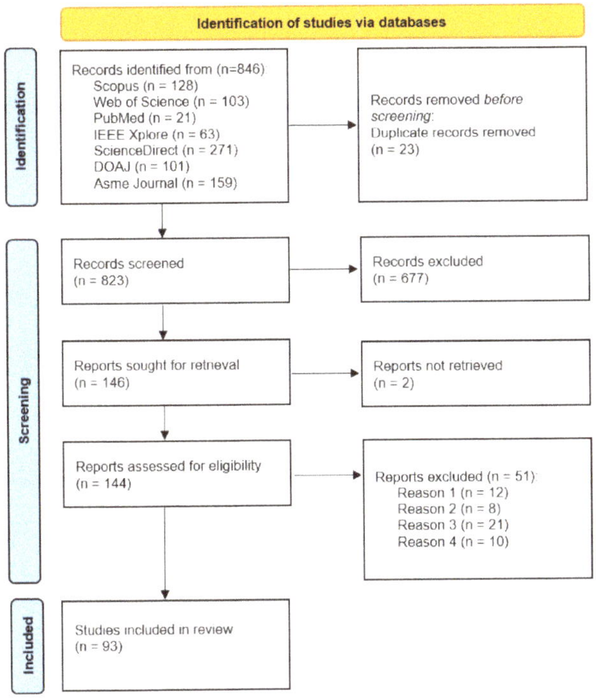

Figure 1. PRISMA flow diagram illustrating the application of inclusion and exclusion criteria to studies selected from 2012 to 2023, focusing on the shoulder, neck, wrist, hip, and ankle joints.

3. Biomechanics of the Human Joints

The relationship between the planes, axes, and ranges of motion in human joints is fundamental to understanding how joints function. These concepts are interrelated and play roles in describing and analyzing joint movement. Planes and axes of motion serve as useful tools for describing the directions and orientations of movements in the human body. The three primary planes, sagittal, frontal, and transverse, are associated with the frontal, sagittal, and vertical axes, respectively [34], as illustrated in Figure 2.

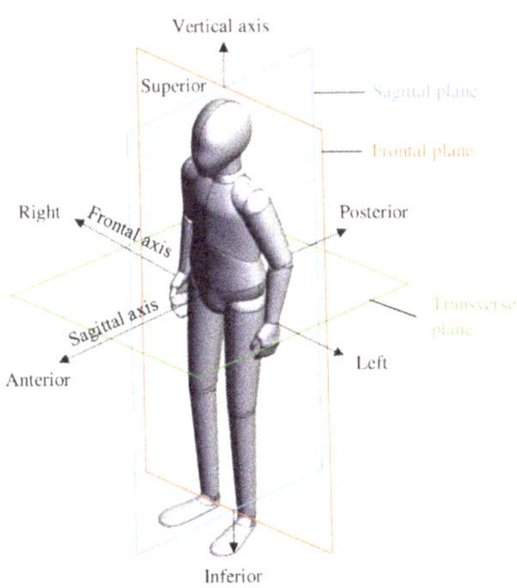

Figure 2. Anatomical sagittal, frontal, and transverse planes and vertical, sagittal, and frontal axes.

The range of motion is fundamental to the movement and functionality of the human body. The term "range of motion" refers to the various directions and amplitudes through which a joint or body segment can move. Biomechanics investigates not only how these movements occur but also how they are generated and influenced by factors such as anatomical structure, muscle strength, and external forces. This field of study also encompasses different types of joint movements, including flexion, extension, abduction, adduction, and rotation, describing these movements in three dimensions through a combination of planes and axes.

- Anatomical planes:
  Sagittal plane (or median plane): Divides the body into left and right halves. Movement in this plane is forward and backward.
  Frontal plane (or coronal plane): Divides the body into anterior (front) and posterior (back) halves. Movement in this plane is side-to-side.
  Transverse plane (or axial/horizontal plane): Divides the body into superior (upper) and inferior (lower) halves. Movement in this plane involves rotation.
- Anatomical axes:
  Sagittal axis (or anteroposterior axis): Extends front to back and is perpendicular to the frontal plane. Movements around this axis include abduction and adduction.
  Frontal axis (or horizontal axis): Extends side to side and is perpendicular to the sagittal plane. Movements around this axis include flexion and extension.
  Vertical axis (or longitudinal axis): Extends top to bottom and is perpendicular to the transverse plane. Movements around this axis include internal and external rotation.
- Range of motion of human joints:
  The range of motion (ROM) varies depending on the specific joint and is influenced by factors such as age, gender, flexibility, and the individual's physical condition. Table 1 presents the planes and axes of motion, types of motion, and ranges of motion for the neck, shoulder, hip, wrist, and ankle joints [34].

Table 1. Dominant movement, planes, axis, and range of motion (ROM) of the neck, shoulder, hip, wrist, and ankle joints.

Joints	Movement	Plane	Axis	ROM (°)
Neck	Flexion/extension	Sagittal	Frontal	0–35/0–45
	Right/left bending	Frontal	Sagittal	0–35/0–45
	Right/left rotation	Transverse	Vertical	0–60/0–80
Shoulder	Flexion/extension	Sagittal	Frontal	0–150/0–170
	Abduction/adduction	Frontal	Sagittal	0–160/0–30
	Internal/external rotation	Transverse	Transverse	0–70/0–70
Hip	Flexion/extension	Sagittal	Frontal	0–140/0–10
	Abduction/adduction	Frontal	Sagittal	0–50/0–30
	Internal/external Rotation	Transverse	Vertical	0–40/0–50
Wrist	Flexion/extension	Sagittal	Frontal	0–50/0–30
	Abduction/adduction	Frontal	Sagittal	0–25/0–30
	Pronation/supination	Transverse	Vertical	0–85/0–90
Ankle	Plantarflexion/dorsiflexion	Sagittal	Frontal	0–50/0–30
	Abduction/adduction,	Frontal	Sagittal	0–10/0–20
	Inversion/eversion	Transverse	Vertical	0–60/0–30

4. Parallel Robots and Applications in the Medical Field

Parallel robots have had a significant impact on the medical field, finding applications in areas such as rehabilitation, assistance, and humanoids.

- Rehabilitation robots are designed to assist individuals in regaining motor skills, functionality, and strength following injury or illness. Deployed in therapeutic settings, these devices aim to enhance the individual's physical capabilities and long-term quality of life.

- Assistive robots are created to help individuals perform activities of daily living, such as dressing, running, or walking, when these activities are restricted due to disability or injury. These devices serve as aids to improve the individual's quality of life, offering autonomy and independence.
- Humanoid robots represent an impressive advancement in technology, designed to mimic and replicate human form and behavior. These robots are built with anatomical features resembling those of humans, including heads, torsos, arms, and legs, enabling them to move and execute tasks in a manner similar to humans.

Based on the preceding discussion, it is crucial to initially distinguish between parallel robots and their serial and hybrid counterparts. Following this differentiation, we can then delve into their applications in rehabilitation, assistance, and humanoid contexts, particularly focusing on specific joints like the neck, wrist, shoulder, hip, and ankle.

4.1. Distinctions Between Serial, Parallel, and Hybrid Robots

Serial robots have links and joints sequentially connected to maneuver the end-effector in relation to a stationary base. Parallel robots, on the other hand, feature several serial chains that link a movable platform to a fixed base using multiple independent kinematic chains. Hybrid robots integrate features from both, blending closed-chain structures with open-chain systems [21]. These distinctions relate directly to the robots' architectural design and intrinsic structure, as depicted in Figure 3.

Parallel robots vary in structure depending on the kinematic chain, which can consist of revolution (R), prismatic (P), universal (U), and spherical (S) type joints. The 3-SPS/S parallel mechanism serves as an example; the "3-SPS" portion of the name indicates the presence of three serial chains, each containing a spherical, prismatic, and then spherical joint. The "S" series chain, although consisting of only one spherical joint, remains parallel to the other three SPS chains [21].

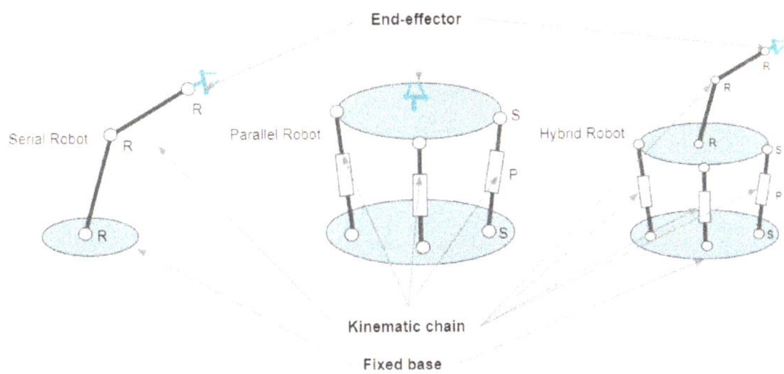

Figure 3. Architectural design of serial, parallel, and hybrid robots.

4.2. Parallel Robot for Neck Joint

Studies related to rehabilitation, assistance, and humanoid applications for the neck joint are summarized in Table 2.

4.2.1. Parallel Robot for Neck Rehabilitation

In 2019, Limgampally et al. introduced a wearable therapy device for treating cervical spine injuries. The device used a three-degrees-of-freedom parallel manipulator and was intended for automated, safe operation under therapist supervision. Figure 4a shows the wearable therapeutic device designed to treat patients with head/neck posture disorders. It employs a 3-RPS parallel mechanism consisting of a movable platform (top) and a fixed platform, connected by three similarly designed supporting links. Each link sequentially connects the top and fixed platforms through a revolute joint (R), a prismatic joint (P), and a spherical joint (S), with the prismatic joint being controlled by an electric linear actuator [1]. Zhang et al. also designed a dynamic neck brace used for characterizing head motion in ALS patients while simultaneously recording the surface electromyography (EMG) of neck muscles. The brace measured muscle activation and could supplement self-reported data to assess head drop and disease progression [2].

In 2022, Lozano et al. developed a closed-chain robotic active neck orthosis featuring four degrees of freedom, based on a four-legged Stewart platform configuration. The orthosis employed a robust control strategy with state restrictions, offering an innovative approach to treating neck ailments. The system was tested on selected volunteer subjects and successfully limited the range of motion within a pre-established area based on the patients' reported range of motion for conditions such as cervicalgia and whiplash syndrome [4].

In 2023, Zhang et al. presented a cable-driven exoskeleton specifically designed for cervical rehabilitation, addressing the urgent need given the increasing prevalence and younger onset of cervical conditions. The research began with an in-depth analysis of the mechanics of neck movement and the specific needs of rehabilitation, setting the foundation for the design criteria and the preliminary sketch of the exoskeleton. Subsequent phases involved kinematic modeling and simulation exercises, which not only confirmed the design's validity but also provided insights into cable adjustments across different rehabilitation paths [35].

4.2.2. Parallel Robot for Neck Assistance

In 2017, Zhang et al. designed a dynamic neck brace to measure and assist head motion in human users. The device offered accurate measurements of head motion and had the potential to improve the neck's range of motion for patients with head/neck posture disorders [36].

In 2018, Zhang et al. developed an active neck brace with three degrees of freedom designed to support patients exhibiting symptoms of head drooping. The device allowed for an enhanced range of motion for both the head and neck and reduced muscle activation when assisted by the brace [37].

In 2019, Liu et al. proposed a rigid–flexible parallel mechanism known as the 3-RXS, designed as a neck brace for patients suffering from head drooping symptoms. The mechanism demonstrated excellent rotational performance and could effectively assist in neck flexion, extension, lateral bending, and axial torsion. As shown in Figure 4b, the device was intended for use in various neurological disorders like amyotrophic lateral sclerosis (ALS), Parkinson's disease (PD), and primary lateral sclerosis (PLS), all of which may lead to head drooping syndrome (HDS). The 3-RXS mechanism facilitated passive neck extension, aiding in the correction of head drooping. The mechanism consisted of a series of links connected to both the top and fixed platforms through a sequence of joints in the following order: revolute joint (R), X-shaped compliant joint (X), and spherical joint (S). This construction was simple and lightweight, enabling smooth flexion/extension, left/right lateral bending, and axial rotational movements [38].

4.2.3. Parallel Robot for Neck Humanoid

In 2013, Gao et al. introduced a cable-driven flexible parallel robot designed to mimic the pitch and roll movements of the human neck. This robot employed three cables and a compression spring to serve as its flexible spine, replicating the motion of the head. Inverse kinematics were addressed using quaternion methods, and workspace analysis was performed under positive cable tension constraints, all validated through simulations [39].

In 2014, Gao et al. compared two different lateral bending models for the compression spring used in their cable-driven parallel robot. The robot aimed to emulate human neck movements by utilizing the spring's bending motion for inverse kinematics [40]. In a related study, Gao et al. presented a cable-driven flexible parallel robot designed to simulate the motion of a human neck while minimizing motion noise. This robot used three cables and a compression spring, with the spring acting as the cervical spine to support the head-like moving platform and cables functioning as muscles around the human neck. Due to the flexible nature of the compression spring, inverse kinematics were not directly solvable. Cable placements were optimized to reduce actuation force, and workspace analysis was conducted under the constraint of positive cable tension. Simulations validated the efficacy of the inverse kinematics and workspace analysis [41].

In 2015, Jiang et al. presented a cable-driven flexible parallel robot featuring a compression spring as the cervical spine. As with previous models, the flexible nature of the spring meant that the inverse kinematics were not directly solvable. To find possible solutions, statical analysis was incorporated. The cable placements were optimized to reduce actuation force, and workspace analysis was conducted under a positive cable tension constraint [42].

In 2017, Gao et al. introduced another cable-driven parallel robot, this one featuring a flexible spine and four cables designed to mimic human neck movements. The lateral bending motion of the spring facilitated pitch and roll movements, while an included bearing enabled yaw motion. Both inverse kinematics and cable placement optimization were investigated through simulations [43].

In 2021, Quevedo et al. applied various linear and non-linear models to design a soft neck mechanism with a central soft link actuated by three motor-driven tendons. The force exerted on the individual tendons allowed the neck to perform motions similar to those of a human neck. The cable-driven parallel mechanism, referred to as 3-CDPM and illustrated in Figure 4c, was constructed from flexible materials and activated by cables, causing the upper platform to tilt. The neck was comprised of a base, movable platform, central soft link, tendons (cables), and motors [3].

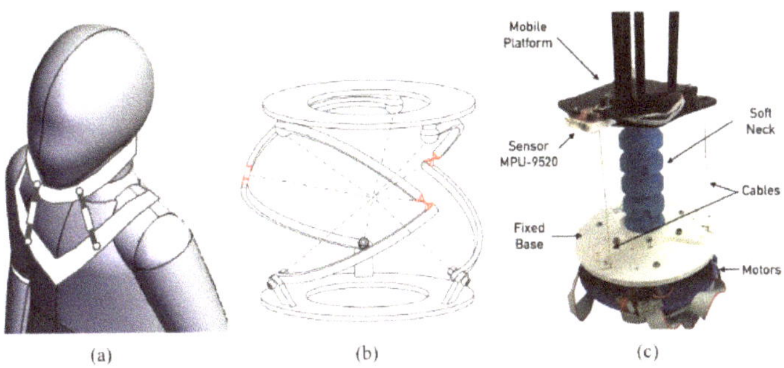

Figure 4. Neck joint mechanisms: (**a**) 3-RPS mechanism by Vellore Institute of Technology, redrawn based on [1]; (**b**) 3-RXS mechanism by the University of Technology Beijing [38], available under a Creative Commons Attribution License; (**c**) 3-CDPM mechanism by University Carlos III of Madrid [3], available under a Creative Commons Attribution License.

Table 2. Parallel robots for rehabilitation, assistance, and humanoids of the neck joint.

Author	Year	Country	Device	TRL	Mechanism	DoF	ToM	Actuator	Model	Tool
Gao et al. [39]	2013	China	NH	1	cable-driven	3	PR	NS	IK	M
Gao et al. [40]	2014	China	NH	2	cable-driven	3	NS	NS	IK	M
Gao et al. [41]	2014	China	NH	1	cable-driven	3	NS	NS	IK	M
Jiang et al. [42]	2015	China	NH	2	cable-driven	3	NS	NS	IK	M
Gao et al. [43]	2017	China	NH	2	cable-driven	3	PR	NS	IK	M
Zhang et al. [36]	2017	USA	NA	3	3-RRS	3	FE, RLB, RLR	NS	FK	M
Zhang et al. [37]	2018	USA	NA	3	3-RRR	3	FE, RLB, RLR	ER	IK	NS
Liu et al. [38]	2019	China	NA	3	3-RXS	3	FE, RLB, RLR	NS	IK	AN

Table 2. *Cont.*

Author	Year	Country	Device	TRL	Mechanisms	DoF	ToM	Actuator	Model	Tool
Lingampally et al. [1]	2019	India	NR	3	3-RPS	3	RLR	EL	IK	M
Zhang et al. [2]	2019	USA	NR	3	3-RRR	3	FE, RLB, RLR	ER	D	M
Quevedo et al. [3]	2021	Spain	NH	3	cable-driven	3	FE, RLR	ER	IK	M
Lozano et al. [4]	2022	Mexico	NR	4	4-SPS	4	NS	EL	D	M
Zhang et al. [35]	2023	China	NR	4	cable-driven	3	FE, RLB, RLR	NS	IK	M

Abbreviations: NH—neck humanoid, NA—neck assistance, NR—neck rehabilitation, 3-RRS—3 (revolute-revolute-spherical), 3-RRR—3 (revolute-revolute-revolute), 3-RXS—3 (revolute-joint X-spherical), 3-RPS—3 (revolute-prismatic-spherical), 4-SPS—4 (spherical-prismatic-spherical), PR—pitch and roll, NS—not specified, FE—flexion-extension, RLB—right–left bending, RLR—right–left rotation, ER—electric rotary, EL—electric linear IK—inverse kinematic, FK—forward kinematic, D—dynamics, M—MATLAB, AN—ANSYS.

4.3. *Parallel Robot for Shoulder Joint*

Studies related to rehabilitation, assistance, and humanoid applications for the shoulder joint are summarized in Table 3.

4.3.1. Parallel Robot for Shoulder Rehabilitation

In 2014, Klein et al. introduced a novel robotic interface to explore the neuromechanical control of redundant planar arm movements. This device featured a 5R closed-loop pantograph design with a wrist flexion/extension cable-actuated mechanism. The interface's characteristics, such as motion range, impedance, friction, and dynamics, were discussed. This lightweight, high-force, and low-impedance device enabled research into redundant motor control in humans [44].

In 2015, Enferadi et al. proposed a new spherical parallel robot design for rehabilitation applications. The robot's full rotational capabilities were highlighted. Dimensional optimization, aimed at maximizing the robot's workspace, was carried out using genetic algorithms. The robot boasted a relatively large workspace and exhibited precision in its kinematics, Jacobian matrices, and workspace analysis [45].

In 2016, Enferadi et al. presented another spherical parallel robot designed for various applications, including rehabilitation (e.g., TV satellite dishes; tracking systems; solar panels; cameras; telescopes; and the rehabilitation of human joints like the ankle, shoulder, and wrist). The robot allowed for complete rotation around an axis. A genetic algorithm optimized its dimensions to maximize the workspace, which was found to be relatively large and free from singularities—a significant advantage [46]. Hunt et al. introduced a low-inertia shoulder exoskeleton with five degrees of freedom (DoF). The first innovation involved a 3DoF spherical parallel manipulator (SPM), which was designed through a new approach that mechanically coupled certain degrees of freedom to constrain the kinematics. The second innovation was a 2DoF passive slip interface that enhanced system mobility and prevented joint misalignment caused by the user's glenohumeral joint motion. Motion capture validated the SPM's kinematics, confirming both its forward and inverse kinematic solutions. Beyond its application for shoulder rehabilitation, the device introduced a novel motion coupling method that was applicable to various parallel architectures. It also showcased the versatility of its passive slip interface in both parallel and serial robotic systems [47].

In 2020, Niketkaliyev et al. shifted the focus to robotic shoulder rehabilitation exoskeletons, which often neglect certain shoulder DoF, leading to discomfort due to joint axis misalignments. They introduced a bio-inspired 5DoF hybrid human–robot mechanism (HRM) that combined serial and parallel manipulators with rigid and cable links. This hybrid mechanism ensured compatibility between human and exoskeleton joint axes and covered the complete range of human shoulder motion in a singularity-free workspace. Numerical simulations and a 3D-printed prototype validated the kinematic model and advantages of the proposed hybrid mechanism [48].

In 2021, Hunt proposed a novel parallel-actuated exoskeleton architecture for rehabilitation. This architecture's stiffness property could be optimized for specific tasks using a stiffness model and bounded nonlinear multi-objective optimization. Figure 5a shows an exoskeleton designed with a four-bar system in a parallel mechanism, denoted as 4B-SPM. This system allowed individuals to modulate their stiffness attribute to optimize activities, such as augmented lifting or impact absorption for the shoulder [49].

4.3.2. Parallel Robot for Shoulder Assistance

In 2013, Sekine et al. presented a systematic approach to designing a shoulder prosthesis with consideration for force and spatial accessibility. Using measurements from Activities of Daily Living (ADLs), the design process entailed evaluating both force and spatial accessibility, followed by optimization based on kinematic and static models. The optimized parallel mechanism was tailored for specific ADL tasks and various spatial specifications, illustrating the potential for individualized shoulder prosthesis design. Figure 5b depicts an optimized, compact, pneumatic-actuator-driven parallel mechanism for a shoulder prosthetic arm [50].

In 2015, Sekine et al. proposed shoulder prostheses designed for transhumeral and shoulder disarticulation amputees. Accessibility and intrinsic viscoelasticity were the focal points of this study. The paper introduced new mechanisms—specifically, an antagonistic mechanism and a soft backbone—to enhance spatial characteristics and responsiveness to disturbances. The evaluation confirmed increased workspace and disturbance responsiveness for the prosthetic arm [20].

In 2018, Leal-Naranjo et al. introduced a synthesis of a spherical parallel manipulator for a seven-degrees-of-freedom (7DoF) prosthetic human arm. The design objectives included workspace, dexterity, and actuator torques. Optimization was performed using genetic algorithms, culminating in a manipulator that met all performance requirements [51]. Hunt et al. unveiled a new parallel-actuated exoskeleton architecture aimed at offering a superior alternative to serial actuation for augmenting multiple-DoF biological joints. This architecture employed a spherical parallel manipulator (SPM) with three four-bar substructures to control three rotational DoF independently. Variants of the four-bar spherical parallel manipulator (4B-SPM) were presented for shoulder, hip, wrist, and ankle exoskeletons. Three actuation methods for the 4B-SPM were explored, each based on different dynamic performance requirements. This work set the stage for advancements in more effective parallel-actuated exoskeletons as opposed to conventional serial-actuated counterparts [52]. Leal-Naranjo presented a low-cost prosthetic device designed for shoulder disarticulation and featuring seven DoF. The mechanisms for shoulder, elbow, and wrist movements were discussed. Dynamic simulations and experimental evaluations confirmed the device's functionality and suitability for daily activities [19]. Figure 5c illustrates a shoulder mechanism integrated into an arm prosthesis. This spherical mechanism, configured as a 3-RRR type, allowed for three degrees of freedom with shoulder motions, using compact motors. Given that the shoulder prosthetic supported the entire structure of the device, this is where the highest joint loads developed [19].

4.3.3. Parallel Robot for Shoulder Humanoid

In 2012, Chen et al. presented a novel homing algorithm for a three-degrees-of-freedom (3DoF) parallel spherical joint in a cable-driven parallel robot. The algorithm utilized incremental encoders and limit switches to identify the home posture automatically, implementing decoupling control for each axis. Simulation results affirmed the algorithm's effectiveness [53]. Both control accuracy and a consistent initial posture are pivotal when evaluating control algorithms. To minimize cumulative errors during control processes or to estimate initial postures, robotic systems need to revert to an approximate home posture.

In 2013, Wang et al. aimed to improve the performance of bionic joints using a five-link parallel mechanism, actuated by two antagonistic artificial pneumatic muscles (PMs). The study examined kinematics, singularity constraints, and joint torque models based on spring-damp dynamics. The joint's compliance, represented by the angle-to-spring torque ratio, was derived. Energy consumption analysis was conducted considering varying PM lengths. The proposed bionic shoulder and elbow joints exhibited an enhanced angular range and decreased PM contraction, contributing to a more humanoid-like design [54].

In 2016, Alfayad et al. undertook a research initiative focused on developing a three-degrees-of-freedom (3DoF) hybrid mechanism suitable for humanoid robotics. This mechanism catered to various modules in humanoid robots and could also be adapted for other legged robots, such as quadrupeds and hexapods. Utilizing kinematic synthesis, the study proposed a novel hip mechanism that combined one rotary and two linear actuators. This approach accommodated the wide motion ranges of the shoulder module and introduced a new perspective on the contributions of linear actuators in both motion and force generation. The research employed kinematic and geometrical models to optimize the hybrid mechanisms, demonstrating their broad applicability in various robotic systems [55].

In 2017, Jiang et al. focused on developing a new hybrid mechanism for humanoid wrist and shoulder joints. A cable-driven parallel robot platform was developed for experimental study. A dynamic model of the mechanism was formulated, considering the coupling theory of flexible body motion and deformation. A nonlinear control method was applied for anti-vibration control. Both simulations and experimental results validated the feasibility and control scheme of the hybrid mechanism [56].

In 2019, Liu et al. introduced a bionic flexible manipulator driven by pneumatic muscle actuators (PMAs). The study outlined the configuration design based on human physiological mapping, established kinematic models, and employed the Lagrange method for dynamic analysis. A fuzzy torque control algorithm, developed using the computed torque method, showed improved trajectory tracking and accuracy compared to traditional methods [57]. In the same year, Bai et al. provided a comprehensive review of state-of-the-art techniques in spherical motion generation via parallel manipulators or spherical motors. The review covered kinematics, dynamics, design optimization, and emerging applications, offering insights into new research challenges and future developments in the field [58].

In 2021, Wang et al. proposed a multi-objective trajectory planning approach for a 7DoF hybrid humanoid robotic arm. The methodology combined kinematic modeling and optimization to achieve faster transit times, lower energy consumption, and higher stability during point-to-point tasks. Simulation results validated the proposed approach [59].

In 2023, Chen et al. introduced an enhanced design and modeling analysis for a 3DoF series-parallel joint module used in humanoid service robots. This module, inspired by human-like 3DoF joints, employed a cable-driven method. Using the shoulder joint as a model, it could execute various motions like arm abduction, back extension, and lifting. The team designed a shared modular connector for easier assembly and disassembly across modules and a tension amplification mechanism for a more compact design. They also developed a transient torque model and a cross-coupling control framework and conducted kinematics analysis based on the anti-parallelogram principle. Prototyping and tests showed the design's potential for realistic humanoid shoulder movements, offering a novel concept for humanoid robots [60].

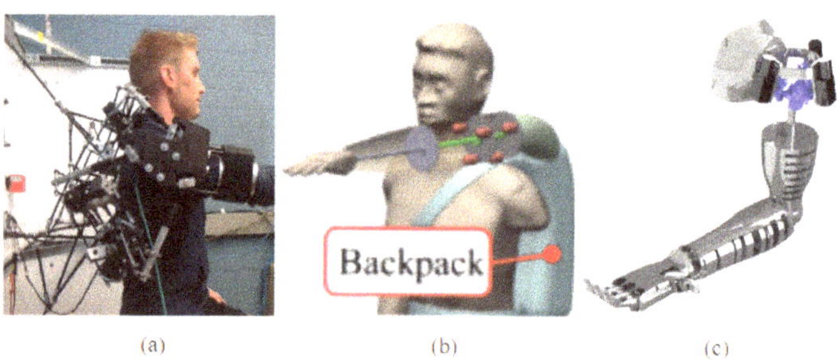

Figure 5. Shoulder joint mechanisms: (**a**) 4B-SPM mechanism by Arizona State University [49]; (**b**) cable-driven mechanism by Chiba University [50]; (**c**) 3-RRR mechanism by Universidad de Guanajuato [19]. All are available under a Creative Commons Attribution License.

Table 3. Parallel robots for rehabilitation, assistance, and humanoids of the shoulder joint.

Author	Year	Country	Device	TRL	Mechanism	DoF	ToM	Actuator	Model	Tool
Chen et al. [53]	2012	China	SH	2	cable-driven	7	NS	ER	IK	AD
Sekine et al. [50]	2013	Japan	SA	2	cable-driven	3	FE, AA, IER	PL	FK	ADLA
Wang et al. [54]	2013	China	SH	2	5-PMA	5	NS	P	FK	MS
Klein et al. [44]	2014	UK	SR	4	5R	3	NS	E	FK	CAD
Sekine et al. [20]	2015	Japan	SA	3	3-SPS/P	3	NS	PL	FIK	NS
Enferadi et al. [45]	2015	Iran	SR	2	3-RSS/S	3	RP	ER	IK	NS
Enferadi et al. [46]	2016	Iran	SR	2	3-RSS/S	3	NS	ER	IK	NS
Hunt et al. [47]	2016	USA	SR	3	SPM	3	NS	EL	FIK	CAD
Alfayad et al. [55]	2016	France	SH	3	2-UPUR/RU	2	PRY	EL	IK	NS
Jiang et al. [56]	2017	China	SH	3	cable-driven	2	NS	ER	NS	NS
Leal-Naranjo et al. [51]	2018	Italy	SA	3	3-RRR	3	FE	ER	IK	NS
Leal-Naranjo et al. [19]	2018	Mexico	SA	2	3-RRR	3	FE	ER	IK	AD
Hunt et al. [52]	2018	USA	SA	3	4B-SPM	3	NS	ER	FIK	M
Lui et al. [57]	2019	China	SH	3	5-PMA	5	NS	PL	FIK	M-AD
Bai et al. [58]	2019	Denmark	SH	3	3-RRR	3	RP	ER	FIK	NS
Niyetkaliyev et al. [48]	2020	Australia	SR	2	cable-driven	3	AA	NS	IK	CAD
Hunt et al. [49]	2021	USA	SR	3	4B-SPM	7	FE, AA	EL	FIK	NS
Wang et al. [59]	2021	China	SH	2	5R	3	FE, AA, IER	ER	FK	AD
Chen et al. [60]	2023	China	SH	4	cable-driven	3	FE, AA, L	ER	IK	NS

Abbreviations: SR—shoulder rehabilitation, SH—shoulder humanoid, SA—shoulder assistance, 5-PMA—5 pneumatic muscle actuator, NS—not specified, 3-SPS/P—3 (spherical-prismatic-spherical)/1 (prismatic), 3-RSS/S—3 (revolute-spherical-spherical)/spherical, 3-RRR—3 (revolute-revolute-revolute), 4B-SPM—4-bar (spherical parallel manipulator), 5R—5 revolute, FE—flexion–extension, AA—abduction–adduction, IER—internal–external rotation, RP—rotational pure movement, L—lifting, PRY—pitch-roll-yaw, ER—electric rotary, EL—electric linear, PL—pneumatic linear, IK—inverse kinematic, FK—forward kinematic, FIK—forward and inverse kinematic, AD—ADAMS, ADLA—(ADL area) is used to evaluate spatial accessibility, MS—MATLAB-SimMechanics, CAD—Computer-Aided Design, M—MATLAB, M-AD—MATLAB-ADAMS.

4.4. *Parallel Robot for Wrist Joint*

Studies related to wrist assistance and rehabilitation and humanoid wrist mechanisms are summarized in Table 4.

4.4.1. Parallel Robot for Wrist Rehabilitation

In 2013, Pehlivan et al. developed an adaptive controller for a robotic mechanism designed for the upper-extremity rehabilitation of the wrist. Compared to a proportional-derivative (PD) controller, the model-based adaptive controller improved trajectory tracking. The adaptive controller achieved similar error performance but used significantly lower feedback gains, offering a more compliant interface for patients during rehabilitation sessions [61].

In 2017, Bian et al. introduced an exoskeleton aimed at rehabilitating the elbow, forearm, and wrist motor functions in stroke patients. The EFW Exo II exoskeleton was based on a hybrid mechanism that combined a parallel 2-URR/RRS mechanism with a serial R mechanism. It included adjustable features to accommodate different arm sizes and utilized force sensors to facilitate patient interaction. The exoskeleton provided ranges of motion that met the requirements of activities of daily living [62].

In 2018, Kitano et al. unveiled a wearable wrist rehabilitation training device that utilized a parallel link mechanism. This innovative approach allowed for the training of both translational and rotational wrist joint motions, filling a gap left by previous techniques that did not address translational motion. The device could enable a range of motion that covered approximately 60% of the wrist's mobility, potentially reducing wrist joint strain [6].

In 2021, Wang et al. introduced a soft parallel robot designed for automated wrist rehabilitation. Figure 6a illustrates the 6-SPS/PS soft parallel robot that employed pneumatic artificial muscles for wrist rehabilitation. This robot merged soft and parallel structures to offer a secure, adaptable, and low-cost personalized rehabilitation solution. Linear actuators, including pneumatic artificial muscles, drove the robot, and an electromyography sensor provided feedback for evaluating rehabilitation progress. Experimental tests confirmed the robot's efficacy in assisting various wrist motions [63].

In 2022, Goyal et al. developed an impedance controller designed to rehabilitate stroke patients with wrist motor impairments. The controller employed a Koopman-operator-based autodidactic system identification model to predict wrist joint stiffness during different rotational motions. It adjusted the applied force based on the subject's joint stiffness, utilizing a parallel-structured end-effector robot equipped with biomimetic muscle actuators. The performance of the controller was validated through tests on healthy subjects [64].

In 2023, Li et al. developed a bionic cable-driven mechanism for forearm-wrist rehabilitation. This device mimics human wrist motions with three degrees of freedom, addressing the full range of wrist and forearm actions. Notably, a spring within a parallel mechanism helps counteract cable slack. The system's design incorporates kinematics and statics for precise movement calculations. Simulations and practical tests confirmed its effectiveness and accuracy for rehabilitation purposes [65]. In the same year, Goyal et al. introduced a robot with a four-link parallel end-effector designed for wrist joint rehabilitation. Using biomimetic muscle actuators (BMAs), the robot had an inherent flexibility, with a fuzzy model developed to recognize the BMAs' nonlinear characteristics. The system's stiffness-observer tailored itself to individual subject stiffness, adjusting the robot's trajectories. An adaptive controller, based on the fuzzy model and stiffness-observer, governed the four BMAs, giving the robot end-effector three rotational degrees of freedom. Initial tests with three healthy individuals confirmed the controller's effectiveness in guiding the robot while accounting for the compliant and nonlinear nature of the BMAs, even adjusting for higher-stiffness areas within the wrist's range of motion [66]. In the same year, Goyal et al. developed a trajectory-tracking controller for a wrist rehabilitation robot with three degrees of freedom. They used the Koopman linear system to address the nonlinearity of human–robot interaction, turning state variables into linear approximations of nonlinear systems through Koopman operators. These operators, learned via linear regression, determined the system dynamics for the robot's trajectory controller. This data-driven method resulted in a clear control-oriented model. Tests with three healthy individuals proved the controller's effectiveness in guiding the wrist along a set trajectory [67].

4.4.2. Parallel Robot for Wrist Assistance

In 2012, Serracin et al. designed a parallel robot to assist in bone milling surgeries. The robot had two active degrees of freedom and was employed for orientation during bone-milling procedures. The paper outlined the kinematic geometry, discussed workspace optimization, and performed force analysis for jawbone reconstruction. The singularities of the mechanism were analyzed, and the motor selection was justified based on torque requirements. The study also presented simulation results and a prototype using linear motors [68].

In 2018, Lee et al. introduced an over-actuated coaxial spherical parallel mechanism optimized for efficient wrist motion. The mechanism's rotation axis was aligned with the wrist's pronation-supination movement. The prototype demonstrated a design that enhanced user comfort and workspace efficiency. Simulation results indicated improved performance compared to similar devices, along with extensive motion coverage [69]. Figure 6b depicts the wrist human–machine interface based on this over-actuated coaxial spherical parallel mechanism. The base was connected to two links that were coaxially coupled, and the rotation axis was specifically designed to align with the wrist's pronation/supination motion, which has the most extensive operational range of wrist movements [69].

In 2021, Lee et al. proposed a robotic exoskeleton interface (REI) for the wrist, based on a fully actuated coaxial spherical parallel mechanism (CSPM). The CSPM-based interface offered pure rotational motion akin to the human wrist, along with a high torque output and low moving-parts inertia. The device's torque output and range of motion aligned closely with human capabilities [70].

In 2022, Lopez-Custodio et al. conducted a stiffness analysis of the Exechon hybrid manipulator, which served as a five-axis machine tool. The study included the consideration of an offset wrist in both kinematic and stiffness analyses. A compliance model was formulated and validated against experimental data, offering a more accurate portrayal of the manipulator's behavior [71]. Sanjuan et al. proposed a cable-driven wrist prosthesis (CDWP) to address challenges related to limb amputation. The anti-parallel-based local transmission index was introduced to optimize the device's dimensions. The CDWP design aimed to offer both high dexterity and aesthetic appeal to a wide range of patients [72].

4.4.3. Parallel Robot for Wrist Humanoid

In 2013, Chaparro-Altamirano et al. proposed a system utilizing a 3SPS-1S parallel manipulator for surveillance and defense applications. The mechanism incorporated a central leg to increase stiffness and provide three pure rotational degrees of freedom. The study covered inverse kinematics, forward kinematics solved through geometry and neural networks, workspace calculations, parameter optimization, and singularity detection via Jacobian matrix analysis [73].

In 2015, Kong et al. focused on synthesizing two-degrees-of-freedom (DoF) parallel mechanisms (PMs) capable of both spherical translation and sphere-on-sphere rolling modes. They introduced a 2DoF 3-4R overconstrained PM, derived from an existing 5DoF US equivalent PM, to serve as the basis for further developments. The study classified 2DoF 3-4R PMs capable of both identified modes by exploring shared conditions between them [74].

In 2017, Lu et al. conducted a comprehensive study analyzing coordinated grasping kinematics for multi-fingered systems and optimizing grasping force in a parallel hybrid hand. The study included conditions for coordinated grasping and formulas for calculating the displacement, velocity, and acceleration of contact points between the fingers and the object [75].

In 2018, Wu et al. addressed the dynamic stability challenges of a tripod parallel robotic wrist using the monodromy matrix method. The wrist demonstrated uninterrupted end-effector rotation across its orientation workspace, making it suitable for machine tool head tasks like drill point grinding. Stability analysis was conducted using Floquet theory, and stability charts were constructed to identify parametric instabilities [76].

In 2019, He et al. introduced a novel concept of a self-insulating joint design, employing a cable-driven parallel-series (CDPS) mechanism coupled with electrical insulation analysis. This design offered mechanical support and electrical insulation during live-line operations, thus facilitating equipment maintenance without manual intervention or power interruption. Figure 6c depicts the wrist joint of the robotic arm as a CDPS mechanism, featuring a lower and an upper platform in concentric positions for mechanical support. Two concentric shafts in a spring, connected by a universal joint, provided additional structural support. Four cables powered the structure [77].

In 2020, Pang et al. proposed a bionic flexible wrist parallel mechanism that mimicked human wrist joints by using a combination of rope-driven and compression spring-supported hybrid mechanisms. A parallel structure controlled by cables emulated wrist muscles. The inverse kinematics were solved using force and torque balance conditions along with the spring bending equation. MATLAB software was used for analysis, verifying the mechanism's feasibility for wrist rehabilitation and promoting the development of rehabilitation robots and rope-driven technology [9].

In 2021, Wang et al. proposed a seven-degrees-of-freedom serial–parallel hybrid humanoid robotic arm. The arm's trajectory planning was optimized for speed, energy consumption, and stability using a multi-objective optimization approach. Simulation results validated the effectiveness of this methodology [59].

In 2022, Bazman et al. introduced a 4DoF parallel forceps mechanism designed for minimally invasive surgery. This mechanism employed a unique center leg to convert thrust into gripping motion. The study included kinematic analysis and workspace assessment and addressed unintended gripper motion. Human-in-the-loop simulations validated the design [78]. Li et al. developed a 3-RPS/US parallel mechanism with two DoF to enhance load-bearing capabilities. The study covered kinematics, Jacobian matrix analysis, workspace calculations, singularity considerations, and static analysis. The potential application of this mechanism in tracking photovoltaic brackets was demonstrated [79].

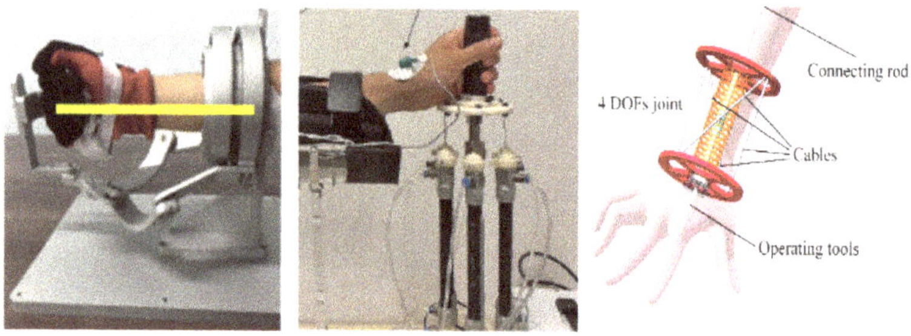

(a) (b) (c)

Figure 6. Wrist joint mechanisms: (**a**) 6-SPS/PS mechanism by the University of Macau [63]; (**b**) 3-CSPM mechanism by Kwangwoon University [69]; (**c**) 4-CDPS mechanism by Southeast University [77]. All are available under a Creative Commons Attribution License.

Table 4. Parallel robots for rehabilitation, assistance, and humanoids of the wrist joint.

Author	Year	Country	Device	TRL	Mechanism	DoF	ToM	Actuator	Model	Tool
Serracin et al. [68]	2012	Spain	WA	3	2-UPS/S	2	NS	EL	IK	M
Chaparro et al. [73]	2013	Mexico	WH	3	3-SPS/S	3	RPY	EL	FK	NS
Pehlivan et al. [61]	2013	USA	WR	3	3-RPS	2	FE, AA	PL	FIK	M
Kong et al. [74]	2015	UK	WH	2	3-4R	2	NS	NS	IK	NS
Lu et al. [75]	2017	China	WH	3	5-SPM	5	NS	EL	FIK	NS
Bian et al. [62]	2017	China	WR	3	2-URR/RRS	3	FE, AA	ER	IK	NS
Lee et al. [69]	2018	South Korea	WA	3	2-RRR	3	FE, AA	ER	FIK	NS
Kitano et al. [6]	2018	Japan	WR	3	3-RRR	3	FE, AA, SP	ER	FIK	NS
Wu et al. [76]	2018	China	WH	3	2-PUU/RPS	2	TR	ER, EL	IK	NS
He et al. [77]	2019	China	WH	2	cable-driven	4	NS	ER	IK	NS
Pang et al. [9]	2020	China	WH	3	cable-driven	2	FE, AA	ER	IK	NS
Wang et al. [63]	2021	China	WR	3	6-SPS/PS	3	FE, AA, SP	PL	FIK	NS
Lee et al. [70]	2021	Korea	WA	3	2-CSPM	2	FE, AA, SP	EL	FIK	NS
Wang et al. [59]	2021	China	WH	2	3-UPS/S	2	FE, SP	EL	FK	AD
López et al. [71]	2022	UK	WA	3	2-RRRPR/RRPR	3	NS	EL	IK	M
Goyal et al. [64]	2022	Australia	WR	3	3-RPR	3	FE, AA, PS	PMA	IK	NS
Bazman et al. [78]	2022	Turkey	WH	2	3-RSR/UUP	2	RPY	NS	FK	MS
Sanjuan et al. [72]	2022	USA	WA	2	cable-driven	2	NS	NS	IK	NS
Li et al. [79]	2022	China	WH	2	3-RPS/US	2	NS	EL	IK	NS
Li et al. [65]	2023	China	WR	4	cable-driven	3	FE, AA, SP	ER	IK	M
Goyal et al. [66]	2023	Australia	WR	4	4-BMA	3	FE, AA, SP	PL	IK	NS
Goyal al. [67]	2023	Australia	WR	4	4-BMA	3	FE, AA, SP	PL	IK	NS

Abbreviations: WR—wrist rehabilitation, WA—wrist assistance, WH—wrist humanoid, 2-UPS/S—2 (universal-prismatic-spherical)/1 (spherical), 3-SPS/S—3 (spherical-prismatic-spherical)/1 (spherical), 3-RPS—3 (revolute-prismatic-spherical), 3-4R—3 (4 revolute), 5-SPM—5 (spherical parallel mechanism), 2-URR/RRS—2 (universal-revolute-revolute)/1 (revolute-revolute-spherical), 2-RRR—2 (revolute-revolute-revolute), 3-RRR—3 (revolute-revolute-revolute), 2-PUU/RPS—2 (prismatic-universal-universal)/1 (revolute-prismatic-spherical), 6-SPS/PS—6 (spherical-prismatic-spherical)/1 (prismatic-spherical), 2-CSPM—2 (coaxial spherical parallel mechanism), 3-UPS/S—3 (universal-prismatic-spherical)/1 (spherical), 2-RRRPR/RRPR—2 (revolute-revolute-revolute-prismatic-revolute)/1 (revolute-revolute-prismatic-revolute), 3-RPR—3 (revolute-prismatic-revolute), 3-RSR/UUP—3 (revolute-spherical-revolute)/1 (universal-universal-prismatic), 3-RPR—3 (revolute-prismatic-revolute), 3-UPU—3 (universal-prismatic-universal), 3-RPS/US—3 (revolute-prismatic-spherical)/3 (universal-spherical), 4-BMA—4 (biomimetic actuators) RPY—roll-pitch-yaw, FE—flexion–extension, AA—abduction–adduction, SP—supination–pronation, TRP—translation-rotation-pure, EL—electric linear, PL—pneumatic linear, ER—electric rotary, IK—inverse kinematic, FK—forward kinematic, FIK—forward and inverse kinematic, NS—not specified, M—MATLAB, AD—ADAMS, MS—MATLAB-SimMechanics.

4.5. Parallel Robot for Hip Joint

Studies related to assistance, rehabilitation, and humanoid applications for the hip are summarized in Table 5.

4.5.1. Parallel Robot for Hip Rehabilitation

In 2016, Rastegarpanah et al. investigated a six-degrees-of-freedom parallel robot developed at the University of Birmingham for use in hip rehabilitation for stroke patients. This robot could replicate foot trajectories associated with three types of rehabilitation exercises: hip flexion/extension, ankle dorsiflexion/plantarflexion, and human gait. The study also emphasized the robot's ability to lift significant loads, indicating its potential for the effective administration of lower-limb rehabilitation exercises [80].

In 2020, Zhang et al. introduced an innovative approach to enhance the performance of a three-degrees-of-freedom parallel mechanism used in hip rehabilitation robots. The objective was to prevent kinematic singularity within the workspace corresponding to human gait and to improve power efficiency. This was achieved by optimizing the geometric parameters of the mechanism and by proposing improved force transmission indices. Optimization was conducted using a multi-objective model and the differential evolution algorithm. The effectiveness of this method was validated by comparing the performance metrics of the optimized mechanism to those of the original model, particularly in terms of Jacobian matrix singularity and output power efficiency [81].

In 2022, Shi et al. explored a serial–parallel lower-limb rehabilitation exoskeleton for its potential to mimic human lower-limb kinematics. To address challenges related to modeling errors and uncertainties, an adaptive control approach was proposed. This approach incorporated sensor data from the parallel mechanism to accurately capture attitude and employed a neural network adaptive controller to compensate for uncertainties and external disturbances. The effectiveness of the dynamic modeling and control system was validated through experimental testing [82].

4.5.2. Parallel Robot for Hip Assistance

In 2019, Ren et al. designed a wearable lower-limb exoskeleton to assist with medical rehabilitation. Traditional exoskeletons often struggle to precisely replicate natural human limb movements. To address this issue, the researchers developed an innovative anthropomorphic lower-limb exoskeleton based on a series-parallel mechanism. Human lower-limb movements were captured using an optical gait-tracking system. By comparing the simulation results from the series-parallel mechanism with the captured human data, a kinematics matching model was established specifically for the hip joint. The study demonstrated that the proposed model effectively reduced kinematic matching errors in multiple directions. This made the anthropomorphic series-parallel mechanism a significant improvement for tracking human hip joint positions. Figure 7a shows a lower-limb exoskeleton based on the series-parallel mechanism. A 6-SPS parallel mechanism was employed to enhance the pelvis's capacity to move, thereby providing the lumbar spine joint with six degrees of freedom (DoF) [83].

In 2020, Song et al. focused on aiding amputees in regaining their daily quality of life through the development of a hip prosthetic mechanism, as shown in Figure 7b. Analyzing human hip motion characteristics, they designed a 2-UPR/URR parallel mechanism with a passive limb. The study delved into the inverse kinematics of this mechanism using a closed-loop vector method. A comprehensive analysis of constrained and kinematic Jacobian matrices was conducted, leading to the construction of a 6×6 fully populated Jacobian matrix. This matrix aided in evaluating kinematic performance. Additionally, a dynamic model based on the virtual work principle was formulated. Its theoretical solution was compared with numerical simulation results, validating the dynamic model's efficacy and the accuracy of the inverse kinematics. The prosthetic mechanism exhibited a larger rotating workspace and superior mechanical performance, closely mimicking the range of motion and bearing capacity of the human hip across various gait modes. The torque change during hip flexion and extension aligned well with human hip behavior, affirming the feasibility and dynamic performance of the prosthetic hip mechanism [84].

In 2023, Wang et al. introduced a hip exoskeleton with a distinctive parallel structure that allowed unrestricted walking and resolved misalignment issues. They proposed a model-based controller rooted in a human–machine integrated dynamic model, enhancing the system's responsiveness to user movements. Unlike most existing exoskeletons, this controller only needed the user's kinematic data, not interaction force, making the system more compact. Tests showcased the proposed hip exoskeleton's kinematic compatibility and assistance capabilities [85].

4.5.3. Parallel Robot for Hip Humanoid

In 2017, Jiang et al. addressed challenges related to mimicking the distribution of human muscles in robots actuated by antagonistic pneumatic artificial muscles. They observed that existing control algorithms often fail to replicate natural muscle patterns accurately. To remedy this, the researchers proposed a humanoid lower-limb parallel mechanism with unevenly distributed muscle representation. They analyzed the kinematics and dynamic movements of the bionic hip joint using an observer-based fuzzy adaptive control algorithm. This algorithm considered the overall movement of the joint rather than focusing on individual pneumatic artificial muscles. The parameters were optimized using a neural network, and experimental results confirmed the effectiveness of this method. The study particularly highlighted the importance of muscle roles in trajectory tracking for specific muscle groups [22]. Wang et al. focused on the dynamic performance of a 4-SPS/CU parallel mechanism that included a spherical joint with clearance. They employed Archard's wear model to analyze the wear properties of the socket. A kinematics model for the spherical joint with clearance was established, and an improved contact force model was introduced. The researchers formulated a dynamic model for the parallel mechanism while taking into account the spherical joint with clearance. Wear analysis involved the decomposition of contact forces and the computation of sliding distances. Simulation results revealed a nonuniform wear depth along the socket surface, which had implications for the mechanism's performance [86].

In 2018, Russo et al. presented the mechanical design of LARMbot 2, an affordable humanoid robot intended for user-centric applications. LARMbot 2 featured parallel architectures for both the torso and legs, as shown in Figure 7c. The design's kinematics for its primary components—including legs, torso, and arms—are detailed and compared with the previous version. A prototype was introduced, showcasing the subsystem construction and technical specifications. Experimental results offered insights into the robot's performance during walking and weight-lifting operations [87].

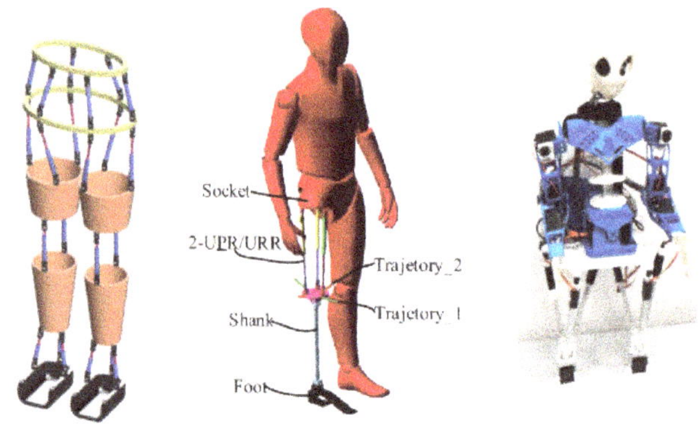

Figure 7. Hip joint mechanisms: (**a**) 6 SPS mechanism by Shangai University [83]; (**b**)4 RPS mechanism by Jiaotong University [84]; (c)3UPR mechanism by the University of Cassino and Southern Latium [87]. All are available under a Creative Commons Attribution License.

Table 5. Parallel robots for rehabilitation, assistance, and humanoids of the hip joint.

Author	Year	Country	Device	TRL	Mechanism	DoF	ToM	Actuator	Model	Tool
Rastegarpanah et al. [80]	2016	UK	HR	4	6-UPS	6	FE	EL	FK	M
Jiang et al. [22]	2017	China	HH	3	PMA	NS	NS	PL	FIK	NS
Wang et al. [86]	2017	China	HH	2	4-SPS/CU	2	NS	EL	NS	NS
Russo et al. [87]	2018	Italy	HH	3	4-SPS	4	NS	EL	IK	NS
Ren et al. [83]	2019	China	HA	3	6-SPS	6	NS	PL	FIK	NS
Zhang et al. [81]	2020	China	HR	3	2-UPS/RRR	3	FE, AA, IER	EL	FK	NS
Song et al. [84]	2020	China	HA	2	4-RPS	3	NS	NS	FK	NS
Shi et al. [82]	2022	China	HR	3	2-UPS/RRR	3	FE-AA, IER	EL	FIK	NS
Wang et al. [85]	2023	China	HA	4	2-UPS+S	2	IER	EL	IK	NS

Abbreviations: HR—hip rehabilitation, HA—hip assistance, HH—hip humanoid, 6-UPS—6 (universal-prismatic-spherical), PMA—pneumatic muscle actuator, 4-SPS/CU—4 (spherical-prismatic-spherical)/1 (cylindrical-universal), 3-UPR—3 (universal-prismatic-revolute), 6-SPS—6 (spherical-prismatic-spherical), 2-UPS-RRR—2 (universal-prismatic-spherical)/1 (revolute-revolute-revolute), 4-RPS—4 (revolute-prismatic-spherical), 2-UPS/RRR—2 (universal-prismatic-spherical)/1 (revolute-revolute-revolute), FE—flexion–extension, NS—not specified, AA— abduction–adduction, IER—internal–external rotation, ER—electric rotary, EL—electric linear, PL—pneumatic linear, IK—inverse kinematic, FK—forward kinematic, FIK—forward and inverse kinematic, M—MATLAB.

4.6. Parallel Robot for Ankle Joint

In the realm of scientific exploration concerning parallel robots, it is paramount to emphasize the scarcity of articles related to assistance and humanoids in the current body of literature. Notably, existing studies tend to focus exclusively on ankle rehabilitation, leaving a significant gap in our understanding of the broader landscape of assistance and humanoid interactions.

Parallel ankle rehabilitation devices aim to improve strength and mobility in patients suffering from ankle injuries, such as sprains or fractures. These devices provide controlled resistance and enable precise ankle motion. Studies related to rehabilitation are summarized in Table 6, each of which is described below.

In 2012, Wang et al. introduced a novel parallel ankle rehabilitation robot and determined its efficacy through kinematic analysis and simulations. The results showcased the robot's versatility under different input scenarios. with single inputs, the robot facilitated a range of ankle movements including dorsiflexion (0–30°), plantar flexion (0–50°), inversion/eversion (0–18°), and adduction/abduction (0–10°), making it suitable for initial ankle rehabilitation training. For double inputs, the robot enabled ankle motions such as dorsiflexion (0–30°), plantar flexion (0–50°), inversion/eversion (0–25°), and adduction/abduction (0–20°), apt for medium-term rehabilitation training. Under three inputs, the robot supported dorsiflexion (0–30°), plantar flexion (0–50°), inversion/eversion (0–40°), and adduction/abduction (0–30°), optimally meeting the needs of comprehensive ankle rehabilitation training. The robot's design effectively accommodated the full range of motion necessary for normal ankle function and provided a versatile platform for various rehabilitation exercises, thereby aiding patients' recovery processes [88].

In 2013, Saglia et al. examined the control architecture and experimental outcomes of the Ankle Rehabilitation Robot (ARBOT), aiming to develop effective control algorithms for aiding ankle joint training and rehabilitation, particularly in the presence of musculoskeletal injuries. They utilized a position control approach for patient-passive exercises and an admittance control technique for patient-active exercises, both with and without motion assistance. The design of the control algorithms was informed by an analysis of the rehabilitation protocol, taking into account system dynamics and human–robot interaction. Experimental assessments involving healthy subjects were conducted to evaluate the performance of the proposed control algorithms [89]. In the same year, Wang et al. presented the design of a novel 3-RUS/RRR redundantly actuated parallel mechanism for ankle rehabilitation, based on conceptual design principles. The mechanism facilitated ankle rotational movements in three directions, aligning the mechanism's center of rotations with the ankle axes. A new actuator redundancy scheme was introduced to enhance system flexibility without compromising inherent advantages. Kinematic attributes such as dexterity, singularity, and stiffness were evaluated using the Jacobian matrix, which was then followed by simulations [90].

In 2014, Jamwal et al. introduced an innovative wearable ankle robot designed for the physical rehabilitation of ankle sprains. Originating from a comprehensive analysis of existing ankle robots, this bioinspired design was adaptable to individuals across different physiological abilities and age groups. The robot was powered by lightweight yet robust pneumatic muscle actuators (PMAs) that emulated skeletal muscles. To address the PMAs' nonlinear characteristics, a fuzzy-based disturbance observer (FBDO) was employed. Additionally, an adaptive fuzzy logic controller—based on Mamdani inference and augmented with FBDO—managed the transient behavior of the PMAs. This control scheme allowed for the simultaneous control of four parallel actuators, achieving three rotational degrees of freedom. Experimental evaluations were conducted with a neurologically intact subject to maintain active–passive robot–human interaction during predefined trajectory tracking. These trials accounted for unforeseen human interventions, nonlinear and compliant actuators, and the parallel kinematic structure of the ankle robot [91].

In 2015, Jamwal et al. delved into the design, analysis, and optimization of a novel wearable parallel robot aimed at ankle rehabilitation. To confront challenges related to parallel mechanisms, flexible actuators, and ankle rehabilitation constraints, a thorough design analysis was executed. Three design stages—kinematic design, actuation design, and structural design—were meticulously investigated, resulting in six critical performance objectives essential for achieving the design goals. Initially, the optimization focused on single objectives; however, due to conflicting objectives, a simultaneous optimization approach was required. The study employed the nondominated sorting algorithm (NSGA II), based on evolutionary algorithms, for multi-objective optimization. NSGA II outperforms single-objective and preference-based optimization methods, providing more comprehensive design solutions. Furthermore, a fuzzy-based ranking method was introduced to select the ultimate design from NSGA II's set of nondominated solutions. This methodology is adaptable for various types of parallel robots [92]. In the same year, Valles et al. from Universitat Politécnica de Valencia developed an economical parallel rehabilitation robot, addressing its design, kinematics, dynamics, and control features. Various position and force controllers were examined to ensure precise tracking performance. The robot was equipped with a force-sensor-integrated orthopedic boot designed for ankle exercises targeting injured areas. It supported passive, active-assistive, and active-resistive exercises for dorsi/plantar flexion, inversion, and eversion ankle movements. Orocos, a component-based middleware, offered a modular and configurable control scheme. Integration with Orocos and ROS enabled real-time teleoperation, represented by a CAD model that mirrored the robot's position. Teleoperated rehabilitation exercises could be conducted using devices like a Wiimote [93].

In 2016, Jamwal et al. developed an ankle rehabilitation robot that employed an interactive training approach based on impedance control. Powered by PMAs, the robot allowed patients to adapt robot-induced motions to their specific limb movements, accommodating for disabilities. Four training modes—position control, zero-impedance control, non-zero-impedance control with high compliance, and non-zero-impedance control with low compliance—were employed to assess the effectiveness of the proposed control scheme. Evaluations involving 10 neurologically intact subjects indicated that increased robotic compliance led to greater participant engagement during training [94]. In the same year, Ruiz-Hidalgo et al. introduced a novel three-degrees-of-freedom (DoF) parallel robot that utilized revolute and spherical joints. The inverse kinematic model incorporate a PID-type controller with tracking capabilities to accurately follow a desired trajectory. Potential applications for this robot include its use as a motion simulator or ankle rehabilitation device. Simulation experiments conducted with a virtual prototype in MD ADAMS software validated the performance of the PID-type controller [95]. Azar et al. introduced an algorithm and an improved rule for controlling a lower-limb rehabilitation system. This system was implemented on a 6DoF Stewart parallel robot. Both impedance control and adaptive control methods were used, and the control parameters were estimated and optimized using artificial neural networks and genetic algorithms. Safety was assured by enabling adaptation under stable conditions. Simulations carried out using MATLAB/SIMULINK demonstrated the effectiveness of this approach when compared to common methods [96].

In 2017, Rosado et al. explored the use of PID controllers for conducting passive rehabilitation exercises. They designed and constructed an ankle rehabilitation prototype that employed a 2-RRSP parallel mechanism. Computer programs for guiding the rehabilitation exercises were developed using open-source software. The prototype facilitated passive exercises that involved path planning and PID control for various ankle movements [97]. Rosado et al. also presented the application of impedance controllers in active rehabilitation exercises. As in the previous study, a parallel ankle rehabilitation prototype using a 2-RRSP mechanism was utilized. They categorized active rehabilitation exercises and user-operation resistance effects into low, medium, and high opposition levels [98]. Du et al. introduced design modeling for a novel 3-RRS spherical parallel mechanism specifically intended for ankle rehabilitation. The study established the kinematics, degree-of-freedom calculations, and inverse kinematics of the mechanism. Multiple inverse solutions were derived, and a forward position analysis method suitable for motor position control was developed. The ankle rehabilitation robot was versatile and applicable in various settings, including homes, hotels, and fitness centers [99]. Zhang et al. highlighted innovative features of a Compliant Ankle Rehabilitation Robot (CARR) that offered an adjustable workspace and torque capacity. The CARR consisted of three rotational degrees of freedom (DoF) and was redundantly actuated by four compliant actuators. Due to the use of a parallel mechanism and compliant actuators, the robot faced the challenge of reconciling conflicting workspace and actuation torque requirements. To address this, the CARR was designed with reconfigurability, allowing it to balance workspace and torque capacity and meet diverse training needs. Theoretical analyses suggested the potential for varying kinematic and dynamic performance by reconfiguring the actuator layout [100].

In 2018, Liao et al. proposed a novel hybrid ankle rehabilitation robot composed of both serial and parallel components. The kinematic performance of this hybrid robot was analyzed, with its parallel part simplified as a constrained 3-PSP mechanism. Mathematical modeling based on screw theory was employed to establish a mathematical model for this component. Inverse kinematics were determined, and factors such as reciprocal twists, Jacobian matrices, and singularities were examined. The study predicted that the workspace of the central point on the moving platform can be expanded while eliminating singularities, making the robot suitable for clinical applications [15]. Rastegarpanah et al. presented a nine-degrees-of-freedom hybrid parallel mechanism designed for ankle rehabilitation, aiming to achieve precise movement in the lower extremities. The methodology for determining stiffness involved calculating the position vectors of each actuator in specific poses using inverse kinematics, thereby obtaining the magnitude and direction of the applied forces. The study leveraged both the stiffness and workspace attributes of parallel robots for ankle rehabilitation. Comparisons were made with standard parallel mechanisms, and the stiffness was evaluated through simulation, which was then compared to a prototype hybrid model in various scenarios [17]. Jamwal et al. investigated the feasibility of a wearable ankle robot for in-home rehabilitation. They began with an analysis of existing technologies and solutions. The complexities of human–robot interactions during rehabilitation were addressed through a fuzzy-logic-based controller designed for ankle treatment. The team proposed visual haptic interfaces to enhance patient engagement and considered web-based communication channels between users and remote physiotherapy staff. The software architecture included patient databases, a graphical user interface, and exercise libraries, thereby ensuring user-friendly operations and offering virtual reality-specific exercises for ankle rehabilitation [101].

In 2019, Wang et al. developed a robot to assist with ankle joint rehabilitation. This robot featured upper and lower platforms connected by a ball–pin pair and driving branches based on an SPS mechanism. Despite having only two degrees of freedom, the upper platform could emulate three types of ankle joint movements. The robot utilized a central ball–pin pair to mimic the natural motions of a patient's foot and ankle joint, while rigid–flexible hybrid driving systems ensured decoupled dorsiflexion/plantar flexion and varus/valgus movements to minimize the risk of secondary injuries [102]. Naruhmi et al. presented a comprehensive analysis of a reconfigurable 3-(rR)PS metamorphic parallel mechanism, exploring its complete workspace and operational modes. The mechanism consisted of three (rR)PS legs, each having an (rR) joint comprising two perpendicular revolute joints. One axis of the (rR) joint could be continuously reconfigured, leading to three distinct configurations. Algebraic geometry techniques were used to derive constraint equations, and primary decomposition computations were carried out for each configuration. The study revealed that the mechanism could exhibit one or two operational modes, depending on the arrangement of the second axes of the (rR) joints. The work characterized the orientation workspaces for both modes and compared the moving platform's instantaneous motion along the same trajectories. An identification approach was also introduced for determining which operational mode corresponded to a given mechanism pose, thereby offering a useful tool for trajectory planning [103].

In 2020, Zuo et al. introduced a wearable parallel mechanism aimed at enhancing the range of equipment options for patients with ankle disabilities. The team conducted kinematic analyses that included inverse position solutions and Jacobian matrices. Performance indices such as reachable workspace, motion isotropy, force transfer capabilities, and maximum torque were developed based on these kinematic solutions. The mechanical structure of the wearable parallel robot was designed to provide motion isotropy, effective force transfer, and high torque capabilities within a broad workspace suitable for ankle rehabilitation [14]. Li et al. concentrated on improving the accuracy and effectiveness of ankle rehabilitation through innovative control strategies. They applied these to a newly developed parallel ankle rehabilitation robot featuring a unique 2-UPS/RRR mechanism, as shown in Figure 8a. The team established the kinematic model for the mechanism and derived both the inverse positional solution and velocity mapping needed for trajectory tracking during passive rehabilitation exercises. Experiments were conducted to determine a torque threshold for detecting the ankle joint's motion intentions, leading to the proposal of an active rehabilitation training strategy. Trials involving healthy subjects showed that the control strategies effectively minimized trajectory-tracking errors during passive exercises and allowed the ankle rehabilitation robot to drive the ankle joint smoothly and flexibly in the intended motion direction. These results confirmed the effectiveness of the proposed control strategies for ankle rehabilitation training [104]. Russo et al. designed the CABLEankle, a lightweight wearable device for ankle motion assistance in both rehabilitation and training, as illustrated in Figure 8b. Employing a cable-driven S-4SPS parallel architecture, the CABLEankle provided motion assistance across a wide range of ankle movements during walking. The design of the mechanism was scrutinized through kinematic and static models, as well as force and workspace analyses. Numerical simulations were conducted to assess the feasibility of the proposed design, specifically characterizing its unique range of ankle motion [105]. Li et al. introduced a new parallel ankle rehabilitation robot with an emphasis on simplicity and nonredundant actuators. They formulated the inverse position solution to calculate the robot's workspace and derived Jacobian matrices for velocity and force mapping. Performance indices such as motion isotropy, force transfer ratio, and force isotropic radius were defined. The inverse dynamic model was developed using the Newton–Euler formulation. The researchers proposed a dynamic evaluation index, called "dynamic uniformity", based on the Jacobian and inertia matrices. A workspace analysis deemed the robot's range suitable for ankle rehabilitation, aligning well with human ankle joint motion. Kinetostatic and dynamic performance analyses confirmed favorable motion properties, high force transfer, and dynamic dexterity, particularly in the central workspace. A simulation validated the rehabilitation process and the inverse dynamic model, underscoring the robot's potential versatility in ankle rehabilitation [106].

In 2021, Pulloquinga et al. discussed the singular configurations of parallel robots (PRs), which can result in control loss. The authors introduced a proximity detection index for type II singularities, aimed at identifying the contributing kinematic chains based on the angle between output twist screws. An experimental benchmark was presented, and PR configurations were analyzed where the proposed index was zero but the forward Jacobian determinant was not. The study illuminated the challenges of handling external actions applied to the PRs in such configurations, highlighting the complexities introduced by manufacturing tolerances [107].

In 2022, Lui et al. introduced a novel ankle rehabilitation robot based on a 2-UPU/RPU parallel mechanism, which they termed the Multi-Locomotion Mode Ankle Rehabilitation Robot (MLMARR). This robot offered various rehabilitation training modes that went beyond basic ankle motion orientation, including up/down and back/forth traction training. The study covered analyses of the degrees of freedom, inverse position solutions, the identification of kinematic singularities, and workspace determination. It also included dynamic simulations aimed at optimizing linear actuators. Experimental data from a prototype MLMARR validated the effectiveness and accuracy of the proposed training modes, showcasing their potential to advance therapies in ankle rehabilitation [108].

In 2023, the field of robotic rehabilitation witnessed significant advancements. Valencia-Segura et al. championed an affordable ankle rehabilitation device, boasting an optimized spherical mechanism with a singular degree of freedom (Figure 8c). This device was tailored for a spectrum of ankle movements, addressing diverse parameters such as motion angle and force transmission [109]. On a parallel note, Wu et al. showcased a groundbreaking 9DoF robotic system aimed at patients with balance disorders. This robot, integrating a series-parallel hybrid motion platform, was constructed from two symmetrically arranged sets of motion platforms, each containing a 6DoF vestibular parallel device and a 3DoF proprioception parallel device. This unique configuration allowed for DoF decoupling and swift responses, leading to a structure optimal for vestibular and proprioceptive simulation. with modularized vestibular and proprioceptive components, the robot's flexibility and capabilities were significantly amplified [110]. Concurrently, Shi et al. enriched the sector with their 5DoF serial–parallel hybrid robot designed for gait and balance training. This robot integrated a 2DoF gait generator employing the H-Bot mechanism and a 3DoF parallel balance module, offering natural balance exercises by aligning the footplate's rotation center with the ankle joint. with embedded sensors, the robot achieved precise closed-loop control for both position and orientation. Post-construction tests demonstrated its prowess in accurate tracking, solidifying its potential as a holistic platform for combined gait and balance rehabilitation. Collectively, these studies illuminate the rapid progress in the domain of robotic rehabilitation, paving the way for sophisticated and patient-centric solutions [111]. Zuo et al. (2023) introduced a groundbreaking design for composite patient-external fixators (PEFs), emphasizing precise correction requirements. By employing multi-site frame mounting and a diverse set of struts, they demonstrated the feasibility of their method, backed by clinical cases and simulations, ensuring that the device could adapt to various deformities and maintain stability [112].

In parallel, Zermane et al. (2023) presented an innovative 3-PRS robotic platform tailored for human balance assessment and rehabilitation. The platform's design, which encompassed rotational and vertical movements, underwent rigorous optimization using multiple algorithms, ultimately pinpointing the optimal geometric parameters [113]. Lastly, Zhang et al. (2023) ventured into the realm of ankle–foot motion, proposing a new evaluation index called the "ankle-foot motion comfort zone". By integrating this index into their optimization method for a 3DoF ankle rehabilitation robot, they not only underscored the interconnection of ankle–foot motion but also showcased the adaptability and efficacy of their method. Collectively, these studies indicate a promising trajectory in the development and optimization of rehabilitation devices, potentially revolutionizing patient care in orthopedics and balance disorders [114].

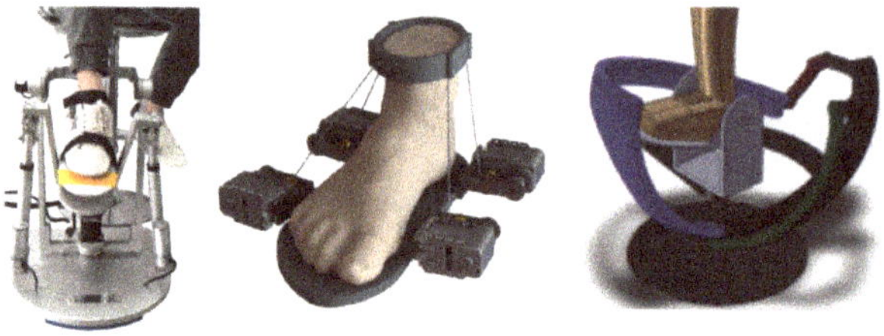

(a) (b) (c)

Figure 8. Ankle joint mechanisms: (**a**) 2-UPS/RRR mechanism by Beijing University [104]; (**b**) S-4SPS mechanism by the University of Nottingham [105]; (**c**) spherical mechanism by Instituto Politécnico Nacional, Mexico [109]. All are available under a Creative Commons Attribution License.

Table 6. Parallel robots for the rehabilitation of the ankle joint.

Author	Year	Country	Device	TRL	Mechanism	DoF	ToM	Actuator	Model	Tool
Wang et al. [88]	2012	China	AR	2	3-SPS/SP	3	PD, IE, AA	NS	FIK	M
Saglia et al. [89]	2013	Italy	AR	4	3-UPS/U	2	PD, IE	EL	FK	NS
Wang et al. [90]	2013	China	AR	2	3-RUS/RRR	3	PD, IE, AA	ER	IK	AD
Jamwal et al. [91]	2014	India	AR	4	PMA	3	PD, IE, AA	PL	FK	M
Jamwal et al. [92]	2015	India	AR	2	PMA	3	NS	PM	FIK	M
Vallés et al. [93]	2015	Spain	AR	3	3-PRS	3	PD, IE	EL	IK	ROS
Jamwal et al. [94]	2016	Kazakhstan	AR	4	PMA	3	PD, IE, AA	PM	FK	NS
Azar et al. [96]	2016	Iran	AR	3	6-UPS	6	NS	PM	IK	NS
Ruiz-Hidalgo et al. [95]	2016	Mexico	AR	3	3-SPR	3	NS	EL	IK	AD
Rosado et al. [97]	2017	Mexico	AR	2	2-RRSP	2	PD, IE	ER	NS	M
Rosado et al. [98]	2017	Mexico	AR	3	2-RRSP	2	PD, IE	ER	NS	AD
Du et al. [99]	2017	China	AR	2	3-RRS	3	PD, IE	EL	IK	NS
Zhang et al. [100]	2017	New Zealand	AR	2	4-PMA	3	PD, IE, AA	PM	IK	NS
Liao et al. [15]	2018	China	AR	2	3-PSP	3	PD, IE	NS	IK	NS
Rastegarpanah et al. [17]	2018	UK	AR	3	6-UPS/3SPR	9	NS	NS	IK	NS
Jamwal et al. [101]	2018	Kazakhstan	AR	4	PMA	3	PD, IE, AA	PM	IK	LV
Wang et al. [102]	2019	China	AR	3	2-SPS	2	PD, IE	EL	FK	NS
Naruhmi et al. [103]	2019	Indonesia	AR	3	3(rR)PS	3	NS	NS	FK	NS
Zuo et al. [14]	2020	China	AR	2	2-UPS/RRR	3	PD, IE, AA	EL	IK	NS
Li et al. [104]	2020	China	AR	3	2-UPS/RRR	3	PD, IE, AA	EL	IK	NS
Russo et al. [105]	2020	UK	AR	2	cable-driven	3	PD, IE, AA	EL	NS	NS
Li et al. [106]	2020	China	AR	3	2-UPS/RRR	3	PD, IE, AA	EL	IK	M
Pulloquinga et al. [107]	2021	Spain	AR	3	3-UPS/RPU	4	NS	EL	IK	M, LV
Liu et al. [108]	2022	China	AR	2	2-UPU/RPU	2	PD, IE	PM	IK	NS
Valencia-Segura et al. [109]	2023	Mexico	AR	3	4-bar	2	PD, IE, AA	NS	IK	CAD
Wu et al. [110]	2023	China	AR	4	6-SSP, 3-RPS	9	PD, IE, AA	EL	IK	NS
Shi et al. [111]	2023	China	AR	4	3-RRR	3	PD, IE, AA	ER	IK	M
Zuo et al. [112]	2023	China	AR	4	RPUR, SPR	5	PD, IE, AA	ER	IK	NS
Zermane et al. [113]	2023	France	AR	4	3-PRS	3	PD, IE, AA	ER	IK	SA
Zhang et al. [114]	2023	China	AR	4	2-UPU/PRPS	3	PD, IE, AA	EL	IK	M

Abbreviations: AR—ankle rehabilitation, 3-SPS/SP—3 (spherical-prismatic-spherical)/ 1(spherical-prismatic), 3-UPS/U—3 (universal-prismatic-spherical)/1 (universal), 3-RUS/RRR—3 (revolute-universal-spherical)/1 (revolute-revolute-revolute), PMA—pneumatic muscle actuator, 3-SPR—3 (spherical-prismatic-revolute), 3-PRS—3 (prismatic-revolute-spherical), 2-RRSP—2 (revolute-revolute-spherical-prismatic), 3-RRS—3 (revolute-revolute-spherical), 6-UPS—6 (universal-prismatic-spherical), 4-PMA—4 (pneumatic muscle actuator), 3-PSP—3 (prismatic-spherical-prismatic), 6-UPS/3SPR—6 (universal-prismatic-spherical)/3 (spherical-prismatic-revolute), 2-SPS—2 (spherical-prismatic-spherical), 2-UPS/RRR—2 (universal-prismatic-spherical)/(revolute-revolute-revolute), 2-UPS/RRR—2 (universal-prismatic-spherical)/1 (revolute-revolute-revolute), 3(rR)PS—two perpendicular revolute joints, 3-UPS/RPU—3 (universal-prismatic-spherical)/(revolute-prismatic-universal), 2-UPU/RPU—2 (universal-prismatic-universal)/1 (revolute-prismatic-universal), 3RRR—3 (revolute-revolute-revolute), RPUR—(revolute- prismatic-universal-revolute), SPR—(spherical- prismatic-revolute), 3-PRS—3 (prismatic-revolute-spherical), 2-UPU/PRPS—2 (universal-prismatic-universal)/(prismatic-revolute-prismatic-spherical), PD—plantarflexion–dorsiflexion, IE—inversion–eversion, AA—abduction–adduction, NS—not specified, ER—electric rotary, EL—electric linear, PL—pneumatic linear, PM—pneumatic muscle, IK—inverse kinematic, FK—forward kinematic, FIK—forward and inverse kinematic, M—MATLAB, AD—Adams, LV—LabVIEW, SA—simulated annealing, ROS—Robot Operating System, CAD—Computer-Aided Design.

5. Discussion

5.1. Timeline of Parallel Robots from 2012 to 2023

This comprehensive timeline (Figure 9) reflects the rapid advancements in robotics tailored to medical and humanoid applications over a decade. The expansive body of research in robotic technology showcases advancements across rehabilitation, assistance, and humanoid applications.

- Rehabilitation focus: Rehabilitative robotics present a variety of versatile solutions, from ankle robots equipped with impedance control and redundant actuation, to teleoperable, modular ankle devices and multi-objective-optimized wearable ankle robots. Shoulder and wrist rehabilitation benefit from adaptive controllers, multi-DoF prosthetics, and anti-vibration control technologies.
  Ankle rehabilitation has been a consistent area of interest every year, with advances in control algorithms [89], wearable designs [91–93], hybrid mechanisms [15], natural motion mimicry [102], and high-torque applications [14,104–106]. As of 2023, rehabilitation techniques have expanded to treat balance disorders [110] and offer solutions for vertigo diagnosis [113].
- Cable-driven and parallel robots: Starting in 2012 with applications like shoulder movement simulation and ankle rehabilitation, there has been a clear focus on cable-driven and parallel mechanisms. These designs emphasize human mimicry, offering smooth and organic movements. Examples include neck movements [39–41], wrist motion [68,74], and shoulder movements [45,53]
  In terms of mimicking human biomechanics, cable-driven robots excel in recreating neck and cervical spine movements, while various exoskeletons and prosthetics, such as 7DoF shoulder and arm mechanisms, offer enhanced spatial characteristics.
- Humanoid design: The timeline indicates the progression in bionic and humanoid designs. From bionic joints in 2013 [54] to more comprehensive hybrid humanoid arms in 2021 [59,59], there has been an evident push towards creating robots that closely emulate human characteristics. The introduction of LARMbot 2 in 2018 [87] demonstrates the practical implementations of these humanoid designs.
  Humanoid robots feature innovations like bionic joints, pneumatic-muscle-driven manipulators, and multi-objective-optimized trajectory controls, designed for speed, stability, and energy efficiency.
- Neck and shoulder rehabilitation: Starting from 2013 with the cable-driven robot mimicking human neck movements [39], there has been a sustained focus on neck and shoulder rehabilitation mechanisms over the years. By 2021, there was a move towards more flexible and soft mechanisms for neck rehabilitation [3], further improving in 2023 with cable-driven exoskeletons for cervical rehabilitation [35].
- Wrist and hand innovations: Over the timeline, wrist innovations have ranged from modular designs [68] to stiffness-optimizable exoskeletons [49] and cable-driven prostheses [72]. The emphasis on wrist rehab and prostheses highlights the importance of restoring hand functions in patients.
- Hybrid designs: The later years (especially post-2020) showcase a strong interest in hybrid systems. For instance, in 2023, highlights include series-parallel hybrid modular joint humanoids [60] and serial–parallel hybrid robots [111].
- Specialized solutions: Some innovations are particularly noteworthy because they address specific medical conditions or challenges, like the EMG-recording dynamic neck brace for amyotrophic lateral sclerosis (ALS) [2] and multi-segment foot and ankle deformity correction in 2023 [112].
- Advanced control and optimization techniques: The use of genetic algorithms [45, 46,51], neural networks [22,82,96], and Koopman controllers [64,67] indicates the integration of sophisticated computational techniques in optimizing and controlling robotic movements.

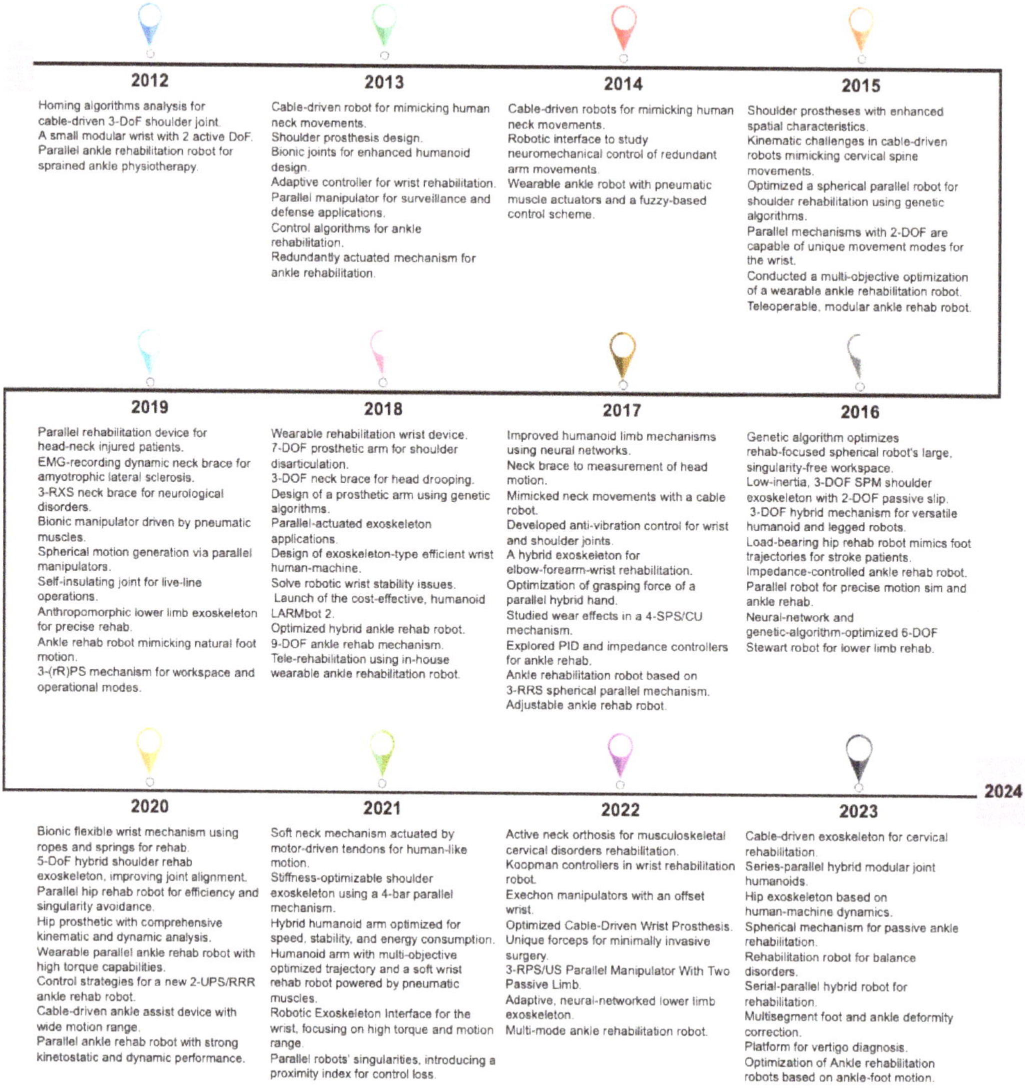

Figure 9. Timeline of parallel robots from 2012 to 2023. References: 2012 [53,68,88], 2013 [39,50,54,61,73,89,90], 2014 [40,41,44,91], 2015 [20,42,45,74,92,93], 2016 [24,46,47,55,80,94–96], 2017 [22,36,43,56,62,75,86,97–100], 2018 [6,15,17,19,31,37,51,52,69,76,87,101], 2019 [1,2,21,27,38,57,58,77,83,102,103], 2020 [9,10,14,33,48,81,84,104–106], 2021 [3,11,16,18,26,49,59,63,70,107], 2022 [4,64,71,72,78,79,82,108], 2023 [35,60,65–67,85,109–114].

5.2. *Advancements in Parallel Robots from 2012 to 2023*

This section presents an analysis of articles published on parallel robots in the fields of rehabilitation, assistance, and humanoid robotics between the years 2012 and 2023, across various countries. Tables 7–9 provide information on the distribution of articles based on the targeted body parts (e.g., neck, shoulder, wrist, hip, and ankle). The "Total" column indicates the sum of articles related to each country, while the "Percentage" column shows the proportion of articles contributed by each country.

- Parallel robots for rehabilitation:
  Firstly, it is evident that China has emerged as a significant contributor to research in this field, accounting for 37.3% of the total articles. This underscores China's robust presence in parallel robot research, particularly in the area of rehabilitation for various joints.
  Mexico accounts for 9.8% of the total articles. This underscores the active participation of researchers from this country in investigating the applications of parallel robots, especially in the neck and ankle areas. On the other hand, each of the following countries: the United States, the United Kingdom (UK), and Australia, accounts for 7.8% of the total articles.
  Moreover, while individual countries like India, Iran, each contributed a smaller percentage, they collectively made up a significant portion of the total articles, at 5.9% each. This fact demonstrates a global interest and investment in advancing the field of parallel robot rehabilitation.
  Conversely, some countries, such as Spain and Kazakhstan, accounting for 3.9%, have demonstrated a more balanced distribution of articles across different joint areas, indicating a diverse research focus.
  It is important to note that there are countries, such as Japan, Italy, New Zealand, Indonesia and France, that each contributed to the total with a smaller percentage, starting from 2.0%. While their contributions may be comparatively modest, they nonetheless play a role in enriching the overall body of knowledge in the field of parallel robot applications for rehabilitation.

Table 7. Analysis of articles published on parallel robots for rehabilitation from 2012 to 2023 across various countries.

Country	Neck	Shoulder	Wrist	Hip	Ankle	Total	Percentage
India	1	-	-	-	2	3	5.9%
USA	1	2	1	-	-	4	7.8%
Mexico	1	-	-	-	4	5	9.8%
UK	-	1	-	1	2	4	7.8%
Iran	-	2	-	-	1	3	5.9%
Australia	-	1	3	-	-	5	7.8%
China	1	-	2	3	13	18	37.3%
Japan	-	-	1	-	-	1	2.0%
Italy	-	-	-	-	1	1	2.0%
Spain	-	-	-	-	2	2	3.9%
Kazakhstan	-	-	-	-	2	2	3.9%
New Zealand	-	-	1	-	-	1	2.0%
Indonesia	-	-	1	-	-	1	2.0%
France	-	-	1	-	-	1	2.0%

- Parallel robots for assistance:
  The United States (USA) leads in contributions to research in this field, accounting for 31.3% of the total articles. This statistic suggests that the USA is vigorously engaged in research on parallel robots for assistance, with particular emphasis on the neck, shoulder, and wrist areas. China follows closely, contributing 25.0% of the articles, with a focus on assistance for the neck and hip.
  Japan also has a significant presence, contributing 12.5% of the articles and concentrating primarily on shoulder assistance. This highlights Japan's active involvement in research concerning parallel robots designed for assistance purposes.
  The analysis revealed that multiple countries, including Italy, Mexico, South Korea, and the United Kingdom (UK), each contributed 6.3% to the total body of articles. These articles were predominantly centered on shoulder and wrist assistance.
  The data indicate a widespread interest in parallel robots for assistance on a global scale. While the USA, China, and Japan emerged as the major contributors, other

countries are also actively participating in research in this field, showing a dispersed research effort.
The lack of contributions focusing on certain body parts, such as the ankle, in some countries may suggest that the research in those nations is primarily oriented toward other specific areas of assistance.

Table 8. Analysis of articles published on parallel robots for assistance from 2012 to 2023 across various countries.

Country	Neck	Shoulder	Wrist	Hip	Ankle	Total	Percentage
USA	2	1	2	-	-	5	31.3%
China	1	-	-	3	-	4	25.0%
Japan	-	2	-	-	-	2	12.5%
Italy	-	1	-	-	-	1	6.3%
Mexico	-	1	-	-	-	1	6.3%
South Korea	-	-	1	-	-	1	6.3%
Korea	-	-	1	-	-	1	6.3%
UK	-	-	1	-	-	1	6.3%

- Parallel robots for humanoids:
  China is the most prominent contributor to research in this field, responsible for 76.9% of the total articles published. This high percentage underscores a significant focus on parallel robots for humanoids within the Chinese research community. In terms of specific body parts, China's contributions span articles related to the neck, shoulder, wrist, and hip.
  Several other countries, including France, Denmark, Mexico, the United Kingdom (UK), Turkey, and Italy, have also made contributions, albeit to a lesser extent. Each of these countries accounts for 3.8% of the total number of articles.

Table 9. Analysis of articles published on parallel robots for humanoids from 2012 to 2023 across various countries.

Country	Neck	Shoulder	Wrist	Hip	Ankle	Total	Percentage
China	6	6	6	2	-	20	76.9%
France	-	1	-	-	-	1	3.8%
Denmark	-	1	-	-	-	1	3.8%
Mexico	-	-	1	-	-	1	3.8%
UK	-	-	1	-	-	1	3.8%
Turkey	-	-	1	-	-	1	3.8%
Italy	-	-	-	1	-	1	3.8%

5.3. *Technology Readiness Levels—TRLs*

Technology readiness levels (TRLs) are used to gauge the maturity level of a technology as it progresses through its research, development, and deployment phases. TRLs are quantified on a scale from 1 to 9, with level 9 representing the most mature technology [115].

Tables 10–12 present an analysis of the technology readiness levels (TRLs) for parallel robots designed for rehabilitation, assistance, and humanoid applications. These tables serve as indicators of a given technology's maturity, spanning from early-stage conceptual models (TRL 1) to fully operational systems (TRL 9). Each table details the distribution of TRLs across various body parts, including the neck, shoulder, wrist, hip, and ankle. The "Total" column lists the cumulative count of articles for each TRL level, while the "Percentage" column indicates the proportion of articles that correspond to each specific TRL.

- Parallel robots for rehabilitation:
  The lack of articles at TRL 1 indicates that most research publications have moved beyond foundational scientific principles to more advanced stages. with 25.49% of the

total articles, TRL 2 serves as a transitional phase from theoretical concepts to practical experimentation. The notable concentration at TRL 2 suggests that researchers are keenly exploring and prototyping new approaches to rehabilitation using parallel robots, especially for the shoulder and ankle regions.
At 45.10%, TRL 3 boasts the largest percentage of articles, indicating that a significant volume of research has progressed to the prototype-building and proof-of-concept stages. The range of body parts covered (neck, shoulder, wrist, hip, and ankle) demonstrates a broad investigation into the applications of parallel robots in various rehabilitation contexts.
Articles at TRL 4 make up 29.41% of the total publications. These studies imply that some research endeavors have reached the level of system integration and functional testing. The existence of articles at this TRL suggests the ongoing refinement of parallel robot systems, potentially bringing them closer to practical implementation in rehabilitation settings.

Table 10. Technology readiness levels of parallel robots for rehabilitation.

TRL	Neck	Shoulder	Wrist	Hip	Ankle	Total	Percentage
TRL 1	-	-	-	-	-	0	0%
TRL 2	-	3	-	-	10	13	25.49%
TRL 3	2	2	6	2	11	23	45.10%
TRL 4	2	1	2	1	9	15	29.41%

- Parallel robots for assistance:
  TRL 1 has no corresponding articles, suggesting that publications are primarily focused on concepts beyond initial scientific principles. Articles at TRL 2 account for 25.0% of the total, indicating a transition in research efforts toward practical experimentation and validation, particularly in the contexts of shoulder and wrist assistance.
  With 68.75% of the total articles classified under TRL 3, the research community has made significant strides in developing prototypes and demonstrating the feasibility of parallel robot systems for assistance across various body parts. The high concentration of articles at TRL 3 signifies a strong emphasis on constructing functional prototypes with the potential for real-world applications.
  Of the total articles, 6.25% are classified under TRL 4. This may indicate that research has advanced to the stages of system integration, advanced testing, or practical implementation.
  The distribution of TRLs across different body parts provides insights into the areas that have garnered more attention in terms of technological development for assistance. While neck, shoulder, and wrist assistance have seen notable progress, ankle assistance has not yet advanced beyond TRL 2.

Table 11. Technology readiness levels of parallel robots for assistance.

TRL	Neck	Shoulder	Wrist	Hip	Ankle	Total	Percentage
TRL 1	-	-	-	-	-	0	0%
TRL 2	-	2	1	1	-	4	25.0%
TRL 3	3	3	4	1	-	11	68.75%
TRL 4	-	-	-	1	-	1	6.25%

- Parallel robots for humanoids:
  Articles at TRL 1 and TRL 2 together account for a collective 50.0% of the total. This suggests a substantial focus on both theoretical groundwork (TRL 1) and practical experimentation (TRL 2). Both neck and shoulder areas have received particular attention at these early stages.

Articles classified under TRL 3 constitute the largest percentage, making up 46.15% of the total. This indicates that researchers have moved beyond initial theoretical concepts to the active development and validation of prototypes. The distribution across various body parts reflects a comprehensive focus on the application of parallel robots in humanoid systems.

There are 3.85% corresponding articles at TRL 4, suggesting that research has advanced to the stage of system integration or operational testing.

The distribution of TRLs across different body parts provides insights into the areas that have received the most attention in terms of technological development for humanoid applications. Notable progress has been observed across TRLs for the shoulder and wrist.

Table 12. Technology readiness levels of parallel robots in the area of humanoids.

TRL	Neck	Shoulder	Wrist	Hip	Ankle	Total	Percentage
TRL 1	2	-	-	-	-	2	7.69%
TRL 2	3	3	4	1	-	11	42.31%
TRL 3	1	4	5	2	-	12	46.15%
TRL 4	-	1	-	-	-	1	3.85%

5.4. Design, Number of Degrees of Freedom, and Kinematics Structure in Parallel Robotics

The intricacies of designing and optimizing parallel robotic systems constitute an area of intense research and development. This arena is particularly significant because the ultimate performance of a robot—be it in rehabilitation, assistance, or humanoid constructs—is deeply rooted in its initial design parameters and subsequent optimization techniques.

5.4.1. Design Principles

The design of parallel robots for use in rehabilitation, assistance, and humanoid applications must account for the specific requirements of each application. Each joint of the human body—such as the neck, shoulder, wrist, hip, and ankle—has unique motion characteristics and specific demands in terms of degrees of freedom, range of motion, required force, and precision [45]. It is, therefore, necessary to tailor the robot's design to each specific joint while considering the patient's needs and rehabilitation goals.

The design process includes structural analysis as well as the selection of actuators and sensors [62]. This allows for the creation of customized designs and a wide variety of configurations, enabling the robots to perform movements like translations, rotations [90], inclinations, or a combination thereof.

Compared to serial robots, parallel robots are designed to be more precise and rigid. This is attributable to the multiple kinematic chains acting in parallel, which offer greater stability [76] and resistance to deformation [98].

Analytically, parallel mechanisms pose additional challenges due to their inherent complexity. The kinematic and dynamic analysis of these mechanisms requires solving nonlinear equations [49,56,91] and constraints and considering interactions between the kinematic chains. Such mechanisms also face limitations in design and control complexity, the need for appropriate actuators and precise sensors [62], associated costs, and technology readiness levels.

5.4.2. Number of Degrees of Freedom and Kinematic Structure

The relationship between the degrees of freedom (DoF) and kinematic structures is foundational in the field of parallel robotics and mechanical design. In essence, the degrees of freedom of a system define the number of independent parameters that determine its configuration. Kinematic structures, on the other hand, describe the arrangement and connectivity of joints and links in a mechanical system. The nature of this arrangement—how parts are interconnected—determines the system's degrees of freedom. For instance, Table 13 presents a comprehensive overview of the degrees of freedom (DoF) and the corresponding kinematic structures associated with various human joints, with a particular emphasis on their utilization in rehabilitation, assistance, and humanoid robot technologies.

- Neck joint:
  For the neck joint, the dominant kinematic structures are primarily based on a 3DoF framework. In the rehabilitation sphere, there are models such as 3-RPS [1] and 3-RRR [2]. In the context of assistance robots, configurations like 3-RRS [36], 3-RRR [37], and 3-RXS [38] are used. Humanoid robotics often favor cable-driven mechanisms [3,39,41–43]. A 4DoF structure, represented by the 4-SPS model [4], is also common in the rehabilitation sector.
- Shoulder joint:
  Given its complex nature, the shoulder joint exhibits a diverse range of kinematic structures. In humanoid robots, 2DoF mechanisms like 2-UPUR/RU [55] and cable-driven [56] systems are prominent. As the DoF expand to 3 and 5, the rehabilitation domain presents an array of models, including 5R [44], 3-RSS/S [45], and SPM [47]. In the assistance robotics realm, structures such as 3-SPS/P [20], 3-RSS/S [46], and 3-RRR [19,51] emerge. The 7DoF mechanisms, including designs like 4B-SPM [49] and cable-driven [50] methods, underscore the sophistication of this joint's modeling.
- Wrist joint:
  Given its essential function in facilitating dexterity, the wrist showcases a broad spectrum of kinematic structures. In rehabilitation for a 2DoF wrist, the 3-RPS [61] configuration stands out. In humanoid applications, models evolve in complexity from 2-PUU/RPS [76] for 2DoF to designs like 5-SPM [75] for 5DoF. The ubiquitous presence of cable-driven [72] solutions across varying DoF emphasizes their role in ensuring flexibility and adaptability.
- Hip joint:
  For the 2DoF hip joint, while the rehabilitation and assistance domains remain relatively unexplored, humanoid designs have proposed models like 4-SPS/CU [86]. A 3DoF hip favors models like 2-UPS/RRR [81,82] in rehabilitation and 4-RPS [84] in assistance robotics. Interestingly, the 6DoF hip primarily features the 6-UPS [80,83] configuration, evident in both the rehabilitation and assistance domains.
- Ankle joint:
  The ankle, crucial for locomotion, has garnered significant attention in the rehabilitation sector. The multitude of designs for a 3DoF ankle, ranging from 3-SPS/SP [88] to 3-PSP [15], attests to this. The frequent inclusion of configurations like PMA in multiple references underscores its significance. Higher-DoF designs, such as the 6-UPS [96] for 6DoF and 6-UPS/3SPR [17] for 9DoF, suggest an emphasis on advancing ankle dexterity in robotics.
  The choice of kinematic structure affects other design parameters like the system's strength, flexibility, control complexity, and efficiency. It becomes evident that the technological advances in rehabilitation, assistance, and humanoid robotics are significantly intertwined with the complexities of human anatomy. The extensive research cited above underscores the relentless endeavors of scientists and engineers to emulate human-like movement and function. As robotics continues its foray into human-centric applications, such insights will become pivotal. They not only reflect the current state of art but also hint at the vast potential awaiting in the future of biomechanics and robotics.

Table 13. Number of degrees of freedom and kinematic structure.

Joint	DoF	Rehabilitation	Assistance	Humanoids
Neck	3	3-RPS [1], 3-RRR [2], cable-driven [35]	3-RRS [36], 3-RRR [37], 3-RXS [38]	cable-driven [3,39,41–43]
	4	4-SPS [4]	-	-
Shoulder	2	-	-	2-UPUR/RU [55], cable-driven [56]
	3	5R [44], 3-RSS/S [45], SPM [47], cable-driven [48]	3-SPS/P [20], 3-RSS/S [46], 3-RRR [19,51], 4B-SPM [52]	3-RRR [58], 5R [59], cable-driven [35]
	5	-	-	5-PMA [54,57]
	7	4B-SPM [49]	cable-driven [50]	cable-driven [53]
Wrist	2	3-RPS [61]	2-UPS/S [68], 2-CSPM [70], cable-driven [72]	3-4R [74], 2-PUU/RPS [76], cable-driven [9], 3-UPS/S [59], 3-RSR/UUP [78], 3-RPS/US [79]
	3	2-URR/RRS [62], 3-RRR [6], 6-SPS/PS [63], 4-BMA [66,67], cable-driven [65]	2-RRR [69], 2-RRRPR/RRPR [71]	3-SPS/S [73]
	4	-	-	cable-driven [77]
	5	-	-	5-SPM [75]
Hip	2	-	2-UPS+S [85]	4-SPS/CU [86]
	3	2-UPS/RRR [81,82]	4-RPS [84]	
	4	-	-	4-SPS [87]
	6	6-UPS [80]	6-UPS [83]	-
Ankle	2	3-UPS/U [89], 2-RRSP [97,98], 2-SPS [102], 2-UPU/RPU [108], 4-bar [109]	-	-
	3	3-SPS/SP [88], 3-RUS/RRR [90], PMA [91,92,94,101], 3-PRS [93], 3-SPR [95], 3-RRS [99], 4-PMA [100], 3-PSP [15], 3(rR)PS [103], 2-UPS/RRR [14], 2-UPS/RRR [104], cable-driven [105], 2-UPS/RRR [106], 3-RRR [111], 3-PRS [113], 2-UPU/PRPS [114]	-	-
	4	3-UPS/RPU [107]	-	-
	5	RPUR, SPR [112]	-	-
	6	6-UPS [96]	-	-
	9	6-UPS/3SPR [17], 6-SSP, 3-RPS [110]	-	-

5.5. Assessment of Workspace, Functional Capabilities, and Performance Methods in Parallel Robotics

The systematic evaluation of workspace dimensions, functional capabilities, and performance indices is critical for the continued advancement of parallel robotics. This appraisal serves to deepen our understanding of the complex operational characteristics of these robots, thereby laying the foundation for future technological innovations and tailored application scenarios.

- Workspace:
  Workspace analysis and optimization are recurring themes in a range of robotic technologies, often serving medical and rehabilitation purposes. Various studies have emphasized different methods and objectives to optimize workspace. For instance, Gao et al. [39,41] and Enferadi et al. [45] used simulations and genetic algorithms, respectively, to assess workspace under positive cable tension constraints and maximize it, with Enferadi's work noting the advantage of a singularity-free workspace. Niyetkaliyev's hybrid mechanism [48] covered a full range of shoulder movements in a singularity-free workspace, while Sekine et al. [20] and Song et al. [84] focused on prosthetic arms and hips that offered increased and rotating workspaces, respectively. Zhang et al. [100] employed geometric parameter optimization to prevent workspace

singularities in an ankle rehabilitation robot. The studies of Lee et al. [69] and Serracin et al. [68] considered enhancing workspace efficiency and user comfort, and Wu and Chaparro-Altamirano [73] addressed orientation workspace and workspace calculations. Additional research by Zhang et al. [81], Liao et al. [15], Rastegarpanah et al. [17], Naruhmi et al. [103], and Zuo et al. [14] delved into adjustable workspaces, the expansion of central points on moving platforms, and broad workspaces for ankle rehabilitation, often validated through simulations, prototypes, or theoretical analyses. Overall, the workspace is considered a critical design factor, evaluated and optimized through various methods to meet different performance requirements.

- Functional capabilities:
  The functional capabilities of parallel robots, such as dexterity, manipulability, and range of motion, are critical for their application in rehabilitation and assistive technologies.
  Dexterity: This is often measured using indices like the "manipulability measure", which is derived from the robot's Jacobian matrix [51,72,90,106].
  Manipulability: This represents the ability of the robot to move in different directions [83,90].
  Range of motion: This typically involves analyzing how well the robot can perform tasks that mimic human joint motions, such as rotations and translations [4,6,36,37,70,84,88].
- Performance evaluation methods:
  The performance evaluation of parallel robots is multi-faceted, typically incorporating metrics such as speed [28,59], accuracy [33,53,57,84,104,108], repeatability, load-carrying capacity [79,80], energy efficiency, and safety [96]. Methods for performance evaluation include computational methods, and simulations for statistical analysis and finite element analysis (FEA) for mechanical property evaluation are popular choices [38].
  Furthermore, various software tools were prevalent for simulation and analysis in the studies reviewed. MATLAB, SimScape Multibody [54,78], ADAMS [19,53,59,59,90,95,98], and ANSYS [38] each offer unique capabilities for evaluating different aspects of robot performance, workspace, and function. Their choice often depends on the specific requirements of the study.
  Based on existing research, MATLAB is commonly used to simulate parallel robots because of its matrix-handling abilities, algorithm implementation, function representation, user interface creation, and interoperability. SimScape Multibody, part of MATLAB's Simulink suite, offers block-diagram-based modeling for rigid-body mechanics, describing kinematics and dynamics through blocks representing bodies and joints. This enables quick validation and offers 3D visualization for system movement comprehension. MSC ADAMS facilitates the study of dynamics, load distribution, and forces in mechanical systems by allowing the creation of virtual prototypes underpinned by real-world physics. Meanwhile, ANSYS specializes in kinematic simulations, providing detailed analyses of motion and structural parameters, including rotation angles and force requirements.

5.6. Material Selection in the Development of Parallel Robotics

The choice of materials in the fabrication of parallel robotic systems plays a role in meeting targeted performance metrics, durability, and functional objectives. Given the expanding applications of parallel robotics in medical rehabilitation, assistive technologies, and humanoid robots, the criteria for material selection have become increasingly stringent.

- Metals
  Aluminum: This metal is highly prized for its advantageous strength-to-weight ratio, thermal and electrical conductivity, and intrinsic resistance to corrosion. It is often the material of choice for constructing lightweight robotic frameworks. This material was used in a hybrid parallel robot designed for foot rehabilitation [17] and a hybrid exoskeleton for elbow–forearm–wrist rehabilitation [62].

Stainless steel: Renowned for its exceptional durability and corrosion resistance, stainless steel finds utility in applications that demand high mechanical strength and rigidity. This material was utilized in a distal arm exoskeleton for stroke and spinal cord injury rehabilitation [7] and a rope-driven mechanism for wrist rehabilitation [9].
Carbon steel: Medium-carbon steels like 1045 are used in applications where a balance of strength and ductility is required. This material was utilized in a novel bio-inspired and cable-driven hybrid parallel shoulder mechanism [48].

- Plastics
These polymers are commonly selected for components that necessitate low mechanical strength but prioritize lightweight characteristics. ABS (acrylonitrile butadiene tyrene) is a type of thermoplastic polymer. It is a copolymer made from three different monomers: acrylonitrile, butadiene, and styrene. Acrylonitrile provides chemical resistance and thermal stability. Butadiene contributes toughness and impact resistance. Styrene offers rigidity and a glossy finish. This material was employed in a parallel robot designed for a prosthetic arm aimed at shoulder disarticulation [19], a rigid–flexible parallel mechanism for a neck brace [38], a prosthetic arm [51], and a humanoid robot with parallel architectures [87].
- Composite materials
Carbon-fiber composites: Valued for their extraordinary strength-to-weight ratios and rigidity, these composites are ideal for crafting structural components that must be both resilient and lightweight. This material was utilized in a hybrid prosthetic mechanism for transfemoral amputees [18].
- Advanced or 'smart' materials
Magnetorheological fluids: These fluids are integrated into adaptive control systems to modulate mechanical properties in real time. Although this material has not yet been used in parallel robots, it is important to note that it can be utilized human prosthesis [116].
Shape-memory alloys (SMAs): Predominantly used in the design of actuators and responsive joints, these alloys enable a broad range of functional capabilities. Although this material has not yet been used in parallel robots, it is important to note that it can be utilized in robotic and biomedical applications [117].
Piezoelectric materials: These are incorporated in systems requiring ultra-precise micromovements and sensing applications. Although this material has not yet been used in parallel robots, it is important to note that it can be utilized in robotic and biomedical applications [118].
- Soft materials
Silicone: Predominantly employed in soft robotics, silicone's deformable yet resilient nature confers flexibility to the robot design. This material was used in a soft robotic neck that employed a parallel robot [3].
Elastomers: Capable of undergoing shape or stiffness alterations in response to electrical or thermal stimuli, elastomers are gaining traction in soft robotic applications. This material was utilized in a parallel lower limb based on pneumatic artificial muscles [22] and a soft parallel robot based on pneumatic artificial muscles for wrist rehabilitation [63].

5.7. Critical Technological Challenges and Future Prospects in Rehabilitation, Assistance, and Humanoids

There are considerable challenges that require keen attention for parallel robot technologies to transcend from research labs to commercial markets.

- Key technological challenges
Precision and reliability: Both rehabilitation and assistive robotics demand highly precise and reliable systems. The challenge extends to humanoid robots, which must interact safely and effectively in human-centric environments. The need for high precision and fail-safe operation in medical rehabilitation sets a challenging bar for

system performance. Addressing issues related to sensor accuracy and the reliability of actuators is of paramount importance.
User interface and experience: As these systems are ultimately designed for human use, developing intuitive, user-friendly interfaces that can cater to a diverse demographic is a significant hurdle. The end-users range from medical professionals to patients and the general public. Designing interfaces that are universally intuitive is an ongoing challenge.
Material constraints: Material choices should balance durability, cost, and biocompatibility, especially for wearable devices. Ensuring that all materials and operational modalities are biocompatible, sterile, and safe for long-term use requires rigorous testing and often leads to increased costs.
Scalability and affordability: Creating systems that are economically viable to manufacture at scale while maintaining high-quality performance presents an ongoing challenge.

- Impediments to commercialization
Regulatory hurdles: Approvals for medical and assistive devices are lengthy and require substantial investment in clinical trials and validation studies.
Cost factor: Advanced materials and smart systems drive up the manufacturing costs, making it difficult to position these products as affordable solutions.
- Future directions
To realize the full potential of parallel robotic systems in the rehabilitation sector, future work needs to focus on solving the technological issues mentioned above. Solutions may lie in material innovations and machine learning algorithms for adaptive control. This would facilitate commercial transitions but also pave the way for more effective and accessible rehabilitation and assistance technologies.

6. Conclusions

- The data presented in Table 8 highlight the global interest and collaborative efforts in researching parallel robots for rehabilitation across various joint areas. China, the USA, and several other countries have actively contributed to this burgeoning field, each with a specific focus on particular joints. This comprehensive effort underscores the significance of parallel robot technology in furthering our understanding of human movement and enabling assistance across various joints.
- As the field of robotics continues to evolve, international collaboration can collectively drive advancements and shape the future of robotic technologies in rehabilitation, assistance, and humanoid applications.
- Technological development is a multifaceted process influenced by numerous factors. In the case of parallel robots, several challenges may have impeded further advancement. These could include technical limitations requiring complex and advanced solutions, as well as the inherent complexity of integrating sophisticated mechanical and electronic systems and control algorithms. Meeting these challenges necessitates ongoing research and development efforts.
- Parallel mechanisms offer multiple advantages over their serial counterparts, including greater rigidity, load capacity, precision, and speed. These characteristics make them well-suited for applications requiring high precision, such as rehabilitation and assistance in the neck, shoulder, wrist, hip, and ankle joints. Due to their design and configuration, parallel robots can distribute loads across their kinematic chains, enabling the manipulation of heavier objects and tasks requiring greater force.
- Parallel robots hold significant potential for rehabilitation, assistance, and humanoid applications, but many challenges and opportunities remain. These include the need for advanced sensors and control systems to improve accuracy and minimize movement errors.
- Looking to the future, continuous technological advancement and collaboration among the scientific community, industry, and healthcare professionals could make parallel robots a mainstay in the market. Their applications could extend to a wide

range of sectors, including but not limited to the automotive, aerospace, and construction industries, as well as military training, medicine, and surgery.

Author Contributions: V.E.A. contributed to the conceptualization, methodology, research, formal analysis, resources, and writing of the original draft. D.A.E. contributed to the conceptualization, review, editing, and supervision. All authors have read and agreed to the published version of the manuscript.

Funding: This research received no external funding.

Conflicts of Interest: The authors declare no conflict of interest.

References

1. Lingampally, P.; Selvakumar, A. A kinematic and workspace analysis of a parallel rehabilitation device for head-neck injured patients. *FME Trans.* **2019**, *47*, 405–411. [CrossRef]
2. Zhang, H.; Chang, B.-C.; Andrews, J.; Mitsumoto, H.; Agrawal, S. A robotic neck brace to characterize head-neck motion and muscle electromyography in subjects with amyotrophic lateral sclerosis. *Ann. Clin. Transl. Neurol.* **2019**, *6*, 1671–1680. [CrossRef] [PubMed]
3. Quevedo, F.; Muñoz, J.; Castano Pena, J.A.; Monje, C.A. 3D model identification of a soft robotic neck. *Mathematics* **2021**, *9*, 1652. [CrossRef]
4. Lozano, A.; Ballesteros, M.; Cruz-Ortiz, D.; Chairez, I. Active neck orthosis for musculoskeletal cervical disorders rehabilitation using a parallel mini-robotic device. *Control Eng. Pract.* **2022**, *128*, 105312. [CrossRef]
5. Keller, U.; Klamroth, V.; van Hedel, H.J.A.; Riener, R. ChARMin: A robot for pediatric arm rehabilitation. In Proceedings of the 2013 IEEE International Conference on Robotics and Automation, Karlsruhe, Germany, 6–10 May 2013; pp. 3908–3913.
6. Kitano, Y.; Tanzawa, T.; Yokota, K. Development of wearable rehabilitation device using parallel link mechanism: Rehabilitation of com pound motion combining palmar/dorsi flexion and radial/ulnar deviation. *Robomech J.* **2018**, *5*, 13. [CrossRef]
7. Pehlivan, A.U.; Celik, O.; O'Malley, M.K. Mechanical design of a distal arm exoskeleton for stroke and spinal cord injury rehabilitation. In Proceedings of the IEEE International Conference on Rehabilitation Robotics, Zurich, Switzerland, 29 June–1 July 2011; pp. 1–5.
8. Pehlivan, A.U.; Sergi, F.; O'Malley, M.K. A subject-adaptive controller for wrist robotic rehabilitation. *IEEE/ASME Trans. Mechatron.* **2015**, *20*, 1338–1350. [CrossRef]
9. Pang, Z.; Wang, T.; Yu, J.; Liu, S.; Zhang, X.; Jiang, D. Design and analysis of a flexible, elastic, and rope-driven parallel mechanism for wrist rehabilitation. *Appl. Bionics Biomech.* **2020**, *2020*, 8841400. [CrossRef]
10. Liu, Q.; Zuo, J.; Zhu, C.H.; Quan, S. Design and control of soft rehabilitation robots actuated by pneumatic muscles: State of the art. *Future Gener. Comput. Syst.* **2020**, *113*, 620–634. [CrossRef]
11. Dong, M.; Zhou, Y.; Li, J.; Rong, X.; Fan, W.; Zhou, X.; Kong, Y. State of the art in parallel ankle rehabilitation robot: A systematic review. *J. Neuroeng. Rehabil.* **2021**, *18*, 52. [CrossRef]
12. Hussain, S.; Jamwal, P.K.; Ghayesh, M.H. State-of-the-art robotic devices for ankle rehabilitation: Mechanism and control review. *Proc. Inst. Mech. Eng. Part H J. Eng. Med.* **2017**, *231*, 1224–1234. [CrossRef]
13. Jamwal, P.K.; Hussain, S.; Xie, S.Q. Review on design and control aspects of ankle rehabilitation robots. *Disabil. Rehabil. Assist. Technol.* **2015**, *10*, 93–101. [CrossRef] [PubMed]
14. Zuo, S.; Li, J.; Dong, M.; Zhou, X.; Fan, W.; Kong, Y. Design and performance evaluation of a novel wearable parallel mechanism for ankle rehabilitation. *Front. Neurorobot.* **2020**, *14*, 9. [CrossRef] [PubMed]
15. Liao, Z.; Yao, L.; Lu, Z.; Zhang, J. Screw theory based mathematical modeling and kinematic analysis of a novel ankle rehabilitation robot with a constrained 3-PSP mechanism topology. *Int. J. Intell. Robot. Appl.* **2018**, *2*, 351–360. [CrossRef] [PubMed]
16. Hussain, S.; Jamwal, P.K.; Vliet, P.V.; Brown, N.A.T. Robot assisted ankle neuro-rehabilitation: State of the art and future challenges. *Expert Rev. Neurother.* **2021**, *21*, 111–121. [CrossRef]
17. Rastegarpanah, A.; Rakhodaei, H.; Saadat, M.; Rastegarpanah, M.; Marturi, N.; Borboni, A.; Loureiro, R.C.V. Path-planning of a hybrid parallel robot using stiffness and workspace for foot rehabilitation. *Adv. Mech. Eng.* **2018**, *10*, 1687814017754159. [CrossRef]
18. Song, M.; Guo, S.; Oliveira, A.; Wang, X.; Qu, H. Design method and verification of a hybrid prosthetic mechanism with energy-damper clutchable device for transfemoral amputees. *Front. Mech. Eng.* **2021**, *16*, 747–764. [CrossRef]
19. Leal-Naranjo, J.-A.; Torres-San Miguel, C.-R.; Ceccarelli, M.; Rostro-Gonzalez, H. Mechanical design and assessment of a low-cost 7-DOF prosthetic arm for shoulder disarticulation. *Appl. Bionics Biomech.* **2018**, *2018*, 4357602. [CrossRef] [PubMed]
20. Sekine, M.; Xie, L.; Kawamura, K.; Yu, W. Improvement and quantification of spatial accessibility and disturbance responsiveness of shoulder prosthesis. *Int. J. Adv. Robot. Syst.* **2015**, *12*, 1–12. [CrossRef]
21. Bajaj, N.M.; Spiers, A.J.; Dollar, A.M. State of the art in artificial wrists: A review of prosthetic and robotic wrist design. *IEEE Trans. Robot.* **2019**, *35*, 261–277. [CrossRef]
22. Jiang, F.; Tao, G.; Li, Q. Analysis and control of a parallel lower limb based on pneumatic artificial muscles. *Adv. Mech. Eng.* **2017**, *9*, 1687814016685002. [CrossRef]

23. Liu, G.; Lou, Y.; Li, Z. Singularities of parallel manipulators: A geometric treatment. *IEEE Trans. Rob. Autom.* **2003**, *19*, 579–594. [CrossRef]
24. Duan, X.; Yang, Y.; Cheng, B. Modeling and analysis of a 2-DOF spherical parallel manipulator. *Sensors* **2016**, *16*, 1485. [CrossRef] [PubMed]
25. Silva, L.A.; Sebastian, J.M.; Saltaren, R.; Aracil, R.; Sanpedro, J. RoboTenis: Optimal design of a parallel robot with high performance. In Proceedings of the 2005 IEEE/RSJ International Conference on Intelligent Robots and Systems, Edmonton, AB, Canada, 2–6 August 2005; pp. 2134–2139.
26. Righettini, P.; Strada, R.; Cortinovis, F. Modal Kinematic Analysis of a Parallel Kinematic Robot with Low-Stiffness Transmissions. *Robotics* **2021**, *10*, 132. [CrossRef]
27. Righettini, P.; Strada, R.; Zappa, B.; Lorenzi, V. Experimental set-up for the investigation of transmissions effects on the dynamic performances of a linear PKM. *Mech. Mach. Sci.* **2019**, *73*, 2511–2520. [CrossRef]
28. Wittenburg, J.; Likins, P. Dynamics of systems of rigid bodies. *J. Appl. Mech.* **1978**, *45*, 458. [CrossRef]
29. Zubizarreta, A.; Marcos, M.; Cabanes, I.; Pinto, C.; Portillo, E. Redundant sensor based control of the 3RRR parallel robot. *Mech. Mach. Theory* **2012**, *54*, 1–17. [CrossRef]
30. Hayes, M.J.D.; Zsombor-Murray, P.J.; Chen, C. Unified kinematic analysis of general planar parallel manipulators. *J. Mech. Des.* **2004**, *126*, 866–874. [CrossRef]
31. Stoeffler, C.; Kumar, S.; Peters, H.; Bruls, O.; Muller, A.; Kirchner, F. Conceptual design of a variable stiffness mechanism in a humanoid ankle using parallel redundant actuation. In Proceedings of the 2018 IEEE-RAS 18th International Conference on Humanoid Robots (Humanoids), Beijing, China, 6–9 November 2018; pp. 462–468.
32. Aracil, R.; Saltarén, R.J.; Sabater, J.M.; Reinoso, Ó. Robots Paralelos: Máquinas con un Pasado para una Robótica del Futuro. *Rev. Iberoam. Autom. Inform. Ind. RIAI* **2006**, *3*, 16–28. [CrossRef]
33. Alvarez-Perez, M.G.; Garcia-Murillo, M.A.; Cervantes-Sánchez, J.J. Robot-assisted ankle rehabilitation: A review. *Disabil. Rehabil. Assist. Technol.* **2020**, *15*, 394–408. [CrossRef] [PubMed]
34. Taboadela, C.H. *Goniometría: Una Herramienta para la Evaluación de las Incapacidades Laborales*, 1st ed.; Asociart ART: Buenos Aires, Argentina, 2007; pp. 65–115, ISBN 978-987-9274-04-0.
35. Zhang, H.; Ma, H.; Ren, J.; Shi, D.; Zhang, W. Design and Kinematic Analysis of a Cable-driven Exoskeleton for Cervical Rehabilitation. In Proceedings of the 2023 IEEE 18th Conference on Industrial Electronics and Applications (ICIEA), Ningbo, China, 18–22 August 2023; pp. 1495–1500.
36. Zhang, H.; Agrawal, S.K. Kinematic design of a dynamic brace for measurement of head/neck motion. *IEEE Robot. Autom. Lett.* **2017**, *2*, 1428–1435. [CrossRef]
37. Zhang, H.; Agrawal, S.K. An active neck brace controlled by a joystick to assist head motion. *IEEE Robot. Autom. Lett.* **2018**, *3*, 37–43. [CrossRef]
38. Liu, J.; Cheng, Y.; Zhang, S.; Lu, Z.; Gao, G. Design and analysis of a rigid-flexible parallel mechanism for a neck brace. *Math. Probl. Eng.* **2019**, *2019*, 9014653. [CrossRef]
39. Gao, B.; Song, H.; Sun, L.; Tang, Y. Inverse kinematics and workspace analysis of a bio-inspired flexible parallel robot. In Proceedings of the 2013 IEEE International Conference on Cyber Technology in Automation, Control and Intelligent Systems, Nanjing, China, 26–29 May 2013; pp. 138–143.
40. Gao, B.; Hu, J.; Guo, S.; Li, W.; Kan, Q. Lateral bending models of spring spine for cable-driven parallel mechanism. In Proceedings of the 11th World Congress on Intelligent Control and Automation, Shenyang, China, 29 June–4 July 2014; pp. 3176–3180.
41. Gao, B.; Song, H.; Zhao, J.; Guo, S.; Sun, L.; Tang, Y. Inverse kinematics and workspace analysis of a cable-driven parallel robot with a spring spine. *Mech. Mach. Theory* **2019**, *76*, 56–69. [CrossRef]
42. Jiang, L.; Gao, B.; Zhao, J. Kinematic and static analysis of a cable-driven parallel robot with a flexible link spine. In Proceedings of the 2015 IEEE International Conference on Robotics and Biomimetics (ROBIO), Zhuhai, China, 6–9 December 2015; pp. 31–36.
43. Gao, B.; Zhu, Z.; Zhao, J.; Jiang, L. Inverse kinematics and workspace analysis of a 3 DOF flexible parallel humanoid neck robot. *J. Intell. Robot. Syst.* **2017**, *87*, 211–229. [CrossRef]
44. Klein, J.; Roach, N.; Burdet, E. 3DOM: A 3 Degree of Freedom Manipulandum to Investigate Redundant Motor Control. *IEEE Trans. Haptics* **2014**, *7*, 229–239. [CrossRef] [PubMed]
45. Enferadi, J.; Haghi, O.R. On the kinematics analysis of a novel spherical robot with unlimited rotation around an axis. In Proceedings of the 2015 3rd RSI International Conference on Robotics and Mechatronics (ICROM), Tehran, Iran, 7–9 October 2015; pp. 158–163.
46. Enferadi, J.; Shahi, A. On the position analysis of a new spherical parallel robot with orientation applications. *Robot. Comput. Integr. Manuf.* **2016**, *37*, 151–161. [CrossRef]
47. Hunt, J.; Lee, H.; Artemiadis, P. A Novel Shoulder Exoskeleton Robot Using Parallel Actuation and a Passive Slip Interface. *J. Mech. Robot.* **2016**, *9*, 011002. [CrossRef]
48. Niyetkaliyev, A.S.; Sariyildiz, E.; Alici, G. Kinematic Modeling and Analysis of a Novel Bio-Inspired and Cable-Driven Hybrid Shoulder Mechanism. *J. Mech. Robot.* **2020**, *13*, 011008. [CrossRef]
49. Hunt, J.; Lee, H. Optimizing the rigid or compliant behavior of a novel parallel-actuated architecture for exoskeleton robot applications. *Front. Robot. AI* **2021**, *8*, 596958. [CrossRef]

50. Sekine, M.; Sugimori, K.; Gonzalez, J.; Yu, W. Optimization-based design of a small pneumatic-actuator-driven parallel mechanism for a shoulder prosthetic arm with statics and spatial accessibility evaluation. *Int. J. Adv. Robot. Syst.* **2013**, *10*, 286. [CrossRef]
51. Leal-Naranjo, J.-A.; Ceccarelli, M.; Torres-San-Miguel, C.-R.; Aguilar-Perez, L.-A.; Urriolagoitia-Sosa, G.; Urriolagoitia-Calderón, G. Multiobjective optimization of a parallel manipulator for the design of a prosthetic arm using genetic algorithms. *Lat. Am. J. Solids Struct.* **2018**, *15*. [CrossRef]
52. Hunt, J.; Lee, H. A New Parallel Actuated Architecture for Exoskeleton Applications Involving Multiple Degree-of-Freedom Biological Joints. *J. Mech. Robot.* **2018**, *10*, 051017. [CrossRef]
53. Chen, W.; Chen, Q.; Liu, R.; Zhang, J. Homing algorithm analysis for a cable-driven 3-DOF shoulder joint. In Proceedings of the IEEE 10th International Conference on Industrial Informatics, Beijing, China, 25–27 July 2012; pp. 1170–1175.
54. Wang, B.; Jin, Y.; Yan, T.; Cheng, M. Mechanism simulation and optimal design of five-link bionic joint driven by two antagonistic pneumatic muscles. *Adv. Mech. Eng.* **2013**, *2013*, 824085. [CrossRef]
55. Alfayad, S.; Tayba, A.M.; Ouezdou, F.B.; Namoun, F. Kinematic Synthesis and Modeling of a Three Degrees-of-Freedom Hybrid Mechanism for Shoulder and Hip Modules of Humanoid Robots. *J. Mech. Robot.* **2016**, *8*, 041017. [CrossRef]
56. Jiang, L.; Gao, B.; Zhu, Z. Design and Nonlinear Control of a 2-DOF Flexible Parallel Humanoid Arm Joint Robot. *Shock Vib.* **2017**, *2017*, 2762169. [CrossRef]
57. Liu, K.; Chen, Y.; Xu, J.; Wu, Y.; Lu, Y.; Zhao, D. Fuzzy torque control of the bionic flexible manipulator actuated by pneumatic muscle actuators. *Chin. J. Mech. Eng.* **2019**, *32*, 79. [CrossRef]
58. Bai, S.; Li, X.; Angeles, J. A review of spherical motion generation using either spherical parallel manipulators or spherical motors. *Mech. Mach. Theory* **2019**, *140*, 377–388. [CrossRef]
59. Wang, Z.; Li, Y.; Sun, P.; Luo, Y.; Chen, B.; Zhu, W. A multiobjective approach for the trajectory planning of a 7-DOF serial-parallel hybrid humanoid arm. *Mech. Mach. Theory* **2021**, *165*, 104423. [CrossRef]
60. Chen, W.; Feng, J.; Guan, Y.; Chen, W.; Wang, M. Improved Design and Modeling of a Series-Parallel Hybrid Modular Joint for Humanoid Arms. In Proceedings of the 2023 IEEE International Conference on Mechatronics and Automation (ICMA), Harbin, Heilongjiang, China, 6–9 August 2023; pp. 2389–2394.
61. Pehlivan, A.U.; Sergi, F.; O'Malley, M.K. Adaptive control of a serial-in-parallel robotic rehabilitation device. In Proceedings of the 2013 IEEE 13th International Conference on Rehabilitation Robotics (ICORR), Seattle, WA, USA, 24–26 June 2013; pp. 1–6.
62. Bian, H.; Chen, Z.; Wang, H.; Zhao, T. Mechanical design of EFW Exo II: A hybrid exoskeleton for elbow-forearm-wrist rehabilitation. In Proceedings of the 2017 International Conference on Rehabilitation Robotics (ICORR), London, UK, 17–20 July 2017; pp. 689–694.
63. Wang, Y.; Xu, Q. Design and testing of a soft parallel robot based on pneumatic artificial muscles for wrist rehabilitation. *Sci. Rep.* **2021**, *11*, 1273. [CrossRef] [PubMed]
64. Goyal, T.; Hussain, S.; Martinez-Marroquin, E.; Brown, N.A.T.; Jamwal, P.K. Impedance Control of a Wrist Rehabilitation Robot Based on Autodidact Stiffness Learning. *IEEE Trans. Med. Robot. Bionics* **2022**, *4*, 796–806. [CrossRef]
65. Li, S.; Wang, Z.; Pang, Z.; Gao, M.; Duan, Z. Design and Analysis of a Hemispherical Parallel Mechanism for Forearm–Wrist Rehabilitation. *Appl. Bionics Biomech.* **2023**, *2023*, 5722499. [CrossRef]
66. Goyal, T.; Hussain, S.; Martinez-Marroquin, E.; Brown, N.A.T.; Jamwal, P.K. Stiffness-Observer-Based Adaptive Control of an Intrinsically Compliant Parallel Wrist Rehabilitation Robot. *IEEE Trans. Hum.-Mach. Syst.* **2023**, *53*, 65–74. [CrossRef]
67. Goyal, T.; Hussain, S.; Martinez-Marroquin, E.; Brown, N.A.T.; Jamwal, P.K. Learning Koopman Embedding Subspaces for System Identification and Optimal Control of a Wrist Rehabilitation Robot. *IEEE Trans. Ind. Electron.* **2023**, *70*, 7092–7101. [CrossRef]
68. Serracín, J.R.; Puglisi, L.J.; Saltaren, R.; Ejarque, G.; Sabater-Navarro, J.M.; Aracil, R. Kinematic analysis of a novel 2-d.o.f. orientation device. *Rob. Auton. Syst.* **2012**, *60*, 852–861. [CrossRef]
69. Lee, J.; Song, B.-W.; Yang, W. Design of exoskeleton-type wrist human–machine interface based on over-actuated coaxial spherical parallel mechanism. *Adv. Mech. Eng.* **2018**, *10*, 1687814017753896. [CrossRef]
70. Lee, J.; Kim, H.; Yang, W. Development of wrist interface based on fully actuated coaxial spherical parallel mechanism for force interaction. *Sensors* **2021**, *21*, 8073. [CrossRef] [PubMed]
71. López-Custodio, P.C.; Fu, R.; Dai, J.S.; Jin, Y. Compliance model of Exechon manipulators with an offset wrist. *Mech. Mach. Theory* **2022**, *167*, 1–18. [CrossRef]
72. Sanjuan, J.D.; Sunny, M.S.H.; Ghommam, J.; Brahmi, B.; Wang, I.; Rahman, M.H. Optimal design of a cable-driven wrist prosthetic device. In Proceedings of the 9th International Conference of Control Systems, and Robotics (CDSR'22), Niagara Falls, ON, Canada, 2–4 June 2022; p. 108.
73. Chaparro-Altamirano, D.; Zavala-Yoe, R.; Ramirez-Mendoza, R. Kinematic and workspace analysis of a parallel robot used in security applications. In Proceedings of the 2013 International Conference on Mechatronics, Electronics and Automotive Engineering, Morelos, Mexico, 19–22 November 2013; pp. 3–8.
74. Kong, X.; Yu, J. Type Synthesis of Two-Degrees-of-Freedom 3-4R Parallel Mechanisms with Both Spherical Translation Mode and Sphere-on-Sphere Rolling Mode. *J. Mech. Robot.* **2015**, *7*, 041018. [CrossRef]
75. Lu, Y.; Zhang, C.; Cao, C.; Liu, Y. Analysis of coordinated grasping kinematics and optimization of grasping force of a parallel hybrid hand. *Int. J. Adv. Robot. Syst.* **2017**, *14*, 1729881417716816. [CrossRef]
76. Wu, G.; Niu, B. Dynamic stability of a tripod parallel robotic wrist featuring continuous end-effector rotation used for drill point grinder. *Mech. Mach. Theory* **2018**, *129*, 36–50. [CrossRef]

77. He, J.; Xu, W.; Gao, B. Self-insulating joint design for live-line operation based on the cable-driven parallel-series mechanism. *Appl. Sci.* **2019**, *10*, 22. [CrossRef]
78. Bazman, M.; Yilmaz, N.; Tumerdem, U. An Articulated Robotic Forceps Design with a Parallel Wrist-Gripper Mechanism and Parasitic Motion Compensation. *J. Mech. Des.* **2022**, *144*, 063303. [CrossRef]
79. Li, X.; Qu, H.; Guo, S. Kinematic Performance and Static Analysis of a Two-Degree-of-Freedom 3-RPS/US Parallel Manipulator with Two Passive Limbs. *J. Mech. Robot.* **2022**, *15*, 021014. [CrossRef]
80. Rastegarpanah, A.; Saadat, M.; Borboni, A.; Stolkin, R. Application of a parallel robot in lower limb rehabilitation: A brief capability study. In Proceedings of the 2016 International Conference on Robotics and Automation for Humanitarian Applications (RAHA), Amritapuri, India, 18–20 December 2016; pp. 1–6.
81. Zhang, W.; Zhang, W.; Ding, X.; Sun, L. Optimization of the Rotational Asymmetric Parallel Mechanism for Hip Rehabilitation with Force Transmission Factors. *J. Mecha. Robot.* **2020**, *12*, 041006. [CrossRef]
82. Shi, D.; Zhang, W.; Wang, L.; Zhang, W.; Feng, Y.; Ding, X. Joint Angle Adaptive Coordination Control of a Serial-Parallel Lower Limb Rehabilitation Exoskeleton. *IEEE Trans. Med. Robot. Bionics* **2022**, *4*, 775–784. [CrossRef]
83. Ren, B.; Liu, J.; Luo, X.; Chen, J. On the kinematic design of anthropomorphic lower limb exoskeletons and their matching movement. *Int. J. Adv. Robot. Syst.* **2019**, *16*, 1–9. [CrossRef]
84. Song, M.; Guo, S.; Wang, X.; Qu, H. Dynamic analysis and performance verification of a novel hip prosthetic mechanism. *Chin. J. Mech. Eng.* **2020**, *33*, 1–17. [CrossRef]
85. Wang, X.; Guo, S.; Qu, B.; Bai, S. Design and Experimental Verification of a Hip Exoskeleton Based on Human–Machine Dynamics for Walking Assistance. *IEEE Trans. Hum.-Mach. Syst.* **2023**, *53*, 85–97. [CrossRef]
86. Wang, G.; Liu, H.; Deng, P.; Yin, K.; Zhang, G. Dynamic Analysis of 4-SPS/CU Parallel Mechanism Considering Three-Dimensional Wear of Spherical Joint with Clearance. *J. Tribol.* **2017**, *139*, 021608. [CrossRef]
87. Russo, M.; Cafolla, D.; Ceccarelli, M. Design and Experiments of a Novel Humanoid Robot with Parallel Architectures. *Robotics* **2018**, *7*, 79. [CrossRef]
88. Wang, Y.; Mei, Z.; Xu, J.; Zhao, G. Kinematic design of a parallel ankle rehabilitation robot for sprained ankle physiotherapy. In Proceedings of the 2012 IEEE International Conference on Robotics and Biomimetics (ROBIO), Guangzhou, China, 11–14 December 2012; pp. 1643–1649.
89. Saglia, J.A.; Tsagarakis, N.G.; Dai, J.S.; Caldwell, D.G. Control Strategies for Patient-Assisted Training Using the Ankle Rehabilitation Robot (ARBOT). *IEEE/ASME Trans. Mechatron.* **2013**, *18*, 1799–1808. [CrossRef]
90. Wang, C.; Fang, Y.; Guo, S.; Chen, Y. Design and Kinematical Performance Analysis of a 3-RUS/RRR Redundantly Actuated Parallel Mechanism for Ankle Rehabilitation. *J. Mechan. Robot.* **2013**, *5*, 041003. [CrossRef]
91. Jamwal, P.K.; Xie, S.Q.; Hussain, S.; Parsons, J.G. An Adaptive Wearable Parallel Robot for the Treatment of Ankle Injuries. *IEEE/ASME Trans. Mechatron.* **2014**, *19*, 64–75. [CrossRef]
92. Jamwal, P.K.; Hussain, S.; Xie, S.Q. Three-Stage Design Analysis and Multicriteria Optimization of a Parallel Ankle Rehabilitation Robot Using Genetic Algorithm. *IEEE Trans. Autom. Sci. Eng.* **2015**, *12*, 1433–1446. [CrossRef]
93. Vallés, M.; Cazalilla, J.; Valera, Á.; Mata, V.; Page, Á.; Díaz-Rodríguez, M. A 3-PRS parallel manipulator for ankle rehabilitation: Towards a low-cost robotic rehabilitation. *Robotica* **2015**, *35*, 1939–1957. [CrossRef]
94. Jamwal, P.K.; Hussain, S.; Ghayesh, M.H.; Rogozina, S.V. Impedance Control of an Intrinsically Compliant Parallel Ankle Rehabilitation Robot. *IEEE Trans. Ind. Electron.* **2016**, *63*, 3638–3647. [CrossRef]
95. Ruiz-Hidalgo, N.C.; Blanco-Ortega, A.; Abúndez-Pliego, A.; Colín Ocampo, J.; Arias-Montiel, M. Design and Control of a Novel 3-DOF Parallel Robot. In Proceedings of the 2016 International Conference on Mechatronics, Electronics and Automotive Engineering (ICMEAE), Cuernavaca, Mexico, 22-25 November 2016; pp. 66–71.
96. Azar, W.A.; Akbarimajd, A.; Parvari, E. Intelligent Control Method of a 6-DOF parallel robot Used for Rehabilitation Treatment in lower limbs. *Automatika* **2016**, *57*, 466–476. [CrossRef]
97. Rosado, W.M.A.; Valdés, L.G.V.; Ortega, A.B, Ascencio, J.R.; Beltrán, C.D.G. Passive Rehabilitation Exercises with an Ankle Rehabilitation Prototype Based in a Robot Parallel Structure. *IEEE Lat. Am. Trans.* **2017**, *15*, 48–56. [CrossRef]
98. Rosado, W.M.A.; Ortega, A.B.; Valdés, L.G.V.; Ascencio, J.R.; Beltrán, C.D.G. Active Rehabilitation Exercises with a Parallel Structure Ankle Rehabilitation Prototype. *IEEE Lat. Am. Trans.* **2017**, *15*, 786–794. [CrossRef]
99. Du, Y.; Li, R.; Li, D.; Bai, S. An ankle rehabilitation robot based on 3-RRS spherical parallel mechanism. *Adv. Mech. Eng.* **2017**, *9*, 1–8. [CrossRef]
100. Zhang, M.; Cao, J.; Zhu, G.; Miao, Q.; Zeng, X.; Xie, S.Q. Re configurable workspace and torque capacity of a compliant ankle rehabilitation robot (CARR). *Rob. Auton. Syst.* **2017**, *98*, 213–221. [CrossRef]
101. Jamwal, P.K.; Hussain, S.; Mir-Nasiri, N.; Ghayesh, M.H.; Xie, S.Q. Tele-rehabilitation using in-house wearable ankle rehabilitation robot. *Assist. Technol.* **2018**, *30*, 24–33. [CrossRef] [PubMed]
102. Wang, C.; Wang, L.; Wang, T.; Li, H.; Du, W.; Meng, F.; Zhang, W. Research on an ankle joint auxiliary rehabilitation robot with a rigid-flexible hybrid drive based on a 2-S'PS' mechanism. *Appl. Bionics Biomech.* **2019**, *2019*, 7071064. [CrossRef] [PubMed]
103. Nurahmi, L.; Gan, D. Reconfiguration of a 3-(rR)PS Metamorphic Parallel Mechanism Based on Complete Workspace and Operation Mode Analysis. *J. Mechan. Robot.* **2020**, *12*, 011002. [CrossRef]
104. Li, J.; Fan, W.; Dong, M.; Rong, X. Research on control strategies for ankle rehabilitation using parallel mechanism. *Cogn. Comput. Syst.* **2020**, *2*, 105–111. [CrossRef]

105. Russo, M.; Ceccarelli, M. Analysis of a wearable robotic system for ankle rehabilitation. *Machines* **2020**, *8*, 48. [CrossRef]
106. Li, J.; Zuo, S.; Zhang, L.; Dong, M.; Zhang, Z.; Tao, C.; Ji, R. Mechanical Design and Performance Analysis of a Novel Parallel Robot for Ankle Rehabilitation. *J. Mech. Robot.* **2020**, *12*, 051007. [CrossRef]
107. Pulloquinga, J.L.; Mata, V.; Valera, Á.; Zamora-Ortiz, P.; Díaz-Rodríguez, M.; Zambrano, I. Experimental analysis of Type II singularities and assembly change points in a 3UPS+RPU parallel robot. *Mech. Mach. Theory* **2021**, *158*, 1–27. [CrossRef]
108. Liu, Y.; Lu, W.; Wu, H.; Xia, Y.; Hu, B.; Zeng, D. Performance analysis and trajectory planning of multi-locomotion mode ankle rehabilitation robot. *Rob. Auton. Syst.* **2021**, *157*, 1–11. [CrossRef]
109. Valencia-Segura, L.E.; Villarreal-Cervantes, M.G.; Corona-Ramírez, L.G.; Cuenca-Jiménez, F.; Muñz-Reina, J.S. Nature-Inspired Intelligent Synthesis of a Spherical Mechanism for Passive Ankle Rehabilitation Using Differential Evolution. *IEEE Access* **2023**, *11*, 51331–51352. [CrossRef]
110. Wu, J.; Liu, Y.; Zhao, J.; Jia, Z. Research on a New Rehabilitation Robot for Balance Disorders. *IEEE Trans. Neural Syst. Rehabil. Eng.* **2023**. [CrossRef] [PubMed]
111. Shi, M.; Hao, L.; Zhou, Z.; Liu, S.; Han, J.; Yu, N. A Novel Serial-Parallel Hybrid Robot for Rehabilitation Training of Gait and Balance. In Proceedings of the 2023 9th International Conference on Electrical Engineering, Control and Robotics (EECR), Wuhan, China, 24–26 February 2023; pp. 5–10.
112. Zuo, S.; Li, J.; Dong, M.; Jiao, R.; Wang, S.; Ju, J.; Wang, Y. Design of Composite Parallel External Fixator for Multisegment Foot and Ankle Deformity Correction: A Correction Requirement- Dominated Approach. *IEEE/ASME Trans. Mechatron.* **2023**, *28*, 2271–2281. [CrossRef]
113. Zermane, A.; Guiatni, M.; Alouane, M.A.; Alloun, F.; Tair, M. Optimal Mechatronics Design and Control of a 3-PRS Robotic Platform for Vertigo Diagnosis. In Proceedings of the 2023 International Conference on Control, Automation and Diagnosis (ICCAD), Rome, Italy, 10–12 May 2023; pp. 1–7.
114. Zhang, J.; Ma, Z.; Wei, J.; Yang, S.; Liu, C.; Guo, S. A Novel Evaluation Index and Optimization Method for Ankle Rehabilitation Robots Based on Ankle-Foot Motion. *ASME. J. Biomech. Eng.* **2023**, *145*, 051006. [CrossRef] [PubMed]
115. Twi-global.Cambridge, UK, 2023. Available online: https://www.twi-global.com/technical-knowledge/faqs/technology-readiness-levels (accessed on 2 July 2023).
116. Hanmant, S.; Surendra, T. Magnetorheological Fluids and Its Applications: Current Scenario and Future Prospects. *Math. Sci. Int. Res. J.* **2021**, *6*, 207–213.
117. Mohd, J.; Leary, M.; Subic, A.; Gibson, M. A review of shape memory alloy research, applications and opportunities. *Mater. Des.* **2014**, *56*, 1078–1113. [CrossRef]
118. Chen-Glasser, M.; Li, P.; Ryu, J.; Hong, S. Piezoelectric Materials for Medical Applications. In *Piezoelectricity-Organic and Inorganic Materiales and Applications*; IntechOpen: London, UK, 2018; Chapter 7, pp. 125–145, ISBN 978-1-78923-647-7.

MDPI AG
Grosspeteranlage 5
4052 Basel
Switzerland
Tel.: +41 61 683 77 34

Robotics Editorial Office
E-mail: robotics@mdpi.com
www.mdpi.com/journal/robotics

www.ingramcontent.com/pod-product-compliance
Lightning Source LLC
Chambersburg PA
CBHW040758250926
51850CB00007BA/1028
9783725852918